Causation and Its Basis in Fundamental Physics

OXFORD STUDIES IN PHILOSOPHY OF SCIENCE
General Editor: Paul Humphreys, University of Virginia

The Book of Evidence
Peter Achinstein

Science, Truth, and Democracy
Philip Kitcher

Inconsistency, Asymmetry, and Non-Locality: A Philosophical Investigation of Classical Electrodynamics
Mathias Frisch

The Devil in the Details: Asymptotic Reasoning in Explanation, Reduction, and Emergence
Robert W. Batterman

Science and Partial Truth: A Unitary Approach to Models and Scientific Reasoning
Newton C.A. da Costa and Steven French

Inventing Temperature: Measurement and Scientific Progress
Hasok Chang

The Reign of Relativity: Philosophy in Physics 1915 – 1925
Thomas Ryckman

Making Things Happen
James Woodward

Mathematics and Scientific Representation
Christopher Pincock

Simulation and Similarity
Michael Weisberg

Causation and Its Basis in Fundamental Physics
Douglas Kutach

Causation and Its Basis in Fundamental Physics

Douglas Kutach

OXFORD
UNIVERSITY PRESS

OXFORD
UNIVERSITY PRESS

Oxford University Press is a department of the University of Oxford. It furthers the University's objective of excellence in research, scholarship, and education by publishing worldwide.

Oxford New York
Auckland Cape Town Dar es Salaam Hong Kong Karachi
Kuala Lumpur Madrid Melbourne Mexico City Nairobi
New Delhi Shanghai Taipei Toronto

With offices in
Argentina Austria Brazil Chile Czech Republic France Greece
Guatemala Hungary Italy Japan Poland Portugal Singapore
South Korea Switzerland Thailand Turkey Ukraine Vietnam

Published in the United States of America by
Oxford University Press
198 Madison Avenue, New York, NY 10016

Library of Congress Cataloging-in-Publication Data
Kutach, Douglas.
Causation and its basis in fundamental physics / Douglas Kutach.
p. cm. — (Oxford studies in philosophy of science)
Includes bibliographical references (pages).
ISBN 978-0-19-993620-5 (hardback : alk. paper) — ISBN 978-0-19-993621-2 (updf)
1. Causality (Physics) 2. Physics—Philosophy. I. Title.
QC6.4.C3K88 2013
530.1—dc23
2012039777
9780199936205

1 3 5 7 9 8 6 4 2

Printed in the United States of America
on acid-free paper

for Lucy

{ CONTENTS }

Preface xi

1. **Empirical Analysis and the Metaphysics of Causation** 1
   - 1.1. Empirical Analysis 2
     - *1.1.1. The Distinctive Features of Empirical Analysis* 9
   - 1.2. Empirical Analysis of the Metaphysics of Causation 13
     - *1.2.1. Effective Strategies* 14
   - 1.3. Empirical Analysis of the Non-metaphysical Aspects of Causation 17
   - 1.4. Causation as Conceptually Tripartite 20
   - 1.5. A Sketch of the Metaphysics of Causation 22
   - 1.6. Fundamental and Derivative 24
     - *1.6.1. The Kinetic Energy Example* 25
     - *1.6.2. Some Constitutive Principles of Fundamentality* 30
   - 1.7. Abstreduction 34
   - 1.8. STRICT Standards and RELAXED Standards 37
   - 1.9. Limitations on the Aspirations of Empirical Analysis 44
   - 1.10. Comparison of Empirical and Orthodox Analysis 45
   - 1.11. Summary 48

PART I **The Bottom Conceptual Layer of Causation**

2. **Fundamental Causation** 53
   - 2.1. Preliminaries 57
     - *2.1.1. Events* 58
     - *2.1.2. Laws* 64
   - 2.2. Terminance 67
     - *2.2.1. Causal Contribution* 75
     - *2.2.2. Trivial Terminance* 76
   - 2.3. The Space-time Arena 79
   - 2.4. Classical Gravitation 82
     - *2.4.1. Galilean Space-time* 82
     - *2.4.2. Terminants in Classical Gravitation* 83
     - *2.4.3. Overdetermination in Classical Gravitation* 85
     - *2.4.4. Instantaneous Causation* 87
   - 2.5. Relativistic Electromagnetism 89
     - *2.5.1. Minkowski Space-time* 89
     - *2.5.2. Terminants in Relativistic Electromagnetism* 91
     - *2.5.3. Classical Unified Field Theory* 92

2.6. Content Independence 94
2.7. Continuity and Shielding 98
2.8. Transitivity 100
2.9. Determinism 101
2.10. Stochastic Indeterminism 104
2.10.1. Stochastic Lattices 104
2.10.2. A Toy Theory of Particle Decay 105
2.11. Non-stochastic Indeterminism 107
2.11.1. Newtonian Indeterminism 107
2.11.2. Contribution Extended 109
2.12. General Relativity 109
2.12.1. Spatio-temporal Indeterminism 110
2.12.2. Closed Time-like Curves 111
2.13. Quantum Mechanics 112
2.13.1. The Quantum Arena and its Contents 112
2.13.2. Bohmian Mechanics 113
2.13.3. Spontaneous Collapse Interpretations 114
2.13.4. Other Interpretations of Quantum Mechanics 116
2.14. Summary 116

PART II **The Middle Conceptual Layer of Causation**

3. Counterfactuals and Difference-making **121**
3.1. General Causation 121
3.2. Counterfactuals 123
3.3. Goodman's Account of Counterfactuals 124
3.4. The Nomic Conditional 128
3.5. Comparison to Ordinary Language Conditionals 134
3.6. Prob-dependence 135
3.7. Contrastive Events 136
3.8. Summary 138

4. Derivative Causation **140**
4.1. Influence 140
4.2. Prob-influence 142
4.3. General Causation 144
4.4. Temporally Extended Events 150
4.5. Idiomatic Differences between Promotion and Causation 152
4.6. Aspect Promotion 154
4.7. Promotion by Omission 155
4.8. Contrastivity 155
4.9. Transitivity 157
4.10. Continuity 160
4.11. Shielding 161
4.12. Partial Influence 162
4.13. Summary 167

5. **The Empirical Content of Promotion** 168
5.1. The Promotion Experiment 169
5.2. Insensitivity Considerations 171
5.3. Thermodynamics and Statistical Mechanics 175
5.4. The Asymmetry of Bizarre Coincidences 184
5.5. The Analogy to Thermal and Mechanical Energy 186
5.6. Broad and Narrow Promotion 187
5.7. Inferences from Empirical Data to Promotion 193
5.7.1. Simpson's Paradox 194
5.8. Why There Are Effective Strategies 197
5.9. Mechanistic Theories of Causation 199

6. **Backtracking Influence** 203
6.1. The Direction of Influence 205
6.2. Proof of Causal Directness 207
6.3. A Search for Empirical Phenomena 214
6.4. 'Past-directed then Future-directed' Influence 219

7. **Causal Asymmetry** 225
7.1. The Empirical Content of the Causal Asymmetry 226
7.2. Causation and Advancement 230
7.3. An Explanation of the Advancement Asymmetry 232
7.3.1. Prob-influence through Backtracking 234
7.3.2. Directly Past-directed Prob-influence 236
7.3.3. Summary 238
7.4. Pseudo-backtracking Prob-influence 239
7.5. The Entropy Asymmetry and Causal Directionality 244
7.6. Recent Alternative Explanations of Causal Asymmetry 249
7.6.1. The Albert-Kutach-Loewer Approach 250
7.6.2. The Price-Weslake Approach 252
7.6.3. The Fork Asymmetry Approach 254
7.7. Fundamental Influence Asymmetry 257
7.7.1. Fundamental Influence Asymmetry by Fiat 258
7.7.2. Fundamental Influence Asymmetry by Happenstance 260
7.8. Summary 262

PART III **The Top Conceptual Layer of Causation**

8. **Culpable Causation** 265
8.1. The Empirical Insignificance of Culpability 270
8.1.1. Part I: Singular Causation 271
8.1.2. Part II: General Causation 272
8.2. Culpability as a Heuristic for Learning about Promotion 275
8.3. Culpability as an Explanatory Device 277
8.4. Culpability as a Proxy for Terminance and Promotion 277
8.5. Commentary 281

9. The Psychology of Culpable Causation 282

9.1. The Toy Theory of Culpable Causation 284
9.2. Culpability$_1$ 287
9.2.1. Salience 287
9.2.2. Irreflexivity 289
9.2.3. Asymmetry 289
9.2.4. Significant Promotion 290
9.3. Shortcomings of Culpability$_1$ 291
9.3.1. Precise Character of the Effect 291
9.3.2. Overlapping Causation 291
9.3.3. Probability-Lowering Causes 292
9.4. Culpability$_2$ 292
9.5. Shortcomings of Culpability$_2$ 297
9.5.1. Saved Fizzles 298
9.5.2. Early Cutting Preëmption 298
9.5.3. Late Cutting Preëmption 298
9.6. Culpability$_3$ 299
9.7. Culpability$_4$ 300
9.8. Summary 305

10. Causation in a Physical World 307

10.1. Summary 307
10.2. Future Directions 312

References 321
Index 329

{ PREFACE }

This work constitutes the first volume of what I hope will become an extended treatise on a philosophical program to be known as Empirical Fundamentalism. This program essentially involves the exploitation of a distinction between fundamental and derivative reality together with a philosophical methodology known as empirical analysis in order to pronounce on a wide range of philosophical problems. Rather than lay out the program at too high level of abstraction, I have chosen to use this first volume to spell out how the program applies to a concept widely believed to have broad philosophical significance: causation. Because its implications for the metaphysics of causation can be adequately explained without detailing the full philosophical system, I will forgo elaboration here. Interested readers can examine a brief summary and sample application to the topic of reductive identities in Kutach (2011b).

The purpose of this volume is to outline a methodology for addressing the topic of causation and to apply this methodology to provide a positive account of the metaphysics of causation. It is not the purpose of this volume to survey the existing literature on causation or even to address the limited subset that pertains to relations between causation and fundamental physics. Furthermore, this book does not attempt to defend the claim that causation is ontologically nothing more than fundamental physics. While I do believe this proposition, its defense requires too much space to fit into a single book. Instead, I will show how some common features found in theories of fundamental physics can help to explain important features of causation in the special sciences, and I will reserve further discussion of whether fundamental physics is all one needs in a fundamental ontology for a future volume. This book's title is intended to signal my intent to discuss how fundamental physics is able to help ground the utility of our talk of causation. It is not a declaration of war against theories that postulate more than physics in their fundamental ontology.

The intended audience for this volume includes readers at roughly a graduate student's level of familiarity with philosophical topics pertaining to causation, but no specific previous knowledge is absolutely required. In particular, no familiarity with advanced physics is assumed, and no physics equations appear in this text. Some prior acquaintance with classical physics and relativity will likely ease digestion of the second chapter, however. More advanced researchers, by contrast, might find this volume to be of limited application to their own work owing to its implementation of a non-standard philosophical methodology. I recognize that a sizable fraction of academics insist that any presentation of a positive view needs to include a thorough discussion of existing alternatives and any views that are nominally similar, but I have three reasons for not doing so in this first volume. First, the economics of print publication forbid a manuscript long enough to include an adequate presentation of my own theory together with enough of the background material needed for a comprehensive comparison with other views. Second, my use of a non-standard methodology makes any direct comparison impossible. If I were to assume my own methodology when measuring the benefits of my own account against others, an obvious and justified rebuttal would be that the other work was never intended to be judged according to the criteria I lay out in this volume. Yet if I were to set aside my own methodology when flagging what I think are suboptimal characteristics of alternative theories, I would have little new to offer. Readers who are interested in discussions of causation using orthodox criteria and assumptions already have more than enough publications to sift through without my adding to the voluminous literature. Third, I personally prefer to maintain a focus on what I see as the positive virtues of my own account as much as possible rather than deficiencies in other theories. I will therefore attempt to do my part for the common good by contributing as little as possible to the relentless negativity, conservatism, and terminological quibbling that pervades contemporary academic philosophy.

That said, I will attempt to comment on a few connections with existing views. My chapter on counterfactuals and difference-making, for example, is longer than strictly necessary because I try to show how my account of difference-making arises naturally from famous precursors in the literature on counterfactuals. Also, the concluding chapter of this volume will offer a few suggestions for how to contrast my account with its distant relatives.

Regardless of length and content, readers will inevitably judge that I discussed some topics too much and others not nearly enough. I welcome any advice from readers on how best to alter my presentation or its substantive content in future editions of this volume and its intended successors. Please consider this book an invitation to engage further on the topic of causation and its basis in fundamental physics. For a reasonable amount of time after the initial publication and my health permitting, I will provide additional summaries, clarify portions that readers find puzzling, and post corrections. I have already made freely available some supplementary material that had to be excluded for the sake of a shorter and more lively text. Many readers may find it useful to locate this material before reading the book in order to print out a handy summary of the more important definitions and to find answers to frequently asked questions. Also, potential readers who are unsure about how much time they can commit to this book can examine shortened presentations of the major ideas.

Although I have made my best effort to discuss compelling topics in an accessible and sensible way, my time and energy are finite and dwindling, and the task of guessing where communication breakdowns will occur is Herculean without further feedback from readers. Because I do not expect any without the book being in print, here it is, warts and all.

I received excellent feedback on my initial drafts from Chiwook Won and Kateryna Samoilova and thorough editing from Sara Szeglowski, Brett Topey, and Vladomir Vlaovic. Steven Sloman improved my chapter on the psychology of causation and the extended version of that chapter I have made available. Thomas Blanchard provided good advice on a more advanced draft. Countless corrections were made as a result of an extremely thorough reading by Chris Hitchcock, whose philosophical ability and knowledge of the causation literature are world class. I was first drawn to the topic of causation by Jon Kvanvig's course in 1991, and I hope this volume repays some of the intellectual debt I incurred back then. I have struggled for a long time in the intervening years to develop a mature view of my own, and I would not have been able to do so without assistance from Paul B. Thompson's Center for Science and Technology Policy and Ethics and Huw Price's Centre for Time. Readers familiar with Price's work should be able to detect some commonalities, but precisely how our views are related has frustratingly remained a mystery to me.

{1}

Empirical Analysis and the Metaphysics of Causation

It is common knowledge that no one understands all the causes and effects that occur in nature, but it may surprise the uninitiated that the idea of causation itself—what causation amounts to—is thoroughly contested among experts without anything remotely approaching a consensus as to the likely form of a satisfactory account. The lack of an accepted theory of the connection between cause and effect coexists with a general agreement that some sort of causation is a critical component of reality. Causation is undoubtedly important in science, but even subjects like ethics, politics, and theology are significantly constrained by the need to make their proclamations compatible with what we know about paradigmatically causal interactions among ordinary physical objects. Causation's central role in linking various components of our overall worldview is evident in the wide range of theories that rely on the coherence of some notion of causation to account for perception, names, time, knowledge, and more.

Satisfying the full range of desiderata for an account of the metaphysics of causation has proven a stiff challenge, illustrated by an expansive technical literature. As I see it, the traditional approaches that dominate philosophical discussion have already succeeded in identifying virtually all the important elements needed for a comprehensive understanding of causation. However, the productive components have not yet been assembled into a convincing systematic metaphysics of causation because the traditional conception of what a metaphysics of causation is supposed to do—provide an informative and principled and consistent regimentation of important truths about causation—virtually ensures failure.

Fortunately, there exists an alternative conception of the task a metaphysical account of causation ought to accomplish: empirical analysis. A successful empirical analysis would vindicate enough of our use of causal concepts in science and philosophy and ordinary life in order for us to claim success in understanding the metaphysics of causation. Empirical analysis redraws the boundaries of the conceptual geography in a way that makes an adequate metaphysics of causation much easier to construct, in effect lowering the bar for success.

For illustration, one can consider the "problem of preëmption" that is believed to plague some prominent theories of causation. Preëmption is when a potential cause is on the way toward producing an effect but is somehow forestalled, like a lit fuse that is severed before it can ignite its rocket. Traditionally, the metaphysics of causation has been understood as needing to provide rules that identify paradigmatic preëmpted would-be causes as not genuine causes. In an empirical analysis of causation, however, it turns out that preëmption does not need to be understood as part of metaphysics, which permits it to be addressed successfully according to more lenient standards. An empirical analysis of the metaphysics of causation can safely count paradigmatic preëmpted would-be causes as bona fide causes.

My task in this book is to explicate this empirical approach and to implement its methodology in constructing a comprehensive metaphysics of causation. Along the way, I will dedicate a great deal of attention to how fundamental physics can serve as a basis for an adequate metaphysics of causation. However, readers are here warned that it is far beyond the scope of this book for me to argue that causation can be reduced to fundamental physics or that causation in the special sciences is merely a bunch of physics. I have much to say on this topic, but to address this important issue properly, I first need to set out my account of causation and the methodology behind it. That alone is a substantial enough task for a single volume.

This introductory chapter consists of two main components: an explanation of the non-standard methodology I will employ and a sketch of the overall structure of causation according to my account. In the first half of this chapter, I will provide a preliminary exposition of empirical analysis and an explanation of how it differs from orthodox conceptual analysis. Then I will illustrate how empirical analysis applies specifically to the metaphysics of causation and to the non-metaphysical aspects of causation. In the second half, I will describe how causation can be distinguished into three stacked conceptual layers. In order to clarify the three layers, I will need to unpack the two distinctions that mark the boundaries between them. One distinction is between fundamental reality and derivative reality, and the other is between STRICT and RELAXED standards of theoretical adequacy. Using the new terminology, I will summarize my account of causation and outline how the remaining chapters will address it in more detail.

1.1 Empirical Analysis

I have chosen 'empirical analysis' as the label for the methodology I will be employing throughout my investigation of causation as a figurative tip of my hat toward Phil Dowe's (2000, Ch. 1) discussion of conceptual analysis and its application to causation. I will soon explain what I mean by 'empirical analysis', but a brief warning is likely warranted for readers familiar with Dowe's work. The version of empirical analysis I will adopt is consistent with what Dowe says about empirical analysis, but I impose further conditions on what constitutes a proper empirical analysis that are substantial enough to make it incorrect to equate my philosophical project with Dowe's. It will turn out, for example, that my metaphysics of causation does not compete with theories where causation is understood in terms of the transfer of physical quantities or the paths of particles bearing conserved quantities. Although such theories are commonly understood as empirical approaches to causation, no existing examples from this tradition count as empirical analyses of causation under my narrower conception of the methodology. I can even go so far as to say I am unaware of any published example of what I would categorize as an empirical analysis.

Because others have not endorsed my construal of empirical analysis, it would be inappropriate for me to contrast other prominent accounts with mine in an effort to assess their relative merits. As I emphasized in the preface, if I were to criticize some account for being inadequate according to the standards of empirical analysis, it would be all too easy for the author to mount an effective defense by simply denying that the targeted account was ever intended to be an empirical analysis (of the form I have defined). Yet, if I were to set aside the method of empirical analysis, I would not have much to say that would still be relevant to the metaphysical project being conducted in this volume. In §5.9, I will clarify why currently existing transference and causal process theories like Dowe's do not count as empirical analyses according the precisification I am invoking.

Let us now attend to a positive characterization of empirical analysis. Uncontroversially, it proves useful in science to employ specialized terminology that is honed to improve precision, simplicity, and generality. To conduct the empirical analysis of some topic *X*, overly broadly speaking, is to identify scientifically improved concepts of *X*. Empirical analysis is a form of conceptual analysis in the broad sense that it provides a link between our ordinary conception of *X* and things in the world, but it is a non-standard form of conceptual analysis by forging the linkage in a manner especially responsive to scientific theorizing and experimental results. Empirical analysis involves not merely setting aside disagreements between theoretically refined terms and the platitudes that characterize *X*, which is common in contemporary versions of conceptual analysis (Jackson 1998), but also adapting the refined terminology to improve explanations of *experiments* that conceptually encapsulate the empirical phenomena that make our concept of *X* worth having.

In my experience, the distinctive features of empirical analysis are surprisingly difficult for experienced philosophers to grasp. Readers are thus cautioned not to be too hasty in concluding that they fully understand what constitutes a proper empirical analysis because some important clarifications cannot be adequately stated until §1.9 after I have introduced some new terminology.

One can acquire a decent preliminary grasp of empirical analysis by reviewing how exemplary sciences engineer their conceptual schemes. For example, food scientists are interested in answering questions about why some foods are healthier than others. A scientific investigation of food provides explanations for the following kinds of empirical phenomena. People who eat a mixture of fruits, vegetables, and nuts are healthier, ceteris paribus, than people who eat sand. The design of concepts for food science ought to be honed toward maximizing the quality of such explanations by having a regimentation of our folk food concept that strikes an adequate balance among various dimensions of explanatory quality such as simplicity, capturing as many empirical phenomena as possible within the scope of the explanation, and fitting properly with related subjects like agronomy, physiology, and chemistry. As it turns out, food science does have an improved concept of food, which we commonly refer to as 'nutrient'. 'Nutrient' serves as an excellent substitute for 'food' when studying the health effects of various ingested substances in part because it is only loosely tethered to our ordinary food concept. Crucially, we do not want to reject a theory of nutrition because it identifies iron crowbars, dust mites, and oxygen as nutrients whereas folk opinion adamantly rejects these as foods. (It ought to go without saying that it is altogether irrelevant that by historical happenstance English has two etymologically distinct words—'food' and 'nutrient'—for the folk and scientific concepts, respectively. There are abundant examples of the same word being used in an informal sense and as a technical term.)

To optimize our native food concept in the service of food science is to make it more precise in a way that achieves an optimal or at least acceptable level of quality according to principles of good conceptual design and according to the needs of food science. A precisification of 'food' or 'nutrient' can be thought of as a stipulation of a maximally precise class of all the possible entities that count as nutrients. I will call each class an 'intension' of the concept.

I will now clarify a few issues relevant to conceptual design and its role in empirical analysis. I will not attempt to provide a complete list of principles for engineering concepts nor a specification of their relative importance, but I will instead assume current scientific practice serves well enough for guidance.

First, philosophers have debated whether we should think of a conceptual analysis of *X* as trying to make an a priori claim about what *X* must be, given how our concept of *X* works or whether we should allow the analysis to incorporate some a posteriori component, for example (Block and Stalnaker 1999, Chalmers and Jackson 2001). From the perspective of empirical analysis, this way of framing the status of conceptual analysis does not elucidate the key issue. The purpose of empirical analysis is not to evaluate our mutually shared folk concept *X* in detail from the armchair, but to take what data science provides and to organize that data from the armchair to arrive at superior surrogates for *X*. The primary task is to balance the generality and specificity of the sought-after scientifically honed concepts. For illustration, consider one of Phil Dowe's glosses on an empirical analysis of causation: that it is intended to "discover what causation is in the objective world" (2000, p. 1). Such a project does not require settling questions about causation in all conceivable worlds, like worlds where magical spells are operative or where time does not exist.

Dowe's approach is rightly criticized by Collins, Hall, and Paul (2004, p. 14) for offering an improperly narrow treatment of the connection between fundamental laws and causal facts. They claim it would be better to "specify the way in which the fundamental laws fix the causal facts in terms that *abstract away* from the gory details of those laws—thereby to produce an account that has a hope of proving to be not merely true, but necessarily so."

I think the correct way to resolve this dispute is to recognize that there are two competing virtues, neither of which should dominate the other. On the one hand, we should prefer our concepts to be insensitive to the details of any empirical phenomenon we have not yet figured out. The nineteenth century conception of energy, for example, is largely insensitive to the details of microscopic interactions. Our later discovery that there is a strong and a weak nuclear interaction did not force a revision of that concept of energy or its central applications—for example, to the feasibility of constructing perpetual motion machines—because the concept of energy was already sufficiently insulated from such details. That counts as a successful application of conceptual engineering.

On the other hand, we should care little about how our concept applies to highly unrealistic possibilities and not at all about whether it applies to absolutely every possibility. It is patently silly, for example, to require a scientific theory of energy to accord with pre-theoretical intuitions we may have concerning the energetic consequences of magic spells. An important rule of good conceptual design is to avoid optimizing concepts to better handle epistemically remote possibilities when it proves costly to the explanation of more realistic possibilities. How we conceive of magic can bear on empirical analysis by helping to clarify how our concepts generalize, but as the imagined possibilities become ever more outlandish, there is less need to fiddle with our concepts in order to accommodate people's gut intuitions. Empirical analysis should not be adapted merely to what we currently believe to be true about the actual world, but neither should it be required to accord with everything we naturally want to say about every conceivable possibility.

Second, a concept can be virtuous by being appropriately insensitive to details that are unimportant in application. For example, whether *S* should count as a nutrient should be insensitive to whether *S* is nutritive in its present condition or only after further chemical changes that will occur during cooking or digestion. An important special case of this principle is that it is virtuous for one's postulated conceptual relations to harmonize with each other and exhibit *graceful degradation* when the applicability of one concept breaks down. For example, an empirical analysis of food ought to be compatible with the observation that there are borderline cases of nutrients and cases where a substance is slightly nutritional in one respect yet slightly poisonous in another. An empirical analysis would be deficient if it required a definite binary fact of the matter about whether *S* is a nutrient even though classical logic requires *S* to be either a nutrient or not.

The importance of engineering the graceful degradation of concepts can be illustrated in terms of the conserved quantity (CQ) account of causation (Dowe 1992a, 1992b, 2000), which postulates that causal interaction requires the transfer of some conserved quantity. Suppose it turns out that the actual fundamental laws ensure that quantities like energy, momentum, and angular momentum are always conserved so that the CQ theory is applicable to the actual world. We can ask about what would have been true about causation if the laws of nature were very slightly adjusted so that conserved quantities became very nearly but not perfectly conserved in a way that preserved all the actual world's macroscopic regularities, the regularities that give us a good reason to believe in causation. It is a consequence of the CQ theory that there would be no causation in such a world. That result by itself may be acceptable, but we have a right to expect the CQ theory to provide some account of why the complete lack of causation coexists with the vast evidence we would have for causation. According to my version of empirical analysis, the provision of such a story needs to be part of the CQ theory's explanation of why the conservation of quantities matters. It is unacceptable for the CQ theorist to balk that because conservation holds in the actual world, consideration of alternative worlds where it does not hold is irrelevant to the study of causation in the actual world. It is relevant to the analysis because if causation requires truly conserved quantities, then it becomes mysterious how we could ever become aware of causal connections, for we are likely not in a position to tell whether nature's quantities are perfectly conserved or just very nearly conserved. The CQ theorist needs to provide some explanation of how we could have epistemological access to causal relations. One schema for a proper explanation would involve demonstrating that our evidence for the existence of causal relations depends on how closely a universe obeys a conservation law. If it could be shown that putative evidence for causation becomes progressively weaker as violations of conservation laws accumulate in number or magnitude, then the CQ theorist could argue that even though worlds with only close approximations to conserved quantities have no genuine causation, we are reasonable to interpret them as having causation when they obey the appropriate conservation laws so far as we can tell. I am not contending that this particular explanation is satisfactory for CQ accounts, only that some story needs to be given about how breakdowns in the applicability of the concepts used in the empirical analysis relate to breakdowns in the applicability of the target concept.

Third, empirical analysis appears to presuppose a distinction between that which is empirical and that which is not. If this distinction is too narrowly construed, problems arise. In a vast array of examples, things we naïvely take to be unproblematically observable turn out to be characterizable only in theoretically loaded language. Also, we can often shift seamlessly between what counts as observed and what counts as inferred. When I claim to see a sheep on the hill, am I seeing a sheep or am I seeing half of a sheepish surface and inferring the rest of the sheep, or am I seeing a colored patch and inferring from that? There appears to be quite a bit of flexibility in how we can answer that question. In order to bracket concerns about how principled the concept 'empirical' is, I intend 'empirical analysis' to be understood not to presuppose a determinate fact of the matter about which items are genuinely empirically accessible. Instead, we should require a successful empirical analysis to make claims that are suitably insensitive to any indeterminacy concerning the empirical. This bracketing will not answer any probing epistemological questions, nor will it ensure the existence of a sufficiently principled empirical basis for adjudicating among competing empirical analyses. However, such deferral is common throughout science. So, to any worries that my empirical analysis of causation requires an unreasonably clear distinction between the empirical and non-empirical, my response is simply that my implementation of the distinction is no different from what is employed throughout science.

Fourth, a principle that is crucially not a part of empirical analysis is a preference for the intension of the analyzed concept to coincide with folk opinions about paradigm cases. It is literally of zero importance for an empirical analysis that paradigm instances of food count as nutrients. If the candidate intension for 'nutrient' happens to count bread as a non-nutrient, that by itself does not count as a deficiency, no matter how strong our pre-theoretical commitment to the proposition that bread is a foodstuff, and no matter how large a fraction of the general public supports the proposition that it is obvious that bread is food, and no matter how many professors one can summon to assert expert testimony that, a priori, bread is food. There does need to be enough of a semantic connection between paradigm cases of food in aggregate and the intension of 'nutrient' so that one cannot pass off an unrelated concept as a theoretical refinement of 'food'.[1] Nevertheless, the connection between the consequences of the theoretical refinement and the original platitudes can permit abundant disagreements without at all detracting from the quality of the conceptual regimentation.

In order to get a better grasp of this contrarian principle, it may help to consider the concept of rotation. Anyone interested in understanding the rotation of material objects is well served by group theory, the branch of mathematics designed to characterize symmetries. There is a group, for example, that represents the relations between all the possible rotations an object can undergo in a two-dimensional plane around a single point. The members of this group can be represented by real numbers. The number θ corresponds to a counter-clockwise rotation by θ radians. Negative numbers correspond to clockwise rotations, and the zero rotation corresponds to no rotation at all. If we were to apply the principle that a conceptual analysis of rotation should make explicitly true those propositions that are analytically true of our folk concept, then we would need to judge that the group-theoretical conception of rotation is in some measure deficient because it counts a rotation of zero as a bona fide rotation. What could be a more paradigmatic non-rotation than something that rotates a zero amount? The reason

[1] One could argue that CQ theories of causation are inadequate for this reason, for contemporary versions do not adequately explain why the transmission of conserved quantities is relevant to the causal principles successfully used in the special sciences and in everyday life.

zero rotations are included in the group-theoretic concept is that it greatly simplifies the definitions and theorems concerning relations among different kinds of rotation. For example, we would like to be able to say that the composition of any two rotations is itself always a rotation, but we cannot state that claim with optimal simplicity if zero rotations are forbidden because a rotation by θ and then by $-\theta$ amounts to a net non-rotation. Mathematicians understand the zero rotation as a trivial rotation rather than something that is not a rotation at all. This respectable attitude is strikingly at odds with the kind of conceptual analysis typically assumed in modern discussions of the metaphysics of causation. It is often taken for granted that events do not cause themselves and that a satisfactory metaphysics of causation needs to accord with this truth by not making it explicitly true that every event causes itself. According to the standards appropriate to empirical analysis, however, it is perfectly acceptable for a metaphysics of causation to ensure that every event causes itself. One can dismiss the importance of this counterintuitive result merely by recognizing that self-causation is a trivial form of causation.

For future reference, I will continue to use the word 'explicitly' in statements of the form, "Theory *T* (or model *M*) makes *P* explicitly true," to communicate that *T* (or *M*) suffices for *P* in the most straightforward interpretation of its claims, setting aside adjustments for language pragmatics. I have just provided two examples of what I mean by 'explicitly' in this context. The mathematician's regimentation of 'rotation' in terms of group theory makes explicitly true that an object at rest is undergoing rotation. One regimentation of 'cause' I will advocate makes explicitly true that every event causes itself. In an empirical analysis, it is no knock on a theory that it makes explicitly true claims we know are false because such discrepancies can be harmlessly explained away in terms of language pragmatics.

Although empirical analysis is largely an activity of regimenting concepts that can be conducted from one's philosophical armchair, the ultimate goal is not the investigation of language or thought but finding the best scientific theory one can. In an empirical analysis of food, the data one seeks to systematize are primarily all the statistical correlations between an animal's biological condition together with what it ingests and its later health condition, but other kinds of data are also relevant. What ought to concern us is learning about robust regularities in these data. The system of concepts provided by an empirical analysis plays a housekeeping role, keeping the conceptual system functioning as efficiently as feasible. Although empirical analysis is subservient to science, that does not trivialize the activity of finding an adequate empirical analysis. For one thing, trying to optimize one's conceptual scheme can play an instrumental role in science. It can raise possibilities that would not otherwise be entertained and can identify some issues as pseudo-problems. My explanation of causal asymmetry in chapter 7 provides an instructive example. For another thing, as Wilfrid Sellars (1962) put it, philosophy aims to find out "how things in the broadest possible sense of the term hang together in the broadest possible sense of the term." Understanding how things hang together is largely a project of conceptual engineering.

1.1.1 THE DISTINCTIVE FEATURES OF EMPIRICAL ANALYSIS

In order to highlight the novel features of empirical analysis, I will now contrast it with what I will call 'orthodox conceptual analysis', or just 'orthodox analysis' for short. Unfortunately, owing to space limitations and the inherent difficulty of criticizing the murky methodology of orthodox analysis in a manner sufficiently resistant to misinterpretation, I can only comment briefly. I have engaged this topic previously in (Kutach 2010), and I have made additional commentary publicly available for interested readers to follow up on this topic in more detail.

A conceptual analysis of *X*, as I will understand it here, is a systematization of the platitudes that constitute our implicit concept of *X*. To conduct a conceptual analysis, one starts with some initial data in the form of uncontroversial truths about the concept, including exemplars of the concept as well as a priori links to other concepts. For example, a conceptual analysis of food would begin with propositions that an orange is food, a hoagie is food, etc., as well as with broader principles that food is the kind of thing people typically like to eat, the kind of thing that relieves hunger, a kind of material substance, a kind that is species-relative, etc. One then attempts to formulate a reasonably small set of principles that (perhaps with some innocuous auxiliary truths) implies a set of claims that comes close enough to matching the initial platitudes. This set constitutes the completed conceptual analysis. It is understood that such a conceptual regimentation can be acceptable and even exemplary even when it rejects the truth of some of the initial platitudes. Indeed there are a wide variety of stances one can take on which kinds of discrepancies between the consequences of the completed conceptual analysis and the initial platitudes are permissible for the conceptual analysis to count as successful.

Some philosophers steeped in the naturalist tradition may think that conceptual analysis has long ago been abandoned, and they would be correct insofar as we understand conceptual analysis narrowly in its old-fashioned forms like Curt John Ducasse's (1926) attempt to define the causal relation. But the more liberal versions described by Collins, Hall, and Paul (2004) are currently in widespread use and have been prominently defended (Jackson 1998).

What makes a conceptual analysis *orthodox*, as I understand it, is the lack of any further systematic method (beyond custom, personal preference, appeasing journal referees, and the like) for adjudicating which discrepancies are acceptable for a satisfactory analysis and assessing the relative merits of competing analyses that differ in how well their accounts match the target platitudes. The implicit conditions of adequacy for orthodox analysis vary quite a bit among those who practice it, but a recurring feature of debates over whose analysis is adequate is the lack of any analysis that perfectly fits the initial platitudes and a proliferation of seeming stalemates among partially successful theories.

There are numerous examples in the philosophical literature on causation where two competing theorists appear to agree on all the relevant facts but disagree on how to incorporate them into an orthodox analysis. One good example is the transitivity of causation. In a scenario to be discussed later, Jane removes her food from a bear box and thus causes her food to be left in the open where bears can get it. Her food being left out causes Jill to recognize the danger and put the food back in the bear box. But Jane's removing the food from the bear box intuitively does not cause the food to be in the bear box later. Some investigators respond by denying the transitivity of causation. Others maintain the transitivity of causation and insist that Jane really does cause the food to be in the bear box but that we do not ordinarily identify such cases as causation because of pragmatic factors. Both sides can agree on all the relevant facts—which interactions occur, what the relevant laws are, which events affect the probabilities of other events—but still disagree about whether Jane's removal of the food was a cause of the food being in the box later. Chris Hitchcock (2003) has conveniently provided a discussion of many such quandaries for orthodox analyses of causation. One upshot is that orthodox conceptual analysis does not prescribe any discernible guidance for adjudicating such disputes and for assessing the relative importance of each initial platitude.

Empirical analysis is crucially different by incorporating specific additional methodology to *guide* movement away from the initial platitudes. Empirical analysis takes the platitudes concerning *X* as a starting point for identifying empirical phenomena, especially by formulating explicit experiments whose results clarify why *X* has some conceptual utility. Then, one's goal is to seek a scientific explanation for the results of such experiments, honing the concepts used in the explanation as much as needed to improve its overall quality, including how it comports with other background theories we accept. Whatever concepts result from this optimization constitute the completed empirical analysis of *X*.

An empirical analysis often results in some of the original platitudes being discarded as irrelevant to the analysis, and the final regimented concept is not to be evaluated in terms of what fraction of the platitudes it makes explicitly true. While orthodox analyses continue to be tethered to some extent to the initial platitudes by always being evaluated in the end in terms of the magnitude and severity of discrepancies with the initial platitudes, the method of empirical analysis encourages us to abandon the platitudes whenever making them explicitly true would result in a suboptimal conceptual scheme.

To encapsulate, empirical analysis may be given the following formal definition. The *empirical analysis* of *X* is the engineering of a conceptual framework optimized in the service of the scientific explanation of whatever empirical phenomena motivate our possession of a concept of *X*, especially insofar as they are characterized in terms of experiments.[2]

[2] In claiming that an empirical analysis of *X* addresses empirical phenomena motivating "our possession of a concept of *X*" I am referring to whatever concept (or perhaps concepts) we possess *before* we improve our conceptual scheme scientifically. Our rough and ready folk conception of *X* can be

It ought to go without saying that scientists have been engaging in the activity of empirical analysis for centuries. In fact, the only argument I offer for empirical analysis being an acceptable form of conceptual analysis is that it has been conducted by scientists in countless instances, and its successful applications have greatly contributed to our understanding of reality. In this sense, there is nothing new about my approach to causation.

It also ought to go without saying that philosophers have long recognized that traditional forms of conceptual analysis include a role for scientific inquiry. Sometimes this idea is expressed as the claim that meanings are not entirely "in the head" (Putnam 1975), an observation illustrated by natural kind terms like 'water'. Not everything that behaves like paradigmatic water is water. Only substances of the same chemical kind as the chemicals that predominate among most paradigmatic instances of water in our local environment count as water. A non-H_2O chemical on the other side of the universe that behaves superficially like water is not water. That we can recognize this feature of our 'water' concept from our philosophical armchairs demonstrates that sometimes the intension of a concept depends on the external environment. Again, this is all well known; the mere incorporation of scientific discoveries into a conceptual analysis is also not a novel feature of empirical analysis.

What is new about empirical analysis—to philosophers, as far as I can tell—is the starring role it casts for explicitly characterized experiments. Later in this volume, I will attempt to construct three experiments in an effort to characterize the empirical phenomena giving us some reason to believe in causation: the promotion experiment, the backtracking experiment, and the asymmetry experiment. Two more examples of how empirical analysis relies on characterizing experiments can be found in (Kutach Forthcoming).

To grasp the crucial role of experiments in an empirical analysis, we can again consider the investigation of food. In an empirical analysis of food, one should attempt to describe a general experiment that captures the empirical phenomena that make 'food' a concept worth having. The following experimental schema, I think, serves reasonably well for a simplistic illustration. One chooses some type of creature, *C*, some type of edible material *M*, and some type of environment *E* for the creature to inhabit during the study. Then, one conducts an experimental run by having a chosen creature ingest a substance and measuring its health outcomes after its stay in the environment. After zillions of such experimental runs for a wide range of creatures, materials, and environments, one will have collected data

understood as having a very low threshold for being motivated or useful or worth possessing. One might say that any concept in regular use very likely has *some* value and is thus worth possessing, for otherwise it probably would have been abandoned. It is certainly possible that scientific investigation or empirical analysis will justify abandoning the folk concept or replacing it with alternatives. Contrary to Peter Godfrey-Smith's (2012) suggestion, empirical analysis does not require that our initial concept *X* retain its utility after we have engineered superior replacements for them.

that can be summarized as a function from these three variables to a set of health outcomes. The results of such experiments presumably verify that humans eating sand and nothing else for a month results in bad health outcomes while eating vegetables, nuts, and fruits mostly results in good health outcomes. The reason our food concept is worth having is that there are robust regularities where certain kinds of ingested materials significantly improve health outcomes relative to other materials. We use 'food' primarily to track these nutrients, and completing the empirical analysis of food requires us to hone 'food' more precisely (using the label 'nutrient', if desired, to avoid the usual connotations of 'food') so that it fits better with everything else we know about physiology, chemistry, disease, and related subjects.

Because there are secondary factors bearing on our use of 'food' like its role in social encounters and its aesthetic qualities, there are mismatches between our judgments about which substances are food and what our science identifies as nutritional. The secondary factors bear on empirical analysis only through a much different set of empirical phenomena: primarily, people's reports about what they consider to be food. These empirical phenomena can be encapsulated in terms of an experiment, a decent first approximation of which would simply involve presenting a sample of material M to a human subject C in environment E and ask, "Is this food?" You could augment the experiment by also testing whether the person eats the substance or serves it to others at dinner, but the basic idea is to test not the bodily effects of the ingested substance but instead how people think of it, talk about it, and use it socially. The data collected from such experiments would constitute empirical phenomena concerning how we conceive of food, and this concept can be made more precise in order to explain why we have the intuition that microbes and humans and aspirin tablets are not food.

The first kind of empirical analysis is typically of much greater philosophical importance because it bears directly on the character of reality generally rather than focusing on how we conceive of it. My main reason for discussing the second kind of empirical analysis is to avoid alarming readers who insist there must be some place in our conceptual scheme for widely shared and strongly held intuitions about important concepts like causation. The intuitions that are properly ignored in an empirical analysis of the metaphysics of causation always have a proper home in this second more psychologically oriented empirical analysis. (Readers who are pressed on time and are untroubled by the fact that my metaphysics of causation does not address several folk intuitions that are traditionally construed as data for a metaphysical investigation of causation should be able to skip chapters 8 and 9.)

An exploration of food using the methodology of empirical analysis thus leads naturally to two somewhat separate investigations: identifying experiments that capture the nutritional aspects of food and identifying experiments that capture the social and psychological aspects of food. This sort of bifurcation happens quite generally when the method of empirical analysis is applied. In effect, empirical analysis attempts to provide with two analyses what an orthodox analysis attempts to accomplish with a single analysis. Typically, an orthodox conceptual analysis of some *X* starts with a set of platitudes concerning *X*, some of which are partly constitutive of the meaning of *X* and some of which are known from empirical investigation. Conducting an orthodox analysis of *X* consists in systematizing all these platitudes concerning *X* together as a single group by identifying a cluster of principles that describe what *X* is in terms of other concepts. In conducting an empirical analysis, by contrast, it usually works out that the original platitudes are best segregated into two groups, those bearing on *X* insofar as it is something "out there in reality" and those bearing on how we think about *X* in ways that go beyond the empirical phenomena in the first group. Then, one conducts two distinct regimentation projects.

Application of the methodology of empirical analysis to causation results in a natural bifurcation into a pair of empirical analyses. The first empirical analysis is more focused on causation insofar as it is something "out there in reality." I will refer to this investigation as the *empirical analysis of the metaphysics of causation*, or sometimes just the *metaphysics of causation*. The second empirical analysis focuses on how we think about causation in ways that go beyond the empirical phenomena addressed by the metaphysics of causation. I will refer to this investigation as the *empirical analysis of the non-metaphysical aspects of causation*. This second empirical analysis will subsume the psychology of causation as well as the subset of epistemology that encompasses the explanatory role of causes and causal modeling. Because this volume is primarily concerned with the metaphysics of causation, this second empirical analysis will receive far less attention from me than the first. My goal will be limited to sketching very briefly a few of its components just to assure readers that the topics it concerns are not being denigrated by my metaphysics of causation or entirely ignored but are merely being reorganized in a way that categorizes them as part of the special sciences themselves rather than as part of metaphysics. This conceptual division is not made for the sake of presentation but for the purpose of assigning to each empirical analysis the standards of theoretical adequacy that are appropriate to it.

Before further clarifying how I think of the proper standards of theoretical adequacy, I will provide some additional detail about the two empirical analyses of causation.

1.2 Empirical Analysis of the Metaphysics of Causation

The purpose of this section is to sketch how the general methodology of empirical analysis will yield my account of the *metaphysics of causation*. Later in §1.10, I will further clarify the character of this empirical analysis by defending an important restriction on the proper scope of metaphysics that will have the crucial consequence of greatly winnowing the sorts of explanation that are appropriate for my investigation of causation. So, I caution readers again not to be hasty in thinking they have fully grasped the essence of empirical analysis based on what I have stated so far.

1.2.1 EFFECTIVE STRATEGIES

A mundane but instructive observation about causation is that it generalizes a wide variety of other concepts like digestion, photosynthesis, rusting, erosion, gravitation, and combustion. One might say these are all species of causation. For each one, there exist conditions or events that are reliably connected to other conditions or events. Where there is abundant dry wood, plenty of oxygen, and a small fire, there will often be a larger fire shortly afterwards. Moreover, such causal regularities are largely insensitive to many other events. There are no remarkable relations between fires and the remote existence of goats or dirt or boron. The lack of notable connections between fire and so many other conditions is partly what makes fuel and oxygen noteworthy vis-à-vis fire.

My particular empirical analysis of the metaphysics of causation attempts to unify our understanding of such connections by concentrating initially on the following entirely unoriginal seed of an idea: *causes are means for bringing about certain kinds of effects*. The label drawn from Nancy Cartwright (1979) for this focus of causal talk is 'effective strategies'. It is empirically verifiable that combining dry wood, oxygen, and some source of heat is a good strategy for creating fire, whereas dunking an ordinary rock in water is demonstrably an ineffective method for starting a fire. The empirical analysis provided by my theory is aimed at facilitating explanations of why there is a regular pattern of events demonstrating that some strategies for affecting the world are better than others. 'Effective strategies' is the name for this empirical content of causation. The effectiveness of a strategy is testable (to a first-order approximation) simply by acting on the strategy a bunch of times, acting on alternative strategies a bunch of times, and observing whether the desired effect occurs more often after using the designated strategy than after using the alternatives.

It will take some work to unpack what 'effective strategies' ultimately amounts to and to ensure that the resulting empirical analysis makes sense of causation that does not involve strategies. This work will not be completed until chapter 5, but I can make a few preliminary comments here.

The attention placed on effective strategies is merely an educated guess about where to begin an exploration of the phenomena we pre-theoretically associate with causation. Nothing about this choice forecloses the possibility that other phenomena associated with causation can be incorporated or prevents an entirely different starting point from leading to a fruitful empirical analysis. Furthermore, nothing ensures that the totality of empirical phenomena relevant to the metaphysics of causation will form a cohesive collection in the end. It might turn out to be conceptually optimal to segregate the empirical phenomena into multiple distinct clusters having little to do with one another. Whether we should explain the empirical phenomena as a cohesive unit or instead as a patchwork of distinct groups of phenomena is not a matter to be decided in advance. What we can say initially is that it makes sense to investigate phenomena that appear to make sense of why we have causal terminology, and that seems to me to be captured in large part by the principle that some happenings are effective at bringing about other happenings of a certain kind. So long as this conception of the empirical focus of causal talk fits comfortably within a suitably broad construal of the empirical phenomena relevant to the metaphysics of causation—including causal regularities having nothing to do with agency—it will not matter that this preliminary choice is somewhat arbitrary.

A key point to keep in mind is that 'effective strategies' is not an expression to which I am attributing any technical meaning. It merely stands for the pre-theoretical idea that some strategies for achieving desired goals are reliably better than others. So one important constraint on the content of 'effective strategies' is that it not take strategies too seriously metaphysically. Naïvely speaking, for a strategy to exist, there needs to be some agent reasoning about how to accomplish a goal, but for the purpose of explaining causation we want to avoid assuming that for causation to exist, there needs to be agency somewhere in the universe. Similarly, we also need our resulting investigation of causation to accommodate the existence of intermediate or borderline cases of agency in a way that exhibits graceful degradation.

A delightful ambiguity in the expression 'effective strategies' is that it suggests enough of a difference between accidental and law-like regularities to substantiate our conviction that causation is more than mere happenstance while not insisting that the reliable effectiveness of some strategies requires some empirically inaccessible non-accidentality. On the one hand, there can be accidental regularities that should not count as causal. On the other hand, if we assume from the beginning that the facts to be explained are precisely the set of non-accidental regularities, that would raise the question of how we could know whether a regularity is accidental or not. We would no longer have uncontroversially empirical phenomena as explananda. A salutary feature of empirical analysis is its compatibility with a flexible distinction between what is accidental and what is enforced by law, avoiding both extremes in an account of the empirical content of causation. This flexibility does not prevent us from invoking a distinction between law-like and accidental in our explanation of the empirical content; it just avoids *requiring* the distinction in order to make sense of the empirical content. To illustrate by analogy, a biologist should not adopt the task of explaining why creatures with souls behave intelligently but instead why creatures that seem to behave intelligently are able to. An explanation for intelligent behavior might postulate a soul, but to assume the soul in the first place would leave unclear whose behavior requires explanation.

'Effective strategies' suggests a flexible distinction between law-like and accidental by encouraging us to think of situations where an agent is selecting or controlling circumstances in order to bring about some desired effect. If a regularity holds even when an agent tests it in numerous circumstances, that potentially counts as strong evidence that the regularity is suitably law-like. No number of test situations will ever ensure that the regularity is law-like, but the fact that we are able to manipulate the world in order to test regularities means that if we account for a pattern of regularities in circumstances where people are trying to test them for potential violations, we in effect account for why there exists a pattern of regularities that look as if they hold by virtue of laws.

One collection of phenomena subsumed under the umbrella of 'effective strategies' is that across a wide range of different kinds of events, materials, and circumstances, there exist exploitable regularities where one type of event *C* is a good means for bringing about an event of type *E*. But there are also important *general* features of such regularities. Two in particular stand out.

The first is that it is seemingly impossible to exert influence in one direction of time to an event and then back in the opposite direction of time to another event. These are *backtracking nomic connections* because they first go in one temporal direction and then backtrack in the opposite temporal direction. (Warning: many philosophers use 'backtracking' misleadingly to refer only to the past-directed half of the backtracking.) The reason someone might hypothesize that nomic connections could backtrack is that frequently the occurrence of one kind of event is correlated with the occurrence of two kinds of effects in its future. Throwing a rock into a pond leads lawfully to a distinctive kerplunk sound and expanding ripples. One might wonder why it is not possible to increase the chance of ripples by making a kerplunk sound. Any such strategy, I think, is demonstrably ineffective except to the extent it exploits a future-directed strategy such as tossing a rock in the pond. *Causal directness* is this (seemingly correct) principle that a backtracking nomic connection between two events never does anything beyond what it already does by virtue of temporally direct nomic connections.

The second (and closely related) general feature of effective strategies is that there are apparently no effective strategies for influencing the past in useful ways. The empirical phenomenon associated with this claim can be roughly characterized as follows. People who are assigned the task of bringing about some future outcome—like writing a haiku or baking bread or establishing a viable human colony on Pluto—are sometimes able to accomplish that task at a significantly higher rate than people who are trying to avoid having that kind of outcome occur. But people who are assigned the task of bringing about some past event of type *E*—no matter what *E* is—never do any better or any worse (on the whole) at having an instance of *E* occur than people who are trying to prevent instances of *E*. Call this phenomenon the *asymmetry of advancement*. We can advance some of our goals for the future but never our goals for the past.

The broad aim of my metaphysics of causation is to provide a conceptual structure optimized for explaining the empirical phenomena associated with effective strategies, understood broadly to include causal relationships that involve no agency. Along the way, in §5.8, I will provide a skeletal explanation of why effective strategies exist across a wide range of activities and an explanation of why causal directness holds and why there is an asymmetry of advancement. There are certainly other features of effective strategies that I will consider, but the upshot of an empirical analysis of the metaphysics of causation is as follows. If an optimal, or at least adequate, set of concepts can be developed that help to explain (in a "complete story" sense of explanation) the empirical phenomena behind effective strategies (broadly construed), then the *metaphysics* of causation will be largely solved. All there is to understanding the metaphysics of causation in the sense relevant to empirical analysis is just understanding how these empirical phenomena are related to fundamental reality. As I will discuss in chapters 5 and 10, no one is currently in a good position to provide an explanation of all the details related to effective strategies, and no one is currently in a good position to provide an adequate comprehensive theory of fundamental physics (much less, fundamental reality). But the *empirical analysis* of the metaphysics of causation does not require that we explain everything about effective strategies; it merely requires a justification of the conceptual architecture that connects the empirical phenomena to fundamental reality. I hope readers will judge that the system of concepts that I will soon introduce are flexible enough to be applicable to a wide range of ways fundamental reality could be structured and to be applicable to any causal regularity, but also inflexible enough to facilitate non-trivial empirical predictions.

1.3 Empirical Analysis of the Non-metaphysical Aspects of Causation

My empirical analysis of the metaphysics of causation addresses causation insofar as we want to tailor our understanding of causation to structures "out there in reality" that are not very closely tied to how we think about causation. But there is a second empirical analysis that is adapted to address further empirical phenomena concerning how we conceive of causation, including causation's role in explanation and the discovery of causal regularities. I will refer to this second investigation as the *empirical analysis of the non-metaphysical aspects of causation*. The purpose of this section is to clarify how the general methodology of empirical analysis applies to the aspects of causation that go beyond the scope of metaphysics, as precisified in this chapter. Its prominent components include (1) the psychology of causation, (2) the role of particular (or token) causes in the explanatory practices of the special sciences, and (3) causal modeling that is sufficiently remote from (or insulated from) the character of fundamental reality.

First, the subfield of psychology dedicated to exploring how people think of causation produces models that attempt to explain uncontroversially empirical data, including people's reactions when they are told stories or shown a sequence of events and asked, "What do you think caused this event?" or "Does event *c* count as one of the causes of event *e*?" This psychology of causation is also meant to be compatible with related phenomena such as how long children look at certain temporal sequences designed to mimic or violate default rules of object behavior, how people conceive of the operation of gadgets, how people attempt to solve mechanical puzzles, etc. Like other scientific theories, these psychological theories can be formulated using a technical vocabulary, distinguishing foreground and background causes, proximate and distal causes, actual and potential causes, etc.

One result we should expect from such a psychologically oriented empirical analysis is that its structures will almost certainly be significantly different from the structures of an empirical analysis aimed at explaining effective strategies. This is easy enough to motivate by virtue of the general pattern whereby a theory of *X* often looks very different from a theory of the psychology of *X*. A scientific theory of space, for example, is aided by having esoteric mathematical structures like manifolds and curvature tensors; a scientific model of how humans naturally think about spatial relations is sure to exclude curvature tensors in favor of structures that better represent the portion of our cognitive processing that manipulates spatial information.

We know enough about our psychology to recognize that humans use various heuristics to understand causal connections in the external world in a simplified way. Because people are poor reasoners about fantastically small probabilities, we should expect them to oversimplify causal relations that involve minute probabilities. People have limited capacity in their working memory, so we should expect them to ignore some causes when a vast multitude of causes are present. There is no a priori reason why the scientific conception of causation *must* differ from our implicit pre-theoretical conception of causation, but it should not be even remotely surprising. More important, there is no reason to assume from the outset that there must be some interesting causation concept simultaneously optimized for both explaining the core phenomena behind causation itself (as some relation out there in reality) and explaining regularities concerning our instinctive causal judgments.

Second, although philosophers do not ordinarily consider causal explanation a topic in psychology, I will discuss in chapter 8 a sense in which disputes about causal explanation—over which individual events explain a particular effect—are psychologically oriented to the extent that they go beyond the "complete story" explanations afforded by the totality of causal relations in my metaphysics. If two people agree on the fully detailed account of how the effect *e* came about by agreeing on all the relevant laws and how they connect the complete arrangement of every last bit of matter, then any further disagreement about which partial causes are *explanatory* cannot be adjudicated by reference to further empirical data about the events leading up to *e* because there would be no further empirical data.

The only extent to which such a debate can be informed by reference to further empirical data would come from investigations of people's explanatory practices, including their revealed explanatory norms. In this limited sense, empirical analysis treats causal explanation like it treats the psychology of causation.

Third, there are many invocations of causation in the special sciences, especially the practice of discerning causal relations from statistical correlations. This includes the scientific and philosophical literature on causal modeling (Spirtes, Glymour, and Scheines 2000; Pearl 2000; Woodward 2003). Much of this scientific activity can be understood without any particular connection being drawn to fundamental reality. As such, these investigations of causation do not count as metaphysical in the framework I have adopted; they count as part of the special sciences, to be addressed by an empirical analysis of the non-metaphysical aspects of causation.

Finally, we should expect an empirical analysis of the metaphysics of causation and an empirical analysis of the non-metaphysical aspects of causation to be related in a fairly straightforward way. The reason people have a concept of causation is that it provides an efficient way to conceptualize those structures responsible for things behaving causally. The metaphysics of causation directly addresses why things "out there in reality" behave causally, why some kinds of events reliably bring about certain other kinds of events. The other empirical analysis addresses how the structures posited in the metaphysics might be simplified for cognitive consumption, paying special attention to people's need to learn about effective strategies and apply them to new circumstances. We ought to suspect our psychology of causation will match, to a first order approximation, the structures that ultimately account for the world behaving causally, but a second order correction would likely take into account our need for an efficient cognitive grasp of these structures. We also ought to suspect that our judgments about which causes are explanatory will fit into the general cognitive system that filters the vast plenitude of partial causes for events that have some cognitive salience. Expressed curtly, we have intuitions and practices for identifying certain partial causes as explanatory as a by-product of their role in our cognition, especially by virtue of our heuristics for learning about effective strategies.

Although I will sketch a theory along these lines in chapter 9, even these suspected connections are not inviolable constraints on the psychology of causation or our theories of causal explanation because the structures that explain the empirical phenomena associated with effective strategies might be so complicated or so remote from our epistemic access to reality that our cognition only makes contact with the metaphysics of causation through roundabout means.

To summarize these last two sections, the application of empirical analysis to causation results in two scientific investigations. The first explores the empirical phenomena related to causation as something "out there in reality," what ultimately becomes the metaphysics of causation. The second explores further aspects of causation that are based on how creatures think about causation. This bifurcation is analogous to how a scientific investigation of food can be divided between an investigation of the nutritional aspects of food and an investigation of food that goes beyond its nutritional aspects to its social role and our usage of the word 'food'. Because the primary goal for this volume is to provide a scientific metaphysics of causation, I will only discuss the empirical analysis of the non-metaphysical aspects of causation in order to show how it relates to the metaphysics of causation and to illustrate how some traditional philosophical problems concerning causation can be resolved when they are properly situated outside the scope of metaphysics.

1.4 Causation as Conceptually Tripartite

Now it is finally time to turn our attention to the structure of my theory of causation. I will initiate the discussion by explaining how the concept of causation should be divided into three stacked layers: bottom, middle, and top. The bottom and middle layers are relevant to the metaphysics of causation whereas the top layer pertains to the non-metaphysical aspects of causation. Then, I will draw a distinction between fundamental reality and derivative reality and describe how the bottom conceptual layer of causation concerns fundamental reality whereas the middle and top layers concern derivative reality. Last, I will draw a distinction between two different sets of standards for evaluating theoretical adequacy, STRICT and RELAXED, and I will defend the thesis that one's metaphysics of causation, the bottom and middle layers, should be evaluated according to STRICT standards whereas an empirical analysis of the non-metaphysical aspects of causation, the top layer, can be entirely adequate even if it only satisfies the more permissive RELAXED standards.

Philosophers who have weighed in with positive theoretical accounts of causation have often focused on a single proffered core aspect of causation—determination by the laws of nature (Mackie 1973), counterfactual dependence (Lewis 1973b), probability-raising (Suppes 1970), transference of some privileged sort of physical quantity (Salmon 1977, Kistler 1999, Dowe 2000)—and have tried to show how legitimate causal claims are vindicated primarily in terms of that one core aspect, whether they involve a magnetic field causing an electron to accelerate or an increase in literacy causing a redistribution of political power. Other theories (Good 1961, 1962, Sober 1985, Eells 1991, Salmon 1993, Hall 2004) are conceptually dual in the sense that they try to make sense of causation in terms of two core causal concepts that operate mostly independently of one another. My own account models causation in terms of three distinct but related conceptual layers.

That my analysis segregates the concepts it constructs for understanding causation into three layers rather than one, two, or forty-seven is not in itself particularly noteworthy. There is no prima facie reason to expect a three-layer account to be superior to a dual or quadripartite account. The tripartite decomposition is merely the result of a natural division of labor concerning what a theory of causation should rightly be expected to accomplish. The two principles that divide the three conceptual layers are these:

1. There is an important metaphysical distinction between that which exists fundamentally and that which does not.
2. There is an important methodological distinction between how one should evaluate a theory of the metaphysics of causation and a theory of the non-metaphysical aspects of causation.

My account of causation thus divides the concepts it employs into three layers corresponding to (1) those appropriate to fundamental reality, (2) those appropriate to derivative reality insofar as it bears on the empirical phenomena associated with the metaphysics of causation, and (3) those appropriate to derivative reality insofar as it bears on the empirical phenomena associated with the non-metaphysical aspects of causation. No single layer, by itself, contains a relation that deserves to be designated as *the* causal relation, but all three layers together constitute a collection of concepts that allow us to make adequate sense of everything regarding causation that needs to be accounted for.

The consequences of this tripartite division are significant and set apart my theory from other existing accounts. Most accounts of causation maintain that there is a cause-effect relation between individual chunks of reality with certain distinctive characteristics. For one thing, the postulated cause-effect relation often holds among mundane objects or events or facts,[3] like a cloud casting a shadow or a virus provoking an immune response, rather than holding only between spatially expansive and microphysically detailed states. For another, the cause-effect relation is normally taken to be irreflexive because it is believed that effects do not cause themselves. Finally, the cause-effect relation is also thought to be non-symmetric because effects do not cause their causes, except perhaps in special circumstances like a time travel scenario. On my account, this crude cause-effect relation has no place in the metaphysics of causation but is suitable for the top layer where the epistemological and psychological roles of causation are properly situated. Relocating the cause-effect relation out of the metaphysics serves to dissolve a large number of problems philosophers routinely assume need to be resolved decisively by any adequate account of causation.

[3] Existing theories vary greatly in their metaphysical account of the causal relata, e.g. whether they are events, property instantiations, aspects, processes, tropes, etc. They also vary in whether they include additional parameters. Causation might not just be a two-place relation between the cause and effect, but a three-place or four-place relation, where the extra parameters can be contrasts, processes, choice of causal variables or choice of causal model, etc. Despite all such differences, most existing accounts of causation are such that when all the additional parameters are filled in, the residual relation between cause and effect shares much of the logical character of folk attributions of causation.

TABLE 1.1 The Three Conceptual Layers of Causation

Layer	Subject	Metaphysical status	Standards of adequacy
Top	Non-metaphysical aspects of causation	Derivative	RELAXED
Middle	Derivative metaphysics of causation	Derivative	STRICT
Bottom	Fundamental metaphysics of causation	Fundamental	STRICT

The three layers are depicted in Table 1.1. One prominent difference among the layers concerns whether they apply to singular or general causation. *Singular* causation applies to cause-effect relations that occur in a single fragment of the world's history. Examples of singular causal claims include, "The collapse of the Tacoma Narrows Bridge was caused by wind," and "The cholera outbreak was caused by a contaminated well." *General* causation addresses the *kinds* of events that can cause some chosen *kind* of effect. Examples of general causal claims include, "Smoking causes cancer," and "Bribes encourage corruption."

The bottom layer addresses an extremely inclusive form of *singular causation* in terms of a theory of fundamental reality employing concepts like determination and probability-fixing. The middle layer addresses *general causation* in a way that abstracts away from the details of fundamental reality. The top layer addresses the less inclusive form of singular causation that people employ in everyday conversations and that scientists employ when giving causal explanations of particular effects. I call these singular causes 'culpable causes'.

These three conceptual layers exhaust the scope of my account of causation. The overall structure of this book—after the methodological issues have been dealt with in this chapter—is simply to fill in the details concerning the bottom, middle, and top layers. The concepts in each layer depend on the resources of the layers underneath, so it is wise to consider the layers from the bottom up.

My goal for the rest of this chapter is to demarcate the three conceptual layers in greater detail. First, I will provide a simplistic overview of my account of causation. Next, I will elaborate on the distinction between fundamental and derivative in order to make clear how the bottom conceptual layer of causation differs from the two layers above it. After that, I will unpack my distinction between STRICT and RELAXED standards of theoretical adequacy, which separates the top layer from the two layers beneath it. Finally, at the end of this chapter, I will return to the three conceptual layers of causation in order to recap how they relate to singular and general causation.

1.5 A Sketch of the Metaphysics of Causation

Before I explicate the more idiosyncratic elements of my overall account, it will likely be useful for me to summarize simplistically how my account will eventually help explain how causes are effective at bringing about their effects. To keep the discussion manageable, I invite readers to accept provisionally that there are some fundamental laws of physics that govern the behavior of all particles and fields and that ordinary macroscopic objects are merely aggregates of these fundamental microscopic parts.

Imagine a magnetic compass lying undisturbed. By moving a lodestone near the compass, one can reliably make the compass needle move. It is uncontroversial and empirically verifiable that events of type *C*, moving a lodestone near a compass, are effective at bringing about events of type *E*, a jostling of the needle. What explains such phenomena, according to my account, is that there exists a fundamental reality that includes extremely detailed facts about how fundamental particles and fields are arranged as well as fundamental laws governing the temporal development of this fundamental stuff. The objective structure behind all causation is located in how the fundamental laws link the fundamental material stuff at different times and places. Specifically, some fundamental happenings *determine* the existence of other fundamental happenings or *fix an objective probability* for their existence, and that is what ultimately grounds all causal relations.

Yet, when we explore the character of plausible fundamental laws, we find good reason to believe that fundamental laws by themselves provide no connection between the highly localized event, *c*, constituting a particular lodestone's motion toward the compass, and the consequent jostling of the needle, *e*. At best, the fundamental laws connect *e* only with a much larger event c' that includes *c* as well as a complete collection of microphysical facts occurring at the time of *c* and occupying a vast expanse of space, perhaps stretching out to infinity. Puny events like *c* are too small to determine or fix any objective probabilities for events like *e*, but gargantuan events like c' can. The localized chunk of reality *c* only plays a fundamental causal role by virtue of its being a part of the much larger c'.

That story is fine as far as fundamental reality goes, but because we humans are unable to perceive microphysical states accurately enough, unable to reckon their nomic consequences accurately enough, and unable to control the world precisely enough, these fundamental relations are by themselves rarely useful to us in practice. Fortunately, our world is amenable to various approximations that allow us to represent aspects of fundamental reality in ways that abstract away from their precise character. Our belief that *c* caused *e* is in part a belief that the lodestone part of the world was somehow a more important part of the vast c' than all the far flung events that seemingly have nothing to do with the motion of the compass needle. What makes *c* the important part of c', I claim, is that the probability that c' fixes for the effect is significantly greater than the probability that would be fixed for the effect by events that are just like c' except that the physics instantiating the movement of the lodestone is hypothetically altered to make the lodestone remain at rest. The motion of the lodestone is causally important to the compass needle because it affects the needle's probability of moving.

The metaphysical picture, boiled down to its essence, is that there is some sort of fundamental reality instantiating relations of determination or probability-fixing among microscopically detailed events and a more abstract or fuzzy construal of reality where events of type *C* raise the probability of events of type *E*. This helps to explain the existence of effective strategies because, to a first approximation, an effective strategy for bringing about some instance of *E* is to bring about an event that raises the probability of *E*. My task in the rest of this book is to provide the resources to facilitate talk of probability-raising, determination, influence, etc. so that the details of this explanation can be specified in an acceptable way.

1.6 Fundamental and Derivative

The account of causation I present in this book crucially relies on a metaphysical distinction between fundamental and derivative. Most people, I think, have at least some intuitive grasp of the difference between fundamental and derivative, and for the purpose of understanding causation, we can mostly rely on that intuitive grasp. However, in order to focus the distinction a bit more, I will list a few guiding principles and then describe an example of how we can think of kinetic and thermal and mechanical energy as derivative properties that reduce (in a sense I will soon clarify) to fundamental attributes, for example, mass and relative speed. (Throughout this book, an *attribute* is a property or relation, broadly construed.) This example will serve as a template for my account of causation, clarifying how derivative aspects of causation can be related to fundamental aspects of causation.

I will attempt to characterize fundamental and derivative reality without introducing undue controversy, but because the distinction bears on broad principles of ontology, truth, and explanation, there will inevitably be plenty of room for disagreement about its ramifications. Because even my best effort to make precise my intended conception of fundamentality will suggest a conclusion that is objectionable to some readers, I want to emphasize from the start that for the purpose of applying the distinction between fundamental and derivative to causation, not everything I say about fundamentality is absolutely essential. My goal here is merely to establish an initial reference. Readers who disagree with me on a few points here and there should be able to grasp the gist of my distinction and translate it into their preferred terminology.

I think the easiest way to get a grip on what is fundamental and what is derivative is to start by thinking about reality in a rather naïve way as *existence*. Just consider everything that exists, including all objects, properties, relations, substances, and whatever else you think needs to be included. The totality of existents, including all their relations with one another, constitutes reality. Then, we can think of reality as subdivided into exactly two disjoint parts: fundamental and derivative. 'Fundamental' and 'derivative' at this point serve as placeholders for a distinction that is filled in by specifying the role that 'fundamental' plays.

I am now just going to present a list of some platitudes that appear to me to capture several constitutive features of fundamental reality:

1. The way things are fundamentally is the way things *really* are.
2. Fundamental reality is the only real basis for how things stand derivatively.
3. Fundamental reality is as determinate as reality ever gets.
4. Fundamental reality is consistent.

While these principles are admittedly vague and subject to philosophical objection, I think they provide a useful starting point for discussing fundamental reality.[4] I will attempt to specify them more precisely by laying out a specific example based on the concept of kinetic energy, which will serve as a reference for further clarification.

For our purposes, it will be helpful to simplify by operating provisionally under the pretense that every existent is either determinately fundamental or determinately derivative. After the basic distinction is clear enough, one can take up the project of evaluating the extent to which there is indeterminacy at the boundary between fundamental and derivative reality.

1.6.1 THE KINETIC ENERGY EXAMPLE

The theory of classical mechanics is a scheme for modeling how material bodies move around according to forces. I will focus on a specific interpretation of classical mechanics whose purpose is to clarify ontological commitments: *the simple theory of classical mechanics*. There are other ways to interpret the content of classical mechanics, but I am not engaging here with technical issues in the philosophy of physics or with historical exposition.

The ingredients of the simple theory of classical mechanics include a space-time inhabited by corpuscles bearing intrinsic properties like mass and charge. A *corpuscle* is by definition a point particle, meaning that it has an identity through time and occupies a single point of space at any given moment so that its history over any span of time is a path in space-time. Corpuscles in classical mechanics rattle around according to exceptionless laws where each corpuscle's acceleration is a relatively simple mathematical function of fundamental attributes like the inverse-square law of gravity and some sort of short-range repulsive interaction that makes corpuscles bounce elastically away from each other when they (nearly) collide. To be more specific, the simple theory posits the following structures: an appropriate space-time, corpuscles, charge and mass properties that adhere to the corpuscles, a distance relation between any two corpuscles at any given time, a relative speed relation between any two corpuscles at any given time, and a law

[4] I provide further discussion of these principles and their role in Empirical Fundamentalism in Kutach (2011b).

governing how these attributes evolve over time. The simple theory posits nothing else. Once all these entities and attributes have been everywhere specified, the entire world has been specified according to the simple theory.

We know that classical mechanics is not an accurate theory of our world, but for pedagogical purposes, it is convenient to consider how we ought to think about reality if it were true that the actual world perfectly matched one of the models of the simple theory of classical mechanics. For the rest of this section, discussion will proceed under the pretense that some model of the simple theory of classical mechanics is the complete and correct account of fundamental reality so that we have a concrete reference for understanding fundamentality. Having adopted the simple theory of classical mechanics, we can distinguish between fundamental and derivative in a fairly intuitive way. The corpuscles and space-time are fundamental entities, their relative distances and speeds are fundamental relations, their masses and charges are fundamental properties, and the laws governing them are fundamental laws. They are all fundamental existents. Poetry, patience, and financial assets, by contrast, are arguably non-fundamental. They do not appear as components or parts of the model nor do the laws of the simple theory make special use of them. It is uncontroversial that poetry, patience, and financial assets exist. Therefore, assuming they are not fundamental existents, they are derivative existents.

In more generality, once we have supposed that some model completely and accurately represents fundamental reality, we can think of derivative entities and properties as existents that are not substructures of the model.

One important group of existents whose status deserves to be considered for illustration are "moderate sized specimens of dry goods" (Austin 1961). Without getting too bogged down in technicalities, I think the most natural method for categorizing macroscopic material objects can be sketched as follows. If fundamental reality includes a space-time containing the corpuscles and fields that instantiate an ordinary material object, then any particular *instance* of this object—that is, a full specification of the complete microscopic content of a maximally determinate region of space-time that includes at least one temporal stage of the object—will be a part of fundamental reality and thus will be a fundamental existent. But insofar as we treat an object as an existent that retains its identity under even the slightest alterations to its boundary or its microscopic instantiation, or its time or place of occurrence, we are treating it as a derivative existent. (There is no fact of the matter in this case as to whether the *object* is fundamental or derivative. There are instances of the object, which are all fundamental, and there are various abstractions or fuzzings or coarse-grainings of the object, which are all derivative.) Alternatively, if fundamental reality is something more esoteric like an entangled quantum field or an eleven-dimensional arena inhabited by strings, then there may well be no parts of fundamental reality that count as an instance of the object, in which case the object is unambiguously derivative.

Insofar as discussion in this volume will be concerned, it will suffice for us to adopt a single sufficient condition for an existent being derivative. I will specify this sufficient condition in terms of the ontological status of *quantities*, but it holds of existents generally.

> A quantity is derivative if its magnitude requires some specification beyond the totality of fundamental reality (and beyond any specification required to locate the quantity in fundamental reality).

For example, the mass of any corpuscle at any time in the simple theory of classical mechanics is a quantity that has a determinate magnitude once we have specified the spatio-temporal location of the corpuscle whose mass we are considering. By contrast, the kinetic energy of any corpuscle (at any time) is derivative. A corpuscle's *kinetic energy* is equal to one-half its mass times its speed squared, $\frac{1}{2}mv^2$. In the simple theory of classical mechanics, no fundamental structures suffice for a given corpuscle's absolute speed; there are only corpuscle speeds relative to other corpuscles. However, if we select an appropriate frame of reference to serve as a universal standard for being at rest, we can say that a corpuscle's speed is its speed relative to this rest frame. Then, because we have associated a determinate speed with each corpuscle, there will be a well-defined value for each corpuscle's kinetic energy. The kinetic energy of a corpuscle is an example of a derivative quantity because there is nothing in fundamental reality that corresponds to a unique correct value for the kinetic energy (at the corpuscle's spatio-temporal location) unless we augment the model with a parameter that doesn't correspond to anything in fundamental reality, namely this stipulation of what counts as being at rest.

Whenever a parameter used for describing reality does not have a unique correct assignment given how fundamental reality is structured, let us say that it is *fundamentally arbitrary*. A choice of rest is one example of a fundamentally arbitrary parameter. More generally, coordinate systems and so-called gauge degrees of freedom are fundamentally arbitrary. Fundamental reality might make some coordinate systems more convenient than others for characterizing the distribution of matter, but fundamental reality itself is independent of our conventions for assigning labels to points of space-time. Any quantity that is coordinate-dependent is derivative.

By convention, we can adopt the policy that the fundamentally arbitrary specification needed to locate a region in space-time (or whatever space is the container of fundamental material stuff) does not by itself make the contents of that region derivative. The locating information should instead be interpreted as merely defining the component of fundamental reality under consideration.

Imagine two solid blocks in an otherwise empty portion of space, each composed of massive corpuscles bound together by short-range forces. Fig. 1.1 provides two different characterizations of the very same fundamental arrangement of corpuscles that constitute the two blocks. By choosing a rest frame, one bestows

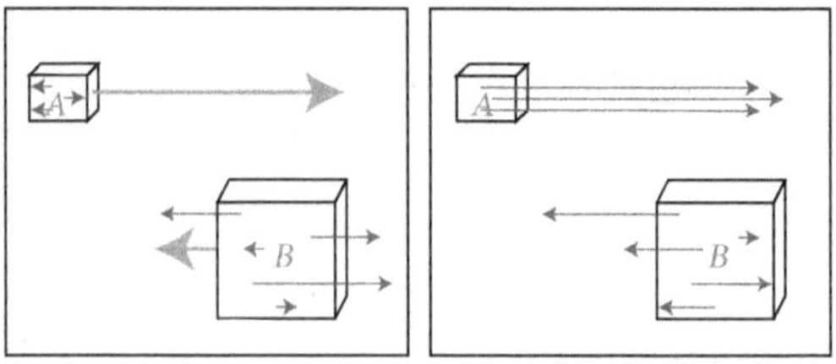

FIGURE 1.1 *Two depictions of the same fundamental reality. On the left, thermal energy is calculated by treating a corpuscle's speed relative to the motion of its block. On the right, thermal energy is calculated by treating every corpuscle's speed as relative to the rest frame.*

on each corpuscle a well-defined (non-relational) velocity. The total kinetic energy, E_T, of the entire system is the sum of each individual $\frac{1}{2}m_i|\vec{v}_i|^2$, where m_i is the mass of the ith corpuscle and $\vec{v}_i$ is the velocity of the ith corpuscle in the rest frame. We can think of this total kinetic energy as divisible into *thermal energy*, and *mechanical energy*, with the relative proportion depending on how we choose to organize the complete collection of corpuscles into groups. Thermal and mechanical energy are both forms of kinetic energy, at least insofar as we are simplifying the physics in this volume for the sake of discussion.

One way to group the corpuscles is to let the A-grouping comprise all the corpuscles in the block marked A, and the B-grouping comprise all the corpuscles in block B. Let $\vec{v}_A$ be the velocity of block A and m_A be the mass of block A, and similarly for $\vec{v}_B$ and m_B. In Fig. 1.1, these two velocities are represented as large gray arrows. The corresponding mechanical energy of block A is $\frac{1}{2}m_A|\vec{v}_A|^2$, and the mechanical energy of block B is $\frac{1}{2}m_B|\vec{v}_B|^2$. The thermal energy of each individual corpuscle can be understood as its kinetic energy *relative to the net motion of its group*. Each corpuscle i on the A-grouping, for example, has thermal energy $\frac{1}{2}m_i|\vec{v}_A - \vec{v}_i|^2$. The thermal energy of block A as a whole is just the sum of the thermal energy of each of its individual members and similarly for block B. The total thermal energy of the whole system is just the sum of the thermal energy of each block.

A second way to group the corpuscles is to put all of them together. Let m_T be the total mass of all the corpuscles and $\vec{v}_T$ be the velocity of the center of mass of the whole system. Then the mechanical energy of the whole system is $\frac{1}{2}m_T|\vec{v}_T|^2$, and the thermal energy is the sum of all the individual terms of the form $\frac{1}{2}m_i|\vec{v}_T - \vec{v}_i|^2$.

The first decomposition of kinetic energy into mechanical and thermal quantities fits our natural inclination to treat the blocks as separate objects and is useful for making predictions about thermodynamic phenomena when each block comes into thermal contact with other objects having their own distinct temperature and composition. If block A is grabbed and put in contact with some ice, and block B is separately placed in a furnace, the thermal energies as calculated in the first decomposition are what figure in predictions of how heat will move between the blocks and their respective environments.

The second decomposition of the kinetic energy into mechanical and thermal quantities is not useful for such calculations. However, there is nothing formally incorrect about it, and there might be some circumstances where that way of breaking apart the kinetic energy into thermal and mechanical components is more useful. The important point here is just that nothing about fundamental reality makes one assignment of mechanical and thermal energy the unique correct assignment, even if fundamental reality makes one assignment especially useful for practical purposes. The allocation of kinetic energy between mechanical and thermal energy is thus fundamentally arbitrary.

Furthermore, there are circumstances where there is no clear best way to distinguish between thermal and mechanical energy even for practical purposes. The oceans have mechanical energy in their currents and thermal energy that plays a role in melting icebergs, but because the ocean is fluid, it can sometimes be unclear how to group the corpuscles. The tiniest eddies in the current might be construed as instantiating purely mechanical energy because they knock pollen grains around, but they could be construed as purely thermal because such a small parcel of energy cannot be extracted by any practical device like a turbine.

Now we are in a position to see why it is reasonable to think of mechanical and thermal energy as metaphysically derivative. First, we know that mechanical and thermal energy can be defined in terms of properties we already accept as fundamental by specifying two fundamentally arbitrary parameters, the choice of rest and the grouping of corpuscles. That leaves us with a choice about how to construe the amounts of thermal energy and mechanical energy:

- One option is to hypothesize that there is a brute fact of the matter about precisely how much thermal and mechanical energy the system has. I interpret such a choice as an addition to what we already accept as part of fundamental reality. This is tantamount to believing there are objective facts about the distribution of thermal and mechanical energy that go beyond how the fundamental masses are arranged in space-time, and a good way to describe such facts is to say that they are fundamental.
- A second option is to declare that there is no ultimate fact of the matter about how much thermal and mechanical energy there is, but that there are still parameter-dependent facts about thermal and mechanical energy. Given that the corpuscles are arranged in such and such a pattern fundamentally, and given that we choose such and such frame of reference for the rest frame, and given that we allocate these corpuscles over here to the *A*-group and those over there to the *B*-group, there is a determinate value for both the thermal and mechanical energy of *A* and *B*. A good way to describe this option is that it treats thermal and mechanical energy as derivative.

There are other interpretational options one could consider, but I will forgo discussion of them because my goal here is just to provide a reference point for how to distinguish fundamental from derivative, not to settle quibbles about how best to understand energy.

I believe the more reasonable stance is to interpret kinetic energy (and thus thermal and mechanical energy) as derivative. There are several reasons to prefer treating them as derivative rather than fundamental. (For concision, I will focus attention on kinetic energy in this paragraph, but everything I say here applies to thermal energy and mechanical energy as special cases.) First, we already have fundamental laws in classical mechanics governing the motions of particles, and if there were some brute (fundamental) fact about precisely how much kinetic energy existed, it would play no essential role in the temporal development of the physics.[5] Second, if there were a fundamental fact about the precise quantity of kinetic energy, we would have no epistemic access to its value. At least, it would be mysterious how we could ever come to know the one true amount of kinetic energy. Third, there is no scientific account of anything that would be defective in any way if we treated kinetic energy as derivative. Nor would any scientific account be improved by treating it as fundamental. These kinds of considerations are standard in scientific practice and provide a practical grip on why we construe some quantities as fundamental and others as derivative. A good way to think about the issue is that if we try to segregate various kinds of properties and relations into fundamental and derivative using scientific methods, we have good reasons to keep the fundamental ontology fairly restricted and to avoid postulating redundancies in fundamental reality. Ceteris paribus, a sparse theory of fundamental reality can provide more reductive explanations, posit fewer epistemically inaccessible facts, posit fewer quantities that do not integrate well with the rest of the fundamental quantities, etc. (A principle of parsimony could conceivably be included in the list of principles associated with the idea of fundamentality, but I think it is ultimately preferable to leave such scientific considerations out of the constitutive principles governing fundamentality in order to accommodate a broader range of approaches toward fundamental reality.) Although I have not argued conclusively that kinetic (and thus thermal and mechanical) energy should be accorded derivative status, the thesis has a lot to recommend it, so from here on, the discussion will assume they are to be understood as metaphysically derivative.

1.6.2 SOME CONSTITUTIVE PRINCIPLES OF FUNDAMENTALITY

With the kinetic energy example in mind, we can now revisit the list of principles I associated with fundamentality.

---

[5] It is possible to formulate classical mechanics so that energy plays a starring role in the temporal development, but the simple theory was constructed to exclude energy from any essential role in the fundamental laws. In any case, kinetic energy by itself plays no role in the fundamental development of the actual world even if energy itself does.

Principle (1) claims that the way things are fundamentally is the way things *really* are. This is a way of assigning privileged metaphysical or ontological status to fundamental reality. Because the topic of ontology is too controversial to take up here, I will only note that philosophical debates over realism and anti-realism can be usefully framed in terms of fundamental reality, and it is one of the main goals of Empirical Fundamentalism to reformulate debates about realism in terms of debates about what is fundamental.

For example, my suggestion that we should interpret kinetic energy as metaphysically derivative accords with Marc Lange's (2002) discussion. Lange cites the frame dependence of kinetic energy as a reason to believe that kinetic energy is not real, unlike a corpuscle's mass, which is well-defined independent of any choice of reference frame or coordinate system. If I have interpreted Lange correctly, what I mean by 'fundamental' is what Lange means by 'real'. This suggests that there is at least one construal of the word 'real' that tracks fundamental existence, though certainly our pedestrian attributions of 'real' apply to both fundamental and derivative existents.

Before going any further, I want to emphasize that a commitment to the existence of fundamental reality, as I construe it, does not impose significant constraints on a theory of reality. Although the simple theory of classical mechanics draws on a familiar conjecture that what is fundamental is the stuff described by theories of fundamental physics, nothing in my account of fundamentality or my account of causation requires that what is fundamental include physics, much less be identified with some instance of fundamental physics. As far as my theory is concerned, one could hold that economic events or geological processes or thoughts are fundamental. Because fundamental reality can in principle comprise just about anything, the mere claim that reality divides into fundamental and derivative has little substantive content. It can only generate rich consequences when conjoined with significant constraints governing fundamental reality.

For that reason, throughout much of this book, I will be exploring the auxiliary hypothesis that fundamental reality resembles models of paradigm theories of fundamental physics. This assumption is not strictly part of the theory of causation; its only purpose is to permit discussion of causation in a concrete context. It would be extremely difficult to say anything interesting about causation without at least exploring some auxiliary hypotheses about the nature of fundamental reality, and a focus on fundamental physics as a preliminary working model for understanding fundamental reality is motivated by the privileged role fundamental physics plays in any scientific investigation of causation that purports to hold for all kinds of causes. Physics has a distinguished role to play because a comprehensive theory of causation is supposed to apply not only to mundane affairs but also to the fantastically small and fantastically large, domains where only physics has provided a rich account of how things operate. Though one's theory of causation should apply to oceans and economies and psychological processes, it is at least a plausible hypothesis that the empirical phenomena that lead us to believe in causal relations described by the special sciences are instantiated by matter that obeys laws of physics. I take it as not even remotely plausible that causation among neutrinos and quarks could be cashed out in terms of oceanic, economic, or mental properties. But the reverse relation—that causation among economic events, say, is a special case of causation among the entities of fundamental physics—is at least a plausible working conjecture.

Although fundamental physics superficially says nothing about economics, it is easy to see skeletally how the laws of physics could impose extremely strong constraints on the physical stuff that instantiates paradigmatic economic activity. If the physical laws are deterministic, for example, a complete specification of the physical state at one time determines the physical arrangement of everything throughout history, including bankers at work, money in people's pockets, merchandise on the store shelves, and just about everything else that is economic in character. It may well be true that even a highly idealized epistemic agent cannot make successful economic predictions knowing the detailed physical state and the deterministic laws, and it is certainly true that the determination does not hold merely by virtue of the economic facts at one time, but it should be easy to understand how there could be non-trivial implications among the physical instances of economic facts.

Another good reason to investigate the auxiliary assumption that fundamental reality resembles existing paradigm theories of fundamental physics is that it is possible to derive remarkable facts about causation from a few relatively uncontroversial hypotheses about the fundamental physical laws. For example, in §6.2, I will derive causal directness, the principle from §1.2.1 that a backtracking nomic connection does nothing beyond what it does by virtue of operating in a single direction of time. In chapter 7, I will do the same to demonstrate that the past cannot be usefully manipulated. Interestingly, none of my arguments defending these principles presupposes a fundamental asymmetry of causation or a fundamental passage of time or any settledness of the past. So, my explanation of the direction of causation will hold even if the fundamental laws are deterministic in both temporal directions with no fundamental temporal asymmetry.

Regardless of the benefits of my focus on fundamental physics as a guide to fundamental reality, the framework I will be constructing is suitable for a wide range of possible views about what should be included in fundamental reality. It is compatible with models of fundamental reality that include phenomenal properties, theological attributes, intentionality, and aesthetic properties. It is compatible with models of fundamental reality where nothing is physical. Emergent properties and dualistic conceptions of the mind can also be represented within the confines of the framework. Nothing I say about causation rules out any of these possibilities. I will just be using physics as a preliminary working model to help guide our thinking about causation.

Principle (2) claims that fundamental reality is the only real basis for how things stand derivatively. Philosophers have tried to make this idea precise in a variety of ways. One way is to say that fundamental reality fixes derivative reality. Another is to think of fundamental reality as a universal truthmaker, something by virtue of which all true claims about reality are true. Yet another option is to think that what is fundamental serves as a supervenience base for derivative reality. Although I believe such options are aimed approximately in the correct direction, I suspect all of these existing approaches will ultimately provide a suboptimal model for understanding the relationship between fundamental and derivative. To avoid unnecessary controversy, however, I will not provide my own account in this first volume of Empirical Fundamentalism. Instead, see Kutach (2011b) for a sketch of my views on this topic and future volumes for more details.

For present purposes, it will suffice to consider one critical feature of the kinetic energy example that needs to be accommodated by any account of how fundamental reality serves as the "real basis" for everything derivative. Remember that in order to derive any specific value for mechanical energy, say, one needs the fundamentally arbitrary choice of rest and corpuscle grouping. These parameters do not represent some additional fact about fundamental reality; they are stipulations. A complete specification of the fundamental attributes of classical mechanics does not by itself suffice for any particular value whatsoever for mechanical energy. How things are situated fundamentally does not fix how much mechanical energy there is. Yet, any specific choice of parameters will imply a precise amount of mechanical energy. So, given a complete characterization of fundamental reality, there exists a complete conditional characterization of mechanical energy, a complete set of conditionals of the form, "If such and such choice of rest and such and such choice of corpuscle groupings are made, the mechanical energy is such and such." Thus, how things are situated fundamentally (assuming the simple theory of classical mechanics) necessitates how things stand with regard to mechanical energy once (and only once) we have chosen the appropriate fundamentally arbitrary parameters.

Although the kinetic energy example shows how a numerically precise quantity can be conditionally implied by fundamental reality, my conception of derivative existents does not require such a conditional implication in order for them to count as bona fide existents.

Principle (3) claims that the way things are fundamentally is as determinate as reality ever gets. The fact that we have to supplement the fundamental attributes of classical mechanics with fundamentally arbitrary parameters in order to acquire determinate values for the distribution of thermal and mechanical energy illustrates the sense in which reality can be thought of as no more determinate than fundamental reality. Put simply, no specific amount of mechanical energy is implied by fundamental reality even though all the fundamental attributes are absolutely precisely defined. Instead, there is only a conditional of the form, "For any choice of rest R and choice of corpuscle groupings A and B, there will be a determinate value for the thermal energy and mechanical energy." If there were some brute fact of the matter about the amount of thermal or mechanical energy that went beyond what was already implied by one's theory of fundamental reality plus any fundamentally arbitrary parameters, that would indicate that this brute fact should count as fundamental.

Discussion of principle (4), which claims that fundamental reality is consistent, is deferred until §1.8.

To conclude, let me note several ideas often associated with 'fundamental' that I believe are best kept separate from our general conception of fundamental reality. For one thing, nothing about fundamental reality, as I conceive of it, requires that there are *levels* of reality or degrees of reality beyond the mere distinction between fundamental and derivative. So, many of the criticisms addressed at the idea of a fundamental *level*, like Schaffer (2003), do not apply to my conception of fundamental reality. Nor does my conception of fundamentality require that what is fundamental be small, like a point-like property instance. Nor is it required that what is fundamental be metaphysically simple; a fundamental entity can have complexity and consist of fundamental parts.

Although much more could be said to delineate fundamental from derivative, I hope I have sketched a clear enough distinction in order to be able to make sense of its primary function in my account of causation: to support a certain kind of reductive relationship, which I will now examine.

1.7 Abstreduction

The relation between thermal and mechanical energy and the fundamental attributes of the simple theory of classical mechanics illustrates an important kind of reduction. Unfortunately, 'reduction' is a famously over-used term with so many different interpretations that it cannot be trusted for secure communication. So, in order to minimize the potential for misleading associations with other people's usage, I hereby introduce a proprietary version of the general idea of reduction: *abstreduction*. A paradigm example of abstreduction is the relation between mechanical (or thermal) energy and the fundamental attributes of the simple theory of classical mechanics. Mechanical energy abstreduces to fundamental reality (under the pretense that fundamental reality answers to the simple theory of classical mechanics).

Reduction is closely associated with reductive explanation. Because explanation is an extremely contentious topic, I want to be clear that although I believe an abstreduction is a legitimate form of reductive explanation, I do not subscribe to any particular theory of explanation nor do I have any ax to grind concerning which explanations count as genuinely reductive. My aim is merely to cite the preceding account of how mechanical and thermal energy are related to the fundamental attributes posited by the simple theory of classical mechanics and then to argue that in whatever sense that account serves as a reductive explanation of mechanical and thermal energy, my metaphysics of causation will incorporate a reductive explanation of the derivative aspects of causation to the fundamental aspects of causation. The only prominent disanalogy is that we know well enough the content of the simple theory of classical mechanics, but at this stage of human history we can only speculate about a correct and complete theory of fundamental physics and thus fundamental reality more generally.

The point of an abstreduction is to abstract away from the details of fundamental reality in a way that allows us to make sense of derivative quantities in terms of fundamental reality and fundamentally arbitrary parameters. It provides a structure for fuzzing fundamental reality. Imagine we start with a particular model of fundamental reality, *F*, which specifies fundamental laws and specifies how all the fundamental attributes are arranged. Suppose further that we believe in the existence of a certain quantity *D* whose value is not implied by *F*. In order to provide an abstreduction of *D* to *F* we engage in the following two stage process.

In the first stage, we supplement *F* with fundamentally arbitrary parameters and provide a function so that the quantity *D* has a specific value in terms of these parameters together with quantities from *F*. The illustration of mechanical energy above was meant to demonstrate that E_M can be derived from a choice of rest and a choice of corpuscle groupings in conjunction with the masses and relative speeds present in any model of the simple theory of classical physics.

The mere fact that we can derive some quantity *D* from a model of fundamental reality *F* (with any extra parameters we choose) shows nothing interesting by itself. It is trivial, for example, to create some function of the fundamental variables of the simple theory of classical mechanics. We could have invented a quantity, quinergy, defined as $\sqrt{m}v^5$ for any corpuscle, given some standard of rest, and defined for collections of corpuscles by summing their individual quinergies. The reason no one takes quinergy seriously as an existing property, I think, is that it is not a particularly useful scientific quantity. It plays no role in systematizing or explaining the behavior of anything anyone cares about; it is not a conserved quantity; it plays no compelling role in any macroscopic phenomena. What it seems like we need to do in order to justify the status of thermal and mechanical energy as derivative properties is to account for their utility. For instance, we might note that the distinction between mechanical and thermal plays a role in our account of how much energy can be extracted from a system. (For a system at one temperature, only mechanical energy can be extracted.) We might also note that the stability of thermal and mechanical energy over appropriate time scales helps to make them useful.[6]

In the second stage of an abstreduction of *D* to *F*, one attempts to explain why the quantity *D* is a useful magnitude to consider. Unfortunately, it is difficult if not

[6] According to the formula for thermal energy, the thermal energy does vary sharply as the corpuscles slow down when they (nearly) collide with one another, but so long as there are many particles integrated within the same physical system, these brief jiggles in the amount of thermal energy are small relative to the total thermal energy and become negligible if one averages them over suitable time scales.

impossible to formalize how one explains the utility of some quantity. There is no general scheme anyone is aware of for measuring and comparing the usefulness of different quantities. So this stage of the abstreduction is going to involve appealing to common sense and our collective wisdom concerning what quantities are worth positing. This makes the boundary between derivative reality and the nonexistent at least as imprecise as our criteria for utility, but I cannot see why this consequence would be problematic. In particular, there is no harm in concluding that quinergy exists (derivatively) but is not worth bothering with. In any case, if we complete both stages, we have completed an abstreduction of *D* to *F*.

A brief terminological note is needed here before I discuss how my metaphysics of causation is abstreductive. Because people's prior commitments about causation are so diverse, I prefer to avoid as much as possible referring to a cause-effect relation in my account. To the extent I refer to causation, that is meant non-technically as a way of speaking with the masses. I will instead use the term 'causation-like' for relations that play some of the roles we ordinarily associate with the cause-effect relation. In particular, 'causal relations' misleadingly suggests irreflexivity, asymmetry, and a discrimination between important causes and negligible background factors. It will be important for my account that there are some fundamental relations that can take over the role traditionally played by token cause-effect relations and thus serve as singular causal relations, but it will not be important whether these fundamental causation-like relations are irreflexive, asymmetric, or suitable for distinguishing the relative importance of partial causes.

My main task in this volume is to conduct an abstreduction of general causation to singular causation. I will define some fundamental singular causation-like relations in my account of the bottom conceptual layer and define some derivative general causation-like relations in my account of the middle conceptual layer using some fundamentally arbitrary parameters. Once the metaphysical structures have been defined, it should be obvious how any derivative causation-like relation with a well-defined value gets its unique determinate value from fundamental reality together with the specified fundamentally arbitrary parameters. I then only need to demonstrate the utility of my derivative causation-like relations, which will be accomplished throughout the middle of this volume by appealing to its simplicity, its flexibility, its generality, and its role in the explanation of causal asymmetry.

In order to carry out my abstreduction, I will stick to the following plan. In chapter 2, I will present an account of fundamental causation-like relations. The most important relations in my account of fundamental reality are determination and (a form of) probability-fixing among events. For example, the complete state of the world at one time might determine a later event, or it might determine that some kind of event has a one-third chance of occurring. This form of singular causation is similar to the models of causation proposed by John Stuart Mill with his "real causes" and by J. L. Mackie with his inus account of causation. It also resembles "productive" notions of causation when the laws propagate states continuously through time.

In chapters 3 and 4, I will present an account of derivative causation-like relations, which inhabit the middle conceptual layer of causation. These will include my own variant of a counterfactual conditional and a corresponding notion of counterfactual dependence or difference-making. Traditionally, difference-making accounts of causation have mostly been competitors to determination accounts of causation, but in my theory, the (derivative) difference-making relations are defined in terms of how fundamental laws propagate hypothetical fundamental states through time, whether deterministically or with fundamental chanciness.

Just as I described parameters that allow one to determine the amount of kinetic and thermal and mechanical energy given the totality of corpuscle attributes, I will provide parameters that allow one to determine the magnitude of difference-making (or counterfactual dependence) using any fundamental laws that determine or fix probabilities. This abstreduction allows my metaphysics of causation to quarantine the shiftiness and vagueness of counterfactuals, which have long plagued difference-making accounts of causation.

Summarizing the important points discussed in this section, to abstreduce some quantity *D* to a model of fundamental reality *F* involves specifying some fundamentally arbitrary parameters and explaining how these parameters (together with *F*) make *D* a determinate quantity with sufficient utility. Abstreduction reveals how an existent *D* is nothing more than a handy way to abstract away from the details of the fundamental existent *F*. The goal for my theory of causation is (1) to show how relations of difference-making (or counterfactual dependence) can be defined in terms of fundamental laws and fundamental events using fundamentally arbitrary parameters, and then (2) to show how these relations of difference-making are useful for abstracting away from the fundamental laws governing the detailed motion of matter.

1.8 STRICT Standards and RELAXED Standards

In this section, my goal is to clarify how the bottom and middle conceptual layers of causation, which I associate with the metaphysics of causation, differ from the top conceptual layer of causation, which I associate with various non-metaphysical aspects of causation, especially including the role of causation in the special sciences.

The distinction that separates the top layer is a methodological one based on how theoretical inconsistency should be evaluated. While it is widely believed that avoiding contradictions is important for any theory, there are systematic practical differences in how threats to consistency are resolved, differences related to whether a theory concerns fundamental reality. There are several case studies that successfully illustrate how a theory can be inconsistent yet hedged in a way that allows it to provide highly non-trivial predictions as well as acceptable explanations. These include the old quantum theory of black body radiation (Norton 1987), relativistic electromagnetism (Frisch 2005a), and more (Meheus 2002). Implementing a system of managed inconsistency by disallowing a restricted class of troublesome inferences allows us to make sense of theories that are strictly speaking inconsistent or incoherent as complete theories. Despite formal inconsistency when construed as complete, such theories can still succeed in their usual roles of predicting, systematizing, and explaining, by being treated as incomplete or imprecise.

When a theory purports to be a complete fundamental theory and proposes inconsistent rules for its components, we rightly reject the theory out of hand as unacceptable. This practice is justified by virtue of our conception of what it is for a theory to be about fundamental reality. Our theories of fundamental reality forbid contradictions, I think, because of a commitment to the thesis that no matter how inscrutable and paradoxical reality may seem, deep down, there is some consistent way reality is. This is the fourth principle guiding my conception of fundamentality from §1.6, which I expressed as, "Fundamental reality is consistent," by which I meant that fundamental reality obeys some metaphysical correlate to the law of non-contradiction, as discussed for example by Tahko (2009).

When seeking a theory of fundamental physics, we often formulate *dynamical laws*, which are laws that constrain how the universe evolves. Imagine a fundamental theory that specifies two dynamical laws and that in the special case where there is a corpuscle at rest by itself, the two laws disagree about what will happen. One law dictates that the corpuscle will remain at rest; the other dictates that the corpuscle will oscillate. If the fundamental theory permits the possibility of a corpuscle being at rest, then the theory in effect provides two conflicting rules for what will happen. Such theories are uncontroversially and correctly regarded as unacceptable theories of fundamental reality.

For a more realistic illustration, we can consider the theory of relativistic electromagnetism, whose laws include Maxwell's laws and the Lorentz force law. Maxwell's laws require that electromagnetic charges be treated as a field-like quantity, a sort of charged fluid. The Lorentz force law requires that charges be treated as (discrete) corpuscles. These two requirements are inconsistent, and it is not clear how to tweak them to remove the inconsistency. If the particles are truly point-like, then the electromagnetic field at every particle location is infinitely strong, which disallows the Lorentz force law from defining a finite force on the particle. If the particles are truly field-like, the internal electromagnetic field forces should make the particle explode. One could postulate additional physics to hold the charged fluid bunched together as a particle, but that force would imply the falsity or incompleteness of the laws of electromagnetism as applied to the charge itself. Fortunately, the inconsistency of electromagnetism is adequately *managed* by not using both laws at the same time for the same material and by not demanding that the theory fully address the question of how charged particles self-interact. In this way, the theory can be technically inconsistent as a complete theory of fundamental reality, but also acceptable as an incomplete theory or as a theory of derivative reality that is only approximate and relies on additional resources in fundamental reality to adjudicate what is fundamentally going on.

In order to explore two approaches to the threat of inconsistency, it is useful to introduce some new terminology. Let us say that a theory's rules *conflict* when, for some realistic circumstance, they make contradictory attributions. The possibility where a corpuscle must both remain motionless and yet oscillate (relative to the same frame of reference) is a paradigmatic example of a conflict. The meaning of 'realistic circumstance' can vary depending on what kind of theory is being offered. Some theories, like those of fundamental physics, are meant to be rich enough to characterize their own conception of nomological possibility. For such theories I mean to count as realistic any circumstance that is nomologically possible according to the theory itself, regardless of whether it is possible according to the actual laws. (This notion of nomological possibility could take into account restrictions on the kinds of matter allowed and the space-time structure, not just restrictions given by equations of motion or conservation laws.) Other theories, like those in anthropology or food science, do not specify what is nomologically possible but implicitly rely on an imprecise antecedent notion of possibility. For such theories, any situation that is possible according to this antecedent notion counts as a realistic possibility regardless of whether there are any actual laws at odds with the implicit notion. For example, it could turn out that the true laws, whether we know it or not, are so severely restrictive that the only nomologically possible world is the actual world. If so, many possible circumstances entertained by geneticists and economists are not nomologically possible, but our standards for evaluating theories of genetics and economics are such that we treat seemingly realistic circumstances as nomologically possible. The purpose of distinguishing realistic from unrealistic circumstances is to mark the fact that some possibilities are so epistemically remote that we should not care about whether our theory's principles conflict there. For example, we would rightly not reject an otherwise splendid theory of physics merely by virtue of its having principles that conflict in models with a 43-dimensional space-time unless there were some reason to think the actual space-time has 43 dimensions. The same goes for any conflicts a theory might have if we were to countenance the possibility of angels or magic spells. 'Realistic circumstances' is not meant to include all circumstances having an appreciable chance of obtaining. A possibility does not count as unrealistic in my terminology merely because it has an extremely low objective chance; to be unrealistic it must be subjectively very improbable (according to the relevant experts) because of the laws it invokes or the types of matter or space-time it posits.

There is a difference one can draw between rules that have *apparent conflicts* and rules that have *genuine conflicts*. If there are additional principles in a theory that specify how to ameliorate apparent conflicts, then the theory's rules do not genuinely conflict. There are several legitimate ways to ameliorate apparent conflicts to show they are not genuine conflicts. If a theory has two rules that seem to conflict, one could supplement the theory with conditions restricting the circumstances under which each applies so that for any specific circumstance only one of the rules is operative. Alternatively, one could weaken the content of the theory with a qualifying clause so that whenever the rules conflict, neither is operative. Another option is to augment the theory with a qualifying clause so that whenever rules one and two conflict, only rule one is operative. Many of these maneuvers make a theory less appealing, but in general, a theory could have conflicts that only occur in restricted circumstances, and the point of the amelioration clauses would be to establish explicitly what the theory says in every potentially conflicting circumstance so that the theory's apparent conflicts are never genuine. If a theory refuses to clarify what to do in cases where its rules superficially conflict, or it merely claims that there is always some further resolution to the conflict but does not specify the additional structure that resolves the discrepancy, then that theory has a genuine conflict.

Let us now say that any intellectual discipline whose theories are required to avoid genuine conflicts obeys STRICT standards and any intellectual discipline that allows theories to possess genuine conflicts obeys RELAXED standards.

For illustration, imagine a crude psychological theory of our implicit food concept that offers us the following two rules of thumb for when something counts as food.

1. Something is food if and only if it is a substance of the kind humans serve to each other as something to be eaten.
2. Something is food if and only if it is nutritious.

These rules conflict because it is easy to imagine a substance that would be routinely served at meals but which has no nutritional value, or something that is nutritional but which people find objectionable to eat. A theory that tries to provide an account of our ordinary food concept (as part of psychology) is not normally pretending to provide rules that are strict necessary and sufficient conditions. Instead, the necessary and sufficient conditions expressed above are meant to characterize informal heuristics or rules of thumb that link our thoughts about food with other concepts. Their purpose is to make sense of the following kinds of regularities. When people are presented with information that some *S* is nutritionally harmful, they tend to think of it as not being food; with information that *S* is nutritionally beneficial, they tend to think of it as food. There is a default expectation in how we interpret such psychological theories that when a test subject is put into a situation where these default heuristics conflict, additional facts can bear on whether the subject identifies *S* as food. A psychologist who wanted to flesh out such a theory would provide a more thorough account of the factors that affect whether *S* is categorized as food, including predictions about which circumstances result in people becoming less certain of their judgments. Yet, we know from experience that the quantity of factors needed to provide precise predictions may be far too large for a practical theory. A much more precise and accurate theory of our folk food concept would presumably need to account for cultural backgrounds, personal differences in gustatory abilities, hunger, how accommodating the subject is to others' judgments, and many other factors. To identify with great precision across a wide range of varying conditions whether a given person will consider *S* as food requires many more facts that are rightly considered outside the scope of psychology. There is also undoubtedly a tradeoff between predictive accuracy and the number of parameters such a theory would need to incorporate. For such practical reasons, the kind of psychology that deals with our concepts in a way that is fairly remote from its neurological implementation could and should be understood as proceeding under RELAXED standards that permit theories to conflict on some assessments of realistic circumstances.

The same considerations apply to so-called ceteris paribus laws appearing in the special sciences. A theory of ecological genetics, for example, might postulate a law that when new islands are formed near populations, the number of species will increase, and another law that when a cataclysm occurs, the number of species will decrease. These laws conflict because there could be a cataclysmic flurry of volcanic eruptions that create new islands. This superficial conflict—that the number of species would both increase and decrease—does not warrant the rejection of a theory that posits both laws. Such laws are not intended as inviolable dictates of nature but as useful rules of thumb that can be overridden in some circumstances. It is also understood that whether the number of species goes up or down depends on the nature and severity of the volcanic activity, the number of islands created, and many other factors that ecological genetics is simply not in the business of accounting for in detail. Because ecological geneticists are not obligated to spell out all the parameters that would ameliorate all apparent conflicts in its models, we can say that ecological genetics obeys RELAXED standards.

When a theory concerns fundamental reality, it is appropriate to hold it to STRICT standards. In most (but not all) cases, when a theory concerns only derivative reality, it is arguably appropriate to hold it only to RELAXED standards. For an example of a case where the subject matter uncontroversially concerns derivative reality but ought to obey STRICT standards, consider the narrow subset of thermodynamics that deals with the distinction between mechanical and thermal energy, again under the pretense that fundamental reality is completely and correctly described by the simple theory of classical mechanics. I know of no reason to think that thermodynamics generally has to hold to STRICT standards, but in the special case of theories that deal only with concepts that have a close enough fit with fundamental reality, it is reasonable to maintain STRICT standards. If we have accepted that the mechanical and thermal energy of macroscopic objects are so closely related to the fundamental attributes of classical mechanics that it does not require the services of any other scientific discipline, we are justified in demanding that a theory rule out any genuine conflicts in its pronouncements regarding thermal and mechanical energy. For example, if a theory claims that very nearly two percent of the energy of any large body of water is thermal, and another part of the same theory claims that very nearly thirty percent of the energy of salt water is thermal, it is appropriate to demand some account of how these two claims are compatible when applied to an ocean, which is both large and salty. An outcome that is uncontroversially unacceptable is for the theory to declare that both percentages are accurate general rules of thumb while remaining silent on what the theory claims about the ocean's thermal energy.

What makes this subset of thermodynamics different from the case of ecological genetics is that the predictions of ecological genetics depend on factors well outside the scope of ecological genetics. We know that conflicts between the law that islands increase the number of species and the law that cataclysms decrease the number of species can be ameliorated by considering a richer set of facts addressed by physics that can ultimately settle whether the number of species increases or decreases in any particular case and over what time scales. But for the physicist whose theory refers to thermal energy, there is no further discipline to which conflicts can be delegated. If we have accepted that classical mechanics is our fundamental theory and that thermal energy is abstreduceable to the fundamental attributes of the corpuscles, the only resources available to ameliorate the apparent conflict are the additional parameters that make thermal energy determinate. The difference between STRICT and RELAXED standards is just that any theory adhering to STRICT standards cannot just hand-wavingly assert that there are further details that ameliorate apparent conflicts. It must explicitly state the parameters that ensure its concepts are being applied consistently. The reason for holding STRICT standards in this special subset of thermodynamics is that we already hold STRICT standards for theories of fundamental reality and we have committed ourselves to the thesis that there are no further facts that some other discipline could supply that would ameliorate the conflict. If we were to abandon belief that there is a fundamental physics or commit ourselves to a different fundamental physics that makes the relation between it and energy more opaque, that could motivate us to adopt RELAXED standards for this portion of thermodynamics.

Throughout the preceding discussion, I have not in any way ruled out the possibility that some special sciences ought to operate under STRICT standards. I am only providing a few examples where I think it is intuitively plausible that the appropriate standards to hold are RELAXED. The kind of psychological theory that tries to model our implicit beliefs about some concept like 'food' or 'causation' is operating at a fairly high level of abstraction, high enough so that its pronouncements can in principle be perfectly acceptable qua high-level theory of our concepts even though it does not provide parameters that guarantee a lack of conflict. This is entirely compatible with the possibility that some other kinds of psychology need to obey STRICT standards. Also, I do not intend my terminology to insinuate that theories operating under RELAXED standards are in any way less respectable than theories that are STRICT or that there is any less rigor in disciplines that employ RELAXED standards. The distinction between STRICT and RELAXED is merely a device to help delegate responsibility among disciplines for ameliorating conflicts.

Although it is difficult to rigorously defend hypotheses about which disciplines ought to hold STRICT rather than RELAXED standards, I do think it is a fair characterization of the intellectual activity known as metaphysics that people engaging in it believe metaphysical theories should obey STRICT standards, and I think they are correct to do so. Although 'metaphysics' is a term with evolving and contentious meanings, metaphysics is uncontroversially the general study of reality. In particular, theories of metaphysics are aimed at an account of reality that is not merely a patchwork of conflicting rules of thumb but a more systematic structure that is ultimately consistent. The motivation for adhering to STRICT standards in metaphysics makes sense given that foundational role of metaphysics does not permit it to delegate conflicts to other disciplines. Conflicts within theories of the special sciences often do not need explicit amelioration because there are virtually always additional physical facts not subsumed by the special science in question that one can plausibly appeal to for amelioration, but metaphysics has no other discipline available to ameliorate its conflicts. Metaphysics does often delegate to other disciplines to fill in some details. For example, a metaphysical theory might pronounce on what kinds of properties are possible and then task biology with discovering which particular biological properties exist. But it is not the role of biology to clarify the conditions under which the metaphysical theory's characterization of properties would be inapplicable or overridden by some alternative. A special science might reveal inadequacies of a metaphysical theory of properties, but it wouldn't provide a richer story about the general nature of properties than what metaphysics itself is expected to provide. In that sense, it is appropriate to hold metaphysical theories accountable to STRICT standards.

The point of this section has been to introduce some theoretical machinery so that I can now state a conclusion that I will eventually defend in chapter 8. Although any theory concerning the metaphysics of causation should obey STRICT standards, there is an activity commonly regarded by philosophers as part of the metaphysics of causation that can be entirely adequate even if it only satisfies the weaker RELAXED standards. This activity is the provision of rules for when a singular event counts as "one of the causes" of some chosen event. Sometimes, such rules are known as theories of *actual causation*, though I forego this terminology because of its incorrect implication that less noteworthy singular causes are not part of the actual world. Weakening the conditions of adequacy for theories governing such singular causes makes it easier to understand the range of intuitions that have long been considered by philosophers as touchstones for identifying the correct metaphysics of causation. What my account does, in effect, is to relocate some aspects of causation that have traditionally been understood as metaphysical, like preëmption, to the non-metaphysical aspects of causation. The purpose of the distinction between STRICT and RELAXED is to mark the boundary between those aspects of causation that need to be systematized in a principled and fully consistent system and those that do not. Chapter 9 will illustrate how our intuitions about singular causation could be systematized in a principled and explanatory theory with genuine conflicts.

1.9 Limitations on the Aspirations of Empirical Analysis

Putting together the new terminology from the previous two sections, I can now complete my presentation of empirical analysis by pointing out an important restriction on what activity is needed to produce an adequate empirical analysis. In §1.1, I initially characterized the goal of an empirical analysis of X as identifying "scientifically improved concepts of X." However, when we are engaged in an empirical analysis that concerns an X that is rightly considered part of *metaphysics*, as elucidated above, the scope of the project is automatically limited in the sense that one is not required to provide a regimentation to serve as an improvement for *all* appearances of X in the sciences. In particular, my empirical analysis of the metaphysics of causation is not required and is not intended to address all the locations where causal terminology is invoked. Quite to the contrary, it is intended to accomplish the much narrower task of connecting *all* causal regularities in the special sciences to fundamental reality (simplified provisionally as fundamental physics) in a STRICT manner. Some readers might have thought on the basis of my earlier discussion that an empirical analysis of causation requires an examination of the many uses of causal terminology or the variety of causal principles invoked in the special sciences, but such thoughts are incorrect. An empirical analysis of the metaphysics of causation only needs to develop concepts needed for the STRICT connection between derivative reality and fundamental reality. The omitted discussion is a task for an empirical analysis of the non-metaphysical aspects of causation.

It ought to go without saying that this methodological division of labor imposed by empirical analysis does not in any way denigrate special sciences or cast doubt on the importance of the full range of causal principles and causal concepts used in the special sciences. Rather, the consequence of this maneuver is in general to insulate the practices of the special sciences from the details of fundamental reality and in particular to grant them wide latitude to use a variety of causal concepts without having to draw any explicit connection to fundamental physics. This is analogous to the division of labor induced by the portion of physics that abstreduces a limited class of energy types, like thermal and mechanical, to fundamental attributes. By making explicit how every invocation of energy within a limited class of energy types can in principle be connected to fundamental reality in a consistent way, other special sciences are thereby freed to mention and build on these energy types without having to provide a maximally detailed account of what these energies consist of.

An ecologist, for example, might want to discuss energy flow through trophic levels by referring to the amount of energy held in plants that is available for appropriation by herbivores. Using the division of labor marked by the STRICT and RELAXED distinction, the ecologist would need to ensure that her energy concepts are well enough managed so that ecological processes she models do not violate conservation of energy or permit perpetual motion machines. But, crucially, she would not be required to be maximally precise in her account of what the boundary is between, say, plant energy and bacterium energy. That is a welcome consequence, given that plants are laden with bacteria. The RELAXED standards enforce enough linkage between ecology and fundamental physics to ensure that ecology does not violate laws of physics but otherwise leaves ecology free to posit forms of energy without having to express them as an explicit function of the attributes of fundamental physics. In general, RELAXED standards permit managed inconsistency.

Similarly, the metaphysics of causation I will provide will not be of any practical use to researchers who seek the causes of cancer. Nor will it directly address how to conduct causal modeling projects. Its purpose instead is to serve as a universal basis to which the special sciences can defer in order to backstop their use of not-maximally-precise causal terminology. With my account in place, the ecologist will be free to attribute the decline of tiger populations to the human appropriation of its prey without such claims hinging on the contentious question of whether fundamental reality includes something more than physics or the contentious question of whether causation is ontologically more than matter evolving according to laws of physics. I fully recognize that such worries are not pressing to practicing scientists, but it is of central concern to the long-standing philosophical question of whether and how there could be a relatively sparse model of fundamental physics that is sufficient for all of reality.

1.10 Comparison of Empirical and Orthodox Analysis

In order to illustrate the practical import of the distinction between STRICT and RELAXED, I will now emphasize how my pair of empirical analyses differ from orthodox analyses in their approach to singular causation.

Although orthodox analyses of causation have varying overall goals, one of the recognized tasks for any orthodox analysis is to identify non-trivial rules for which events count as causes, given not-too-causally-loaded information about the laws of nature and the history of occurrent facts. When we cite instances of causation—a whale breach causing a splash, for example—we intend to draw special attention to a small portion of the universe as being important to the effect. These events are called "the causes" of the effect or, more recently, the "actual causes" of the effect.

Orthodox theorizing about causation is expected to provide rules for what makes something count as one of these causes.

The singular causes sought by orthodox analyses are typically not fantastically detailed physical states but are intended to be the kinds of events people cite when asked about the causes of some particular event. For example, they might mention the launching of a ship, the loss of one tooth, or an increase in the gross domestic product during the fourth quarter of 1968. From here on, I will refer to such events as *mundane events*. Orthodox accounts of singular causation focus on relations among mundane events even when they allow that causal relations can exist among events that are physically sophisticated like the complete microphysical state existing on an infinitely extended time slice. Because these sophisticated kinds of events might play a role in singular causation, it is valuable to distinguish the kind of singular causes that are typically mundane events. Let us say that a *culpable cause* of some event *e* is an event that counts as "one of the causes of *e*" in the sense employed by metaphysicians who study causation. 'Culpable cause' is not a technical term but merely a label for the "egalitarian" (Hall 2004) notion of cause that orthodox metaphysicians seek when they ask, "What are the causes of (the singular event) *e*?" I emphasize that 'culpable cause' is my proprietary expression[7] introduced to reduce confusion about what 'cause' by itself connotes. Two further qualifications can be made at this point. First, culpable causes are so named because they are events that are blameworthy for the effect, but the terminology is not meant to imply that our intuitions about the relevant notion of singular cause absolutely perfectly matches our intuitions about how to attribute causal blame. Second, there is perhaps an ambiguity in the expression "a cause of *e*." It could mean "one of the causes of *e*" or it could mean "something that caused *e*." These are not always recognized as equivalent. When Guy won the lottery, his purchase of the ticket was one of the causes of his winning, but people would not normally say that purchasing the ticket caused Guy to win. 'Culpable cause' refers to the 'one of the causes' disambiguation.[8]

One feature that makes the orthodox analysis of causation a project in metaphysics rather than armchair psychology is that a proper analysis is required to provide a principled account of what is common to all cases of causation. Imagine that a psychologist offers a theory of causation consisting of a list of eight exemplars of the cause-effect relation and thirteen exemplars of the lack of a cause-effect relation. The theory says *c* is a cause of *e* if and only if the situation where *c* and *e* happen is closer to one of the positive exemplars than to any of the negative exemplars, closeness being judged by one's own intuitive off-the-cuff assessment of

[7] The term 'culpable cause' has been used previously by Mark Alicke (1992) to designate something altogether different: the psychological effect of perceived moral blameworthiness on judgments of causal impact.

[8] Some professionals report detecting no ambiguity here. I am fairly certain that if 'a cause' does not strike you as ambiguous, it is already picking out the intended conception of culpable cause.

similarity. A theory of this form might make for an interesting psychological theory and might even accrue empirical support if our causal reasoning is based less on rules than on some sort of pattern-matching capacity. But from the perspective of metaphysics, it would fail to capture what is similar in all cases of causation in an appropriate way. Such theories are merely fitting the data, whereas the metaphysician is interested in a theory based on principles that would connect causation with laws, chance, time, and would more closely resemble necessary and sufficient conditions.

Another feature that distinguishes an orthodox analysis of causation is the expectation that it is to be held to STRICT standards of consistency. For illustration, consider the following crude theory of our concept of causation, which is meant to parallel the crude theory of our concept of food from §1.8.

1. An event *c* is a cause of *e* iff *c* raises the probability of *e*.
2. An event *c* is a cause of *e* iff there exists a chain of probability-raising relations going from *c* to *e*.

This conjunction of rules might be faulty for multiple reasons, but let us focus just on realistic possibilities where the rules conflict. In an example (Suppes 1970) attributed to Deborah Rosen, a golfer's slice, *c*, lowered the probability of a good shot, *e*, and so was not a cause of *e* according to the first rule. But the slice did raise the probability of hitting a tree, which in turn raised the probability of the ball bouncing back in a better position, thus making *c* a cause of *e* according to the second rule.

By the RELAXED standards appropriate to most special sciences, including the kind of psychology concerned with modeling people's responses to questions about what caused what, it is acceptable for a theory to claim that people employ both biconditionals as rough-and-ready heuristics for assessing the existence of a cause-effect relation. Under RELAXED standards, having multiple conflicting rules for what events count as causes can be acceptable even if there is no further account in the theory of how to resolve (for all realistic circumstances) which heuristic is operative.

From a metaphysical point of view, such conflicting rules are unsatisfactory as an account of causation. In metaphysics, one is thinking of the causes as some element of external reality. A theory that provides conflicting pronouncements about whether *c* is a cause and provides no further resources to settle which rule is applicable and fails to relativize the incompatible facts to parameters that would remove the conflict, is in effect stating that its model of the actual world is inconsistent, which is uncontroversially unacceptable. One of the crucial standards by which orthodox metaphysical theories are to be judged is that their rules for causation need to be consistent. Furthermore, one is not allowed to save the inconsistent rules merely by adding a hand-waving qualifier that says, "In some cases, the first rule holds, and in other cases, the second rules holds." For metaphysical theories, room is typically permitted for vagueness by allowing a theory to avoid issuing a determinate judgment in all cases, but there is an obligation to ensure the theory does not judge that in a single scenario, *c* is both a cause of *e* and not a cause of *e*.

Orthodox accounts of causation attempt to find rules for attributing causation that on the one hand are principled and obey STRICT standards of adequacy, and on the other hand closely fit the relevant psychological data, including informed judgments about which partial causes should count as explaining the effect. What my account does is to replace this project with two empirical analyses. The empirical analysis of the metaphysics of causation is supposed to be principled and STRICT but is not supposed to fit any psychological data in the sense of rendering people's judgments about culpable causation explicitly true. The empirical analysis of the non-metaphysical aspects of causation is intended to be principled and fit the psychological data in the sense of systematizing common judgments about culpable causation, but it only needs to satisfy RELAXED standards to count as adequate for its intended purpose. This pair of empirical analyses accomplishes what the orthodox approach attempts to do in a unified treatment, but because it segregates the needed concepts into a metaphysical part and a non-metaphysical part, it is able to optimize metaphysical concepts in accord with the demands of fundamental reality and non-metaphysical concepts in accord with the demands of folk psychology or epistemology or whatever practices in the special sciences one wishes to consider. It is thereby able to achieve greater optimization without significant loss.

1.11 Summary

To conclude this introductory chapter, I will now return attention to the three conceptual layers of causation described in §1.4 and summarize how each layer of my tripartite account of causation is intended to work. Each of the three conceptual layers of causation, depicted in Table 1.2, contains its own causation-like concepts, none of which needs to match what we pre-theoretically think of as causation. Yet, all three layers together allow us to make sense of everything concerning causation that we need to make sense of.

The bottom layer contains concepts that are not tailored to match our everyday causal talk but are supposed to provide just enough structure to support the work of the middle and top layers in explaining the utility of folk causal talk and

TABLE 1.2 The Three Conceptual Layers of Causation

Layer	Subject	Metaphysical status	Standards of adequacy
Top	Non-metaphysical aspects of causation	Derivative	RELAXED
Middle	Derivative metaphysics of causation	Derivative	STRICT
Bottom	Fundamental metaphysics of causation	Fundamental	STRICT

vindicating causal principles in the special sciences. It does so mainly by guaranteeing the existence of a consistent basis of facts to which the special sciences can delegate the amelioration of conflicts, thus freeing the special sciences to employ concepts that are not completely, explicitly, consistently systematized and connected to fundamental reality.

The fundamental causation-like relations serve as the foundation for all causal relations and are as objective as causation ever gets. In chapter 2, I will lay out the details of these fundamental causation-like relations, which are based on the hypothesis that some chunks of fundamental reality—events that instantiate fundamental particles or fields of some sort—fix objective probabilities for (or determine) other chunks of fundamental reality. These relations of probability-fixing and determination serve as singular causal relations in my metaphysics of causation.

The middle layer abstracts away from the fundamental relations by incorporating parameters that fuzz the fundamental details in order to represent the sort of behavior we humans deal with in the special sciences and in everyday life. This helps to account for why some kinds of events reliably bring about characteristic effects. The generality that smoking causes cancer, for example, will be understood on my account in terms of derivative metaphysical relations. Most relevant smoking events fix a higher probability of acquiring cancer than the probability fixed by relevant non-smoking events. Relations of probability-raising and related forms of probabilistic influence exist only relative to parameters that are fundamentally arbitrary. These parameters characterize how to fuzz the microphysics and specify counterfactual scenarios to help contrast how things actually evolve with how things could have evolved. Because there is no unique correct way to set the values of these parameters, the relations that incorporate them do not correspond to fundamental reality but involve some degree of arbitrariness, just like mechanical and thermal energy. Nevertheless, the difference-making relations holding in the middle layer are not independent of the bottom layer. The determination and probability-fixing relations from the bottom layer constitute the basic materials for quantifying difference-making, which acquires determinate values once the designated fundamentally arbitrary parameters are assigned values.

In the end, my metaphysics of causation incorporates several common themes in the causation literature: difference-making, nomic dependence, and production.

Together, the middle and bottom layers support a scientific account of the empirical phenomena associated with effective strategies. How they do so is the subject of chapter 5. I will follow the general explanation of effective strategies with a proof of causal directness in chapter 6, and then an account of causal asymmetry in chapter 7. These chapters will exploit the technical terminology developed in chapter 2 to demonstrate important characteristics of causation that hold by virtue of fundamental laws of physics and thus bolster the hypothesis that causation is at least partly based on fundamental physics.

The purpose of the top layer is to provide an account of those aspects of our causal concepts that are inessential to the explanation of the metaphysics of causation but are important components of causation and causal explanation. The main concept present in this layer is the notion of a culpable cause expressed in statements like "The intruder caused the dog to bark," and "Oppressive heat was one of the causes of the traffic jam." I believe the primary (though not sole) reason we have this kind of causal concept is that it allows us to grasp the important metaphysical relations in a cognitively convenient form. Ideally, we could figure out what kinds of strategies are effective by running lots of controlled studies with a large sample of initial conditions that are tailored to expose how much difference each aspect of reality makes in bringing about any desired effects. But because humans need to gain knowledge of effective strategies even when such studies are not feasible and because our ancestors needed causal knowledge when they did not know how to run controlled studies, we have evolved cognitive shortcuts that allow us to make good guesses about general causal relations from an impoverished data set. This, I contend, is one good reason for our having strong intuitions concerning relations of singular causation. Our concept of culpable causation on the whole does a good enough job of tracking the causation-like relations from the bottom and middle layers for practical purposes, but because the folk conception of causation incorporates additional epistemic features that play no essential role in accounting for effective strategies, there are some significant mismatches. Our instinctive judgments of causation will often enough identify *c* as a non-cause of *e* when *c* is generally useful for bringing about events of the same chosen type as *e*. Yet, the main reason we have a concept of causation is that we need to distinguish between the kinds of events that are generally effective at achieving desired results. In chapter 8, I will attempt to explain why this folk causal notion is unneeded in an account of the metaphysics of causation, and then in chapter 9, I will provide a simplistic psychological theory in order to demonstrate why it makes sense for us to have this folk conception of causation given that the metaphysics of causation obeys the system I lay out in the rest of the book.

In the end, the conceptual system I advocate proves to be an extremely revisionary model of our ordinary understanding of causation. It involves reconfiguring the relation between singular and general causation, abandoning traditional models of counterfactual dependence, modifying the accepted distinction between causal and non-causal statistical correlations, and even dispensing with the dogma that we are unable to influence the past. A radical architecture for causation is hardly surprising, though, since the account focuses on explaining empirical phenomena and refuses to be held captive to common sense.

{ PART I }

The Bottom Conceptual Layer of Causation

{2}

Fundamental Causation

According to the scheme laid out in the introductory chapter, causation is conceptually divisible into three layers of causation-like concepts that together explain everything about causation that needs explaining. This chapter is dedicated to elucidating causal concepts insofar as they apply to fundamental metaphysics. The goal here is not to identify the relation among bits of fundamental reality which best fits all the important platitudes associated with our intuitive conception of causation. It is to find some structure that—when combined with a further story about how everything derivative relates to fundamental reality—facilitates explanations of all the empirical phenomena that make our causal concepts worth having.

Ever since Bertrand Russell issued his famous (1913) rejection of the "law of causality" and causal talk more generally, debate has simmered[9] over whether causation exists in fundamental physics and even whether causation exists at all. One position sympathetic to Russell—causal eliminativism—can be based on the observation that theories of fundamental physics standardly posit particles or fields constrained by laws expressed as differential equations, sometimes with a probabilistic component. Such connections among worldly happenings, the eliminativist would argue, do not match the logical structure of neuron diagrams, where mundane events frequently instantiate binary cause-effect relations and where generalities are cashed out in terms of such singular causal relations. A fully fleshed out version of the eliminativist's argument would conclude that fundamental reality is so poorly described by the terms 'causation' and 'causal law' that it is fair to say that causation does not exist.

An antagonist of Russell, however, could simply point to the tremendous utility of causal claims in everyday contexts and throughout unimpeachable scientific practice and argue that causation of some sort must exist regardless of the character of the fundamental laws and regardless of whether there even are fundamental laws. "If causation does not exist," the antagonist can complain, "how is

[9] See (Field 2003; Norton 2003, 2007; Price and Corry 2007) for recent discussion.

it that some ways of influencing the world are reliably more effective at producing desired outcomes?"

The tension between these two positions is easy to reconcile. On the one hand, if fundamental physics is a good guide to fundamental reality, the eliminativist is almost certainly correct that what is fundamentally driving the development of the world does not match philosophers' model of singular causation as directed graphs linking mundane events. On the other hand, Russell's antagonists are certainly correct that there are objective structures in the world that account for why reality behaves in paradigmatically causal ways, structures that call out for explanation. To frame the debate as a struggle over whether causation exists is to draw the discussion into a pointless verbal quibble about where to attach the label 'causation'. The substantive issue is to explain why the world seems to behave so causally, given the plausible hypothesis that the underlying rules for how the world develops do not closely match the platitudes we ordinarily associate with the word 'causation'. Russell himself never proposed a thorough explanation of the utility of causal talk in terms of physics and so did not engage much in the kind of positive philosophical project that would advance our understanding of causation beyond deriding simplistic slogans about causality. (He did make a brief foray into this project by constructing the concept of a causal line (1948), a forerunner of contemporary causal process theories of causation.)

My metaphysics of causation is intended to supply a much more thorough positive account of how a fundamental reality that is similar to models of paradigm theories of fundamental physics can help to explain the utility of our causal concepts. It does so primarily by formulating an all-purpose representation for any causal regularity expressed using components of fundamental reality. The primary virtue of my metaphysical system is its comprehensive applicability to the causation-relevant empirical phenomena in all scientific disciplines. A secondary justification for the metaphysics comes from its identification of several plausible principles obeyed by the most famous theories of fundamental physics and its use of these principles in a derivation of several prominent general features of derivative causation such as causal asymmetry. It does not matter ultimately whether 'causation' is a term appropriately applied to the fundamental causation-like relations or whether there are "causes" in physics. The important task is to uncover which fundamental structures serve as an adequate foundation for the scientific explanation of empirical phenomena that motivate causal terminology.

In order to keep the discussion concrete, I will operate under the working assumption that our present grasp of fundamental physics is a decent preliminary guide to a complete account of fundamental metaphysics, at least insofar as it bears on causation. I certainly have many doubts about the veracity of contemporary theories of fundamental physics and suspicions about what their shortcomings indicate for the metaphysics of causation, but short of revolutionizing physics, I will have to adopt a rather conventional stance toward well-established physical principles, taking for granted that prominent features of empirically successful theories are clues to the structure of fundamental reality.

Fundamental physics provides an excellent starting point for investigating fundamental metaphysics because it is prima facie plausible that whatever is common among the many instances of causation—on the moon, between protein molecules, and so on—has nothing specifically to do with the normative, intentional, or phenomenal. Even if there is some kind of pan-psychism or omnipresent influence by sundry deities, paradigmatic causation among purely physical entities will presumably need to accord with what we know of physics, for example by not routinely violating known conservation laws at noticeable levels. Indeed, given that theories of physics are far richer in the kind of details relevant to causation than any theory of the mental, normative, or theological, it is difficult to believe a comprehensive understanding of causation could ever be achieved without positing enough fundamental structure to make sense of space-time and the microphysical interactions it seemingly contains. Because developing a plausible comprehensive theory of fundamental reality is too grandiose a task for me to take up here, I will simplify matters by presuming that the kinds of things physicists posit when they investigate what they call 'fundamental physics' are among the kinds of things that are part of fundamental reality, as opposed to quarks and space-time and forces being derivative by virtue of some non-physical fundamental reality.

Whether fundamental reality includes anything beyond fundamental physics is far more controversial, but as I noted in §1.6, much of the discussion in this book will have to operate under this further simplifying assumption that fundamental reality includes *only* the kinds of structures we would readily recognize as components of fundamental physics and not any fundamental volitional or normative or economic or biological attributes. I recognize that setting aside these possibilities will forestall a proper engagement with the literature on emergence and non-reductive models of causation in the special sciences. As I emphasized in the introductory chapter, this omission exists because (1) the economics of print publication make a longer book prohibitive and there are already too many other topics that demand immediate attention, and (2) what I have to say about reduction, physicalism, and related topics requires prior familiarity with my theory of causation, my distinction between fundamental and derivative, and the method of empirical analysis, all of which need to be clarified with the expansive discussion and illustrations and justifications included in this volume. So, remember that the auxiliary hypothesis that fundamental reality consists only of fundamental physics is held merely to facilitate discussion of causation; it is not constitutive of my theory of causation, nor am I arguing for its truth in this volume.

One notable problem with using fundamental physics as a guide to fundamental reality is that no one currently has a plausible complete theory of fundamental physics, and it is famously difficult to integrate the partial insights of quantum mechanics and general relativity into even a skeletal framework for a complete theory.

Nevertheless, if we examine empirically successful theories that superficially pass as theories of fundamental reality, we can make progress in understanding causation despite vast ignorance of how the world behaves fundamentally. By authorial fiat, I hereby declare the following to be paradigm theories of fundamental physics: *classical gravitation*, *relativistic electromagnetism*, *general relativity*, and *non-relativistic quantum mechanics*.[10] The goal of this chapter is to identify some commonalities among these paradigm fundamental theories to serve as a guide for what fundamental reality is like and to identify some causation-like relations that will hold for all such theories. Then, in the following chapters, we can use these fundamental causation-like relations to construct models of influence, probability-raising, causal regularities, and more.

By labeling these theories as paradigmatic, I do not intend to suggest that every one of them is flawless as a theory of fundamental reality, but only that each comes close enough to being a coherent theory of fundamental reality that we should try to postulate fundamental causation-like concepts that fit well with them. I think that most of what needs to be said about fundamental causation-like relations is applicable to all the paradigm theories, but allowances can be made here and there when one of them departs in some significant respect from the others. We ought be on guard against the possibility that these four paradigm theories have some shared features that are less the signature of some notable feature of fundamental reality and more the result of these theories being simple enough for humans to invent. To encourage robustness against accidentally shared features of the paradigm theories, it helps to check our conceptual design against toy models that are currently recognized as far-fetched, like ones that incorporate magic spells or mimic the computational structure of video games. But we should not take these outlandish theories too seriously; if accommodating one of them would unduly clutter the conceptual landscape, we should set it aside. Finally, if we are confident of the existence of some particular kind of macroscopic causal behavior, our metaphysics of causation should avoid requiring the existence of fundamental structures incompatible with that behavior.

The concepts in this chapter are certainly not intended to apply to all conceivable models of fundamental reality. Although my definitions accommodate a wide range of fairly realistic possibilities, there are plenty of conceivable worlds with what looks like causation but which instantiate none of the causation-like relations I define. The point of empirical analysis, recall, is not to characterize what causation is in all possible worlds. It is to optimize causal concepts so that

[10] Although quantum field theory (QFT) could be included among the paradigm fundamental theories, because of the difficulty of interpreting it properly and because many of its causal features are already present in non-relativistic quantum mechanics and special relativity, I will set QFT aside, remembering to keep an eye out for any part of my metaphysics of causation that stands an appreciable chance of conflicting with it. QFT primarily differs empirically from other paradigm theories in its accommodation of particle creation and annihilation, and I will partially address this possibility in §2.10.2 with a toy theory of particle decay.

they are flexible enough to accommodate our ignorance about fundamental reality yet inflexible enough to capitalize on plausible contingent truths about the actual world.

This chapter includes some technical details that might intimidate the uninitiated, but the concepts and principles relevant to causation are quite simple. To convey the basic ideas without making the formalism appear artificially daunting, I have front-loaded as much as possible the definitions and assumptions essential to later discussion. Arguments in the first third of the chapter should be understandable without any knowledge of the paradigm theories themselves. More advanced readers can burrow into the details in the remaining two-thirds of the chapter to review how my definitions accommodate the technical challenges each paradigm fundamental theory poses. The majority of the discussion in this remaining portion is only aimed at establishing some terminology that will be needed for arguments in later chapters, so readers should be patient about seeing the identified principles put into action. For readers who do not need to slog through the details, §2.14 summarizes the key features of the paradigm theories that will help us understand causation among derivative entities.

As we work our way through the discussion, keep in mind that the point of examining the fundamental causation-like concepts is not to resolve debates in the foundations of physics but to assume that the fundamental theories are not too problematic and then to draw from them a common structure that will serve as a foundation for explaining some important features of causation that appear in the special sciences and in folk applications.

2.1 Preliminaries

The paradigm theories of fundamental physics all represent reality in terms of models described using mathematics. To construe one of these theories, T, as complete and correct about fundamental reality is in effect to claim that fundamental reality matches some model of T. As it turns out, all the paradigm theories have a common structure. They are all described using functions defined over some mathematical space intended to represent an arena with material contents. The *material contents* of a theory are whatever force fields, particle paths, and other attributes the fundamental theory postulates. We can also refer to material contents as *matter* or *stuff*. An *arena* is whatever space contains the material contents. In many fundamental theories, the arena is space-time, and its material contents consist of fields like the electromagnetic or gravitational or consist of the existence in some locations (and the non-existence in others) of point-like particles known as *corpuscles*. Some more advanced theories, though, have postulated higher-dimensional arenas, and some interpretations of quantum mechanics use a two-component arena that includes space-time plus a separate higher-dimensional space to be discussed in §2.13.

My discussion of causation does not require any particular metaphysical categories. Whether we think of fundamental stuff in terms of events, properties, tropes, facts, states of affairs, or whatever categories future philosophers invent, will largely be a matter of terminological convenience for present purposes. I believe the optimal conceptual categories for understanding causation are best found by first developing a solid enough theory of the fundamental causation-like structures and only later sorting out whether they ought to be described ultimately in terms of tropes or holistic state descriptions or local property instances or something else. It will not help to be picky about such metaphysical details at this stage because ceteris paribus we want a theory of causation to be insensitive to the precise character of the fundamental metaphysics. In practice, I will formulate my theory in terms of events, but that decision is meant to be provisional, broadly inclusive, and translatable into other terminology.

Although the distinction between an arena and its material contents is useful for an intuitive grasp of the physics, nothing requires that the difference between them be determinate or unrevisable. In general relativity, for example, some aspects of space-time are linked by laws to the distribution of matter, which makes space-time seem to count in some respects as a container for the world's material contents and in other respects as a form of matter itself. One should understand my distinction between arena and contents to be flexible enough so that we do not have to take sides on such debates here. It also might turn out that what we think of as paradigmatic material particles turn out to be geometric features of the arena like knots in space-time. Or it could turn out that relationism is true: that the alleged container of matter is derivative (existing largely by virtue of fundamental relations among material entities only). In my terminology, such relationist theories claim in effect that there is no (fundamental) arena and that space-time is derivative. Again, in order for us to cut to the chase, I will bracket debates about relationism and work provisionally with the hypothesis that when a fundamental theory seems to posit a space-time or some other arena, we take for granted that arena as part of fundamental reality. Relationist theories can be made compatible with my overall theory of causation simply by having adequate causation-like relations among bits of matter without using an arena at all. The task of translating the details of my talk of the arena into a form acceptable to relationists is one I will leave as an exercise for the reader.

2.1.1 EVENTS

The building blocks for my account of causation are *events*. At the most general level of discussion, there are two kinds of events: fundamental and derivative. I will provide a quick initial characterization, followed by some important qualifications about how I am modeling events.

A *region* is any portion of a possible arena. There is no restriction on the shape or size of regions events can occupy. They can be disconnected, highly convoluted, extend off to infinity in some directions but not others, or occupy a single point.

Stated simply, a *fundamental event* is an arrangement of fundamental quantities instantiated in some region. In many fundamental theories, the postulated arena is a space-time, and in that case a fundamental event is a nomologically possible space-time region R together with some material contents occupying R, like the value of the electromagnetic field throughout R. A fundamental event need not specify all fundamental attributes existing within its region. It could comprise a space-time region with the full electromagnetic field but without specifying locations of any corpuscle paths, or it could comprise the electromagnetic field over selected subsets of its region and in the remaining regions specify the presence or absence of corpuscle paths. What is needed for the event to count as fundamental is for it to consist solely of components from fundamental reality with no fuzzing of the fundamental quantities. For example, when a fundamental theory requires that masses can only take determinate real-numbered values (in metric units), a fundamental event (according to that theory) cannot consist of a particle at point *p* with mass somewhere in between 1 and 2 picograms. The event must be as determinate as fundamental reality allows. Another way to put this is that fundamental events are maximally *fine-grained* wherever they define their region and material content. When I speak of events being maximally fine-grained, I do not mean to deny or ignore the possibility of ontic vagueness; I just mean that fundamental events are as fine-grained as they can be, compatible with whatever ontic vagueness is being assumed.

The following is a formal definition framed in terms of nomological possibility:

> A (nomologically) *possible fundamental event e* is a (nomologically) possible arena-region R, together with a (nomologically) possible arrangement of fundamental attributes throughout all of R.

This definition is not meant to imply that all categories of existing fundamental attributes need to be included in all parts of R, but it is meant to imply that no subregion of R may be altogether without any fundamental attributes. For example, if a fundamental theory postulates a electric field and a gravitational field, there are possible fundamental events according to that theory that include a specification of the electric field throughout one-third of the event's region and a specification of the gravitational field throughout the other two-thirds. Also, when I speak of an event as occupying R, that means an event whose region is *exactly* R.

One can define an actual event in a straightforward way.

> An *actual fundamental event e* is an instance of a (nomologically) possible fundamental event in the actual arena.

All actual fundamental events occur at some determinate location in the arena, but when discussing fundamental events, it often proves convenient to conceive of them in a way that is independent of their location relative to fundamental events in other locations. So construed, a fundamental event's region includes all the arena relations among its parts but not any external arena relations. For example, according to relativistic electromagnetism, there is a possible fundamental event *e* comprising a fragment of space-time that extends throughout a spherical spatial region with a radius of one meter and no temporal duration together with a complete specification of an electromagnetic field configuration throughout this sphere. The space-time points in *e*'s region are related to each another spatially (and spatio-temporally) along an infinite variety of paths but nothing about *e* implies anything about what happens outside *e*'s region. One can think of *e* in this way as a freestanding entity, a complete (though very small) spherical universe that lasts only an instant. One might also understand *e* as a *type* of event in the sense that the actual universe might include two or three identically shaped regions all satisfying the defining criteria of *e*, namely, the electromagnetic field values at various points throughout the spherical region together with their spatial relations to each other. In order to help maintain a sufficiently clear distinction between type and token events, though, I will from this point on avoid using any fundamental events to represent event types.

In contrast to fundamental events, derivative events permit us to fuzz the physics. The purpose of postulating derivative events is to abstract away from fundamental reality in order to better represent our ordinary conception of events as imprecise happenings. When we refer to the first moon landing, we often have in mind a conception of reality that is insensitive both to the spatial extent of the event and to its many microscopic details. For illustration, let *e* be some actual fundamental event that instantiates one moment of the first moon landing. A hypothetical alteration of *e* that slightly shifts just one of its molecules will result in a possible fundamental event that is numerically distinct from *e*. Yet, for practical purposes we can often conceive of this moment of the first moon landing as being insensitive to the precise position of a single molecule. When we think of a set of multiple possible instances of whatever counts as (near enough) this moment of the first moon landing, we are thinking of it as a coarse-grained event.

My use of 'coarse-grained' and 'fine-grained' is meant to correspond to terminology that is standard in the branch of physics known as statistical mechanics. Unfortunately, some philosophers who model events in a more language-based way use these terms in another sense, so caution is advised.

In full generality, a *derivative event* is by definition any event that is not fundamental. However, for the purpose of discussing causation, the only derivative events we will need to consider are coarse-grained versions of fundamental events:

> A *coarse-grained event* is a set of (nomologically) possible fundamental events.

The kind of possibility invoked throughout this volume is nomological possibility, as dictated by the actual laws or by whatever theory of fundamental reality is under consideration. It is easy to widen the modality by redefining fundamental events to include a specification of whatever metaphysically or logically possible fundamental laws one chooses. Coarse-grained events can then include some members with one set of fundamental laws operating on them and other members with different fundamental laws. We will have no need in this volume to consider such derivative events, so I will leave this extension for readers to explore on their own.

To facilitate communication from here on, lowercase letters will represent fundamental events, and capital letters will be reserved for coarse-grained (derivative) events. Readers should feel free to assume that when the same letter is used in both lowercase and uppercase form in the same context, I intend the coarse-grained event to contain the fundamental event as one of its members. Although a coarse-grained event typically has more than one member, there is no need to enforce that as a strict requirement. A *trivially coarse-grained event* has, by definition, a single (possible) fundamental event as its only member. Technically, all coarse-grained events should be thought of as derivative events because in general they abstract away from fundamental reality by collecting some number of fundamental events to serve as a single entity. But because trivially coarse-grained events have only one fundamental event in the set, they are in a practical sense equivalent to the fundamental event itself. For the sake of convenience and simplicity, I will apply the convention that all coarse-grained events are derivative and rely on readers to remember that trivially coarse-grained events are a special case where their derivativeness does not amount to anything significant.

In practice, it proves extremely convenient (and does no harm) to restrict discussion to coarse-grained events whose members occupy regions that are the same size and shape. This allows us to speak meaningfully of a coarse-grained event *E* having a single size and shape, and that in turn allows us to interpret each of *E*'s members as a different way to instantiate stuff in the same arena location. Once we get to the discussion of general relativity in §2.12, our construal of "same location" will need to be massaged, but without being overly technical at this early stage, I will just note that everything that needs to be said about causation can be addressed using events that are the same size and shape. There is no significant loss of generality because for all fundamental events we will have occasion to consider, one could always consider larger events, in effect expanding any of one's originally chosen events until they all reach the same size. Throughout the rest of this volume, context ought to be sufficient to settle when a given coarse-grained event *E* should be understood as occupying a particular location in the actual arena and when it should be understood more abstractly as a set of possible ways to instantiate stuff in any region of the shape and size defined by *E*.

A coarse-grained event allows us to speak of happenings without being maximally precise about the fundamental details, but any coarse-grained event has built into it a precise specification of the boundary between the fundamental events that instantiate it and the fundamental events that do not. Let us again attend to the first moon landing. For any precisification R of the space-time region occupied by the first moon landing, there is exactly one fundamental event *e* that occupies R and fully specifies all the fundamental stuff in R. Each coarse-graining E_i of *e* is a set of fundamental events that includes *e* as one member. The various E_i correspond to different ways one can fuzz the fine details of the moon landing. One coarse-graining, E_1, might allow for very slight microscopic alterations of *e* to count as the same moon landing by including fundamental events that are just like *e* except for a small shift of a few atoms. Another coarse-graining, E_2, might allow for there to be an extra screw attached to the lunar lander by including such a fundamental event. Another coarse-graining, E_3, might count the possibility of Buzz Aldrin stepping on the moon as the same moon landing event but not the possibility where Elvis steps on the moon. Each E_i defines a precise and complete set of fine-grained possibilities that one chooses to count as the same coarse-grained event.

Note that there is nothing in the definition of a coarse-grained event requiring any of its members to be actual. A coarse-grained event could consist of a bunch of possible fundamental events, each of which specifies the same shape and size of space-time together with different specifications of how a unicorn could be instantiated in that region. Because there are no actual unicorns, the coarse-grained event is uninstantiated and thus does not occur, but it still exists as a collection of possible fundamental events.

> An *actual coarse-grained event* is a coarse-grained event one of whose members is an actual fundamental event (usually instantiated in some designated region of the arena).

I am now going to emphasize a point that is critical to a proper understanding of everything else in the book. My way of modeling events is crucially different from the conception of events used in many standard theories of causation. Standard accounts often construe the events that are metaphysically basic as at least potentially coarse-grained, and furthermore the prototypical events that stand in singular (token) causal relations are understood as non-trivially coarse-grained. In my account, the events that count as basic are always maximally fine-grained. For example, in many accounts, the claim that the first moon landing caused the president to cry, is modeled as a cause-effect relation holding between a single token coarse-grained event, the first moon landing, and another single token coarse-grained event, the president crying. It is usually understood in such theories that had one of the electrons instantiating the lunar lander been shifted slightly, that would count as the very same token cause. On my account, any hypothetical shift of a single electron results in a numerically different (possible) fundamental event.

Coarse-grained events, according to my account, play the role of event types, not event tokens. A coarse-grained event representing the first moon landing would be a set of all the possible fine-grained token events that one chooses to count as satisfying the description "the first moon landing."[11] Because a coarse-grained event in general can be any set of possible fundamental events, the types that are thus represented are fully general; they are not restricted to types that are "natural" or ungerrymandered.

Singular (or token) events can be discussed using the language of coarse-grained events, but all such talk is to be understood elliptically as conveying information about a fundamental event qua member of some type. To say that the coarse-grained event *E* occurred is just to say that one of *E*'s members occurred. That construal corresponds to fine-grained events being fundamental and coarse-grained events being merely a way of abstracting away from the fine-grained details by clustering a lot of possibilities together under a single type. Much confusion will likely ensue if my account is conjoined with the standard construal of events where token events are understood as coarse-grained. On my account, only fundamental events (which are as fine-grained as reality allows) can be part of the actual world.

Because the focus of this chapter is on fundamental reality, references to events in this chapter are to possible fundamental events unless context indicates otherwise.

It is convenient at this point to introduce a bit of additional terminology to streamline later discussion. Let us define a *subevent* to match the way we speak of subsets.

> An event e' is a *subevent* of an event e iff e''s region is a subset of e's region and the arrangement of e''s fundamental attributes are included in e's arrangement of fundamental attributes. An event e is a *superevent* of e' iff e' is a subevent of e.

A proper subevent of e may occupy the same region as e but specify fewer fundamental attributes.

Although an event does not need to specify all fundamental attributes and entities in its region, such events are particularly useful. These are known as *full events*.

> A fundamental event is *full* iff it comprises fundamental attributes of every type throughout its entire region. (A coarse-grained event is full iff all its members are full.)

For the purpose of identifying full events, a fundamental theory is expected to designate which types of attributes must be specified for an event to count as full according to that theory even if, strictly speaking, the theory's fundamental laws permit more types of attributes than exist in the models one is considering. An actual full event is an event that includes every attribute instantiated in its region;

[11] Experts should avoid tripping themselves up here by worrying about the semantics of names and descriptions. See (Kutach 2011b) for my account of related issues.

it does not need to instantiate types of attributes that do not exist in the actual arena.

For example, in relativistic electromagnetism, the postulated types of attributes are electromagnetic field values and the presence or absence of corpuscle paths bearing charge and mass properties. A full event in region R, according to relativistic electromagnetism, would include a determinate value for the electromagnetic field throughout R as well as a complete specification of everywhere there is (and everywhere there is not!) a corpuscle path and what its mass and charge is. In a theory that also included the strong and weak nuclear interactions, *e* would not be full because it does not comprise the strong and weak charges or fields. By not including any weak or strong attributes, *e* simply remains silent about these types of attributes. Events do not come pre-equipped with any default values. For example, if the complete specification of an event is silent about the magnitude of the electromagnetic field *F* in some region R, then *F* is undefined in R rather than having a value of zero.

The arena and all its contents together count as the largest actual full fundamental event, to be known as the *world event*. Any actual fundamental event is a subevent of the world event.

Note that there are no restrictions on how coarsely grained a coarse-grained event can be. For an extreme example, even the set of all nomologically possible worlds counts as a legitimate coarse-grained event. (Its occurrence is guaranteed by law no matter how the universe's corpuscles and fields are arranged.)

2.1.2 LAWS

All the paradigm fundamental theories postulate laws. What constitutes a law in general is difficult to say, but we have enough exemplars to give us a firm enough practical grip on the concept. A rough initial characterization of laws that is well suited to the study of causation is to think of one's fundamental theory as specifying some sort of arena and various kinds of material contents to inhabit it. The set *S* of all logically or metaphysically possible worlds that instantiate an allowed arena with its material contents constitutes an overly broad set of physical possibilities. The laws are merely additional constraints on *S*. What we get when we restrict *S* to the worlds that obey the designated fundamental laws is the set of worlds that count as the nomologically possible worlds.

One often thinks of the laws and the material content as corresponding to two importantly different kinds of facts. Because there is often an intuitive difference in how these facts are handled in the paradigm theories, I will refer to them as *law facts* and *material facts*. However, none of the conclusions I will draw about causation depend on there being a deep or principled difference. Sometimes interpretations of fundamental theories can differ in what they regard as law and what as material. For example, relativistic electromagnetism can be interpreted as having a fundamental electromagnetic field, in which case all four equations that govern the evolution of the electromagnetic field—traditionally known as Maxwell's laws—count as genuine laws. Alternatively, it can be interpreted as having a fundamental electromagnetic gauge field, in which case two of Maxwell's equations do not count as laws but instead as constraints that obtain merely by virtue of what the gauge field is. For the purpose of understanding causation, there is no need to decide whether the electromagnetic field is fundamental or the gauge potential is fundamental, nor is there a need to insist that there is a definitive answer. One can leave open whether only the two equations count as laws or whether all four equations count as laws or whether some other option holds. Though I speak of law facts and material facts as though they are different, I accept as a condition of adequacy that to the extent that no significant difference exists between material facts and law facts in fundamental reality, none of my theory's claims should depend on the existence of a difference.

My definition of laws in terms of a privileged class of possible worlds is friendly to a reading where physical modality is distinguished metaphysically. A longstanding feud exists over whether laws of nature are to be accorded some sort of robust ontological status (Tooley 1977, Dretske 1977, Swoyer 1982, Armstrong 1983, Carroll 1994, Maudlin 2007a) or whether laws are merely some sort of privileged summary of the actual layout of material contents that are notable for their epistemic value in explanation, prediction, etc. (Mill 1858, Ramsey 1928, Lewis 1973a, Loewer 2004). Unsurprisingly, some alternatives are also in play (Ward 2001, Roberts 2009).

In my terminology, this dispute should be framed as a disagreement about whether fundamental reality includes laws or whether instead all laws are derivative, presumably existing by virtue of the arena and its material contents. As far as I can tell, one does not need to take a side in this debate in order to answer the important questions about causation, so I will avoid engaging in the dispute as much as possible. My references to fundamental laws may be interpreted with either a robust or deflationary conception of the modality inherent in fundamental laws, and whatever robustness exists in one's account of fundamental laws will carry over to the modality inherent in causation. If the deflationary conception is ultimately found to be superior, one presumably would want to adjust the terminology of my theory so as not to mention fundamental laws at all and to instead focus on some privileged class of derivative laws that play a role akin to what are ordinarily recognized as fundamental laws.

One sense in which I take the distinction between law facts and material facts to be important is that in thinking about the hypothetical possibilities relevant to causation, one should in practice consider only those worlds with the same fundamental laws. Some theories of causation, for example (Lewis 1973b, 1986), entertain the idea that we can understand causation by way of miracles, violations of the fundamental laws of the actual world. Although I will not argue that it is incorrect to invoke miracles, they play no role in my theory. When evaluating what would have happened if *C* had happened, we hold the fundamental laws fixed, fill in some supposition about how C is instantiated by fundamental stuff, and then evaluate what would have happened using only the fundamental laws. So, although I do not reject out of hand the thesis that all laws are derivative, I do assume the distinction between law facts and material facts is clear enough to ground the practice of keeping the laws fixed and letting the material facts vary when considering how history could have unfolded. This practice is compatible with a fuzzy boundary between laws and material facts. The fuzziness can be handled by first relativizing claims about hypothetical happenings to various specifications of a precise boundary and then showing that differences among the resulting precisifications are largely irrelevant to the explanation of the target empirical phenomena.

If the actual world does not possess the kind of fundamental laws my theory postulates, there are two substantively equivalent (though verbally different) ways to represent the failure. One option is to accept that the world instantiates no causation, only faux-causation. The other option is to reject theories like mine, interpreting the reality of causation as a falsification of my account. Either way of thinking about causation is acceptable. Such a choice amounts merely to a terminological decision about whether to identify 'causation' with whatever it is that ultimately accounts for putative instances of causation or to incorporate into the concept of causation contingent hypotheses about how paradigmatic causes are related to things like time, laws, and chance. All of this is just another way of stating that there is no fact of the matter about whether my account of causation is falsifiable.

One of the consequences of the privileged ontological status fundamental reality has over derivative reality is that there is a corresponding privilege that fundamental laws have over derivative laws. Derivative laws, as I understand them, are nothing more than rules of thumb that are useful by virtue of fundamental reality. Just as the mere distinction between fundamental and derivative has few consequences by itself, so too does the distinction between fundamental and derivative laws. However, once supplemented with additional theses about the character of fundamental reality, the difference can become much more controversial and interesting. If it turns out that fundamental reality is nothing more than fundamental physics of a kind not too remote from what is posited in our paradigm theories of fundamental physics, then the laws of the special sciences will be derivative laws that are handy rules of thumb merely by virtue of the fundamental existence of the arena and the layout of its material contents and any fundamental laws. Many experts reject this role for special science laws, and within the framework I am advocating, anyone who holds that a given special science law L is more than a rule of thumb is in effect treating L as fundamental in the sense I described in §1.6. Because of space limitations, I cannot engage in any debates here over whether special science laws are best construed as fundamental or derivative. I will instead offer just two brief comments. First, both conceptions of special science laws are compatible with my account of causation and compatible with the broader conceptual framework of Empirical Fundamentalism. Second, an important argument in favor of my account of causation is that—when conjoined with the auxiliary hypothesis that fundamental reality resembles models of the paradigm theories of fundamental physics—one can explain at least two important general facts about special science laws, namely (1) that they are hedged to accommodate the possibility of disruptive external interferences or exceptional initial conditions and (2) that they exhibit the same temporal directionality. Demonstrating such features is hardly a decisive argument that special science laws are derivative, but they count for something.

Throughout this chapter, 'law' will refer only to fundamental laws, and because fundamental physics will be used as a preliminary guide to fundamental reality, the laws may be assumed to resemble laws of paradigm theories of fundamental physics.

2.2 Terminance

The simplest fundamental causation-like relation is determination:

> A fundamental event *c determines* (and is a *determinant* of) a fundamental event *e* iff the occurrence of *c* nomologically suffices[12] for the occurrence of *e* (with *e*'s location relative to *c* being built into this relation). A *determination* relation exists from *c* to *e* iff *c* determines *e*.

The definitions here apply to (nomologically) possible fundamental events, so we can speak of an actual event determining an actual event, or a possible event determining a possible event.

Traditionally, the expressions 'full cause' and 'real cause' have been used to express something in the same conceptual neighborhood as 'determinant' but it is wise to avoid the word 'cause' due to several undesired connotations.

An important technicality in the definition of determination is that *e* is intended to be specified in terms of its size and shape and material content as well as where *e* is located in the arena with respect to *c*. The reason for including information about the location of *e* relative to *c* is illustrated by cases where *c* determines that some *e* occurs one second later but where another event *e*′ that is qualitatively identical to *e* exists somewhere else in the universe. In such cases *c* should not automatically count as a determinant of *e*′. The specification of *e*'s location relative to *c* is additional structure beyond the information needed to define *e* itself (in terms of the arena relations among its subevents). This locating information is defined in terms of the geometrical properties of the arena. The kind

[12] It is already part of the conception of fundamental reality I outlined in §1.6 that fundamental reality as a whole cannot violate the metaphysical correlate of the law of non-contradiction. Thus, no fundamental event will ever suffice for another event by implying a contradiction. Similarly, it is impossible for a fundamental event that determines *e* to determine another event whose existence is incompatible with *e*, that is, that cannot be instantiated in the same possible fundamental reality as *e*.

of relation needed can vary depending on the fundamental theory, but in space-time theories at least, one typically uses some combination of spatial, temporal, or spatio-temporal distances and angles among all the subevents of *c* and *e*, including the geometrical structure between them.

The reason determination is important is that—together with its extension into fundamentally chancy theories—it provides a basis for all other kinds of causation. The relation of some event *c* determining an event *e* somewhat resembles a causal structure because it plays many of the constitutive roles of causation. The event *c* can *make e* happen in the sense of nomologically sufficing for it. Knowing enough about the laws and about *c* and *e* often justifies an accurate *prediction* that *e* will follow with certainty from *c*. If a person can manipulate things so that *c* occurs, she can thereby reliably *bring about e*. The event *e depends on c* in the sense that hypothetical modifications to *c* determine modifications of *e* with systematic covariance. Also, under the right conditions, *c*'s occurrence can *explain e*'s occurrence. Determination relations, one might say, cement various chunks of the universe together fundamentally.

Although determinants are causal in this sense, there are several important features of determination that are in tension with the constitutive features ordinarily thought to apply to causation among mundane events. For example, determination is both reflexive and transitive, and it is compatible with being symmetric. Yet, philosophers tend to think of causation as irreflexive, non-symmetric, and not necessarily transitive. Because of such differences, I will continue to describe determination (and other soon-to-be-defined concepts) as merely causation-like relations.

Despite my hesitancy to use the word 'causation' to describe determination relations, I must emphasize that actual determination relations—that is, determination among actual fundamental events—are genuine relations of singular (and actual) causation in the sense that they are the components of the actual world that bind various happenings causally. The determinants (and their chancy counterparts) are "doing all the causal work," all of the pushing and pulling in the actual world.[13]

Although determination captures one aspect of fundamental reality that is causation-like, some fundamental theories have no determinants subsuming mundane instances of what we intuitively think of as causation. Instead, these theories are *stochastic*, which means that they include a fundamentally chancy

[13] It is sometimes (Russell 1913) claimed that scientific laws (of physics, at least) do not invoke causal relations but only "functional relations between certain events at certain times." The expression 'functional relations' provides a poor description, though, for the causation-like relations in fundamental physics. First, there are many different kinds of functional relations between events, only a few of which serve as an adequate surrogate for causal relations. Second, there can be accidental functional relations between two physical variables without any corresponding interaction. Third, the concept of determination can be generalized into concepts that play the causation-like role needed to make sense of causal regularities but without any functional relations, as discussed in §2.11.2.

component. Stochastic theories posit fundamental events that (together with the fundamental laws) suffice for a probability distribution over other possible fundamental events.

To formalize this idea, let us first introduce a new kind of derivative (coarse-grained) event.

> A *contextualized event* $\overline{E}$ is a coarse-grained event with a probability distribution assigned over all its members.

Throughout, I will use a bar over the capital letter to designate contextualized events. The bar does not represent a function or operator that converts a plain coarse-grained event into a contextualized event; it is merely part of the label of any contextualized event. When the same capital letter appears both unadorned and with a bar over it, as *C* and $\overline{C}$ for example, that is meant to carry the default connotation that every member of *C* is a subevent (in *C*'s designated location) of some member of $\overline{C}$ and every member of $\overline{C}$ has some member of *C* as a subevent (in *C*'s designated location). This captures the idea that if we start with any plain coarse-grained event *C*, we can *contextualize* it by expanding all of its members in the arena as far as needed to incorporate a specification of *C*'s environment and then supplying a probability distribution over these members. Also, one can define a *trivial contextualization* of some *c* in the obvious way as a contextualized event with a single member (that is a superevent of *c*) carrying all of the probabilistic weight).

Probability distributions include both probability mass functions, which are applicable to non-continuous sets and probability density functions, which are applicable to continuous sets by assigning non-negative values to its measurable subsets. Nothing of substance in my account will depend on the technicalities concerning the representation of probability distributions, and readers should feel free to substitute some fancier formalism if desired. For concision, I will not stray into a discussion of non-standard probability theory, but I see no reason why my account cannot be adapted for more general probability-like concepts.

Let us now go beyond the concept of determination to incorporate probability.

> A fundamental event *c* *fixes* a contextualized event $\overline{E}$ iff *c* nomologically suffices for $\overline{E}$.

In other words, *c* (which can be actual or merely possible) together with all the fundamental laws is logically sufficient for $\overline{E}$'s probability distribution over all the possible fundamental events that are members of $\overline{E}$. Like determination relations, we should assume that a fixing relation incorporates a specification of the region for $\overline{E}$ relative to *c*.

An event *c* can nomologically suffice for the occurrence of $\overline{E}$ without fixing it. This can occur, for example, when an actual event *c* determines an event *e* that is a member of $\overline{E}$, thereby sufficing for $\overline{E}$'s occurrence. Fixing, however, requires *c* to suffice nomologically for the probability distribution over all the members of $\overline{E}$, not just for the occurrence of one of $\overline{E}$'s members.

An event $\overline{E}$ fixed by c is allowed to be a proper subevent of a larger or more richly specified event $\overline{E}'$ that c fixes, but such subevents must have the same relative probabilistic weight as the superevents they are drawn from. When representing a subevent of a contextualized event, there is an opportunity to reduce the probability distribution to something equivalent but simpler, but such reductions will not be exploited by any arguments in this volume, so I will leave the details aside.

All of the paradigm theories are such that when an event c fixes an event or probability by way of dynamical laws that propagate an event continuously through time in one direction, c fixes a *unique* event or probability. An argument that the uniqueness holds more generally is postponed until chapter 6, but the definitions of fixing and probability-fixing are not intended to rule out by fiat the possibility that multiple conflicting events or probabilities can be fixed by the same event. To incorporate such a possibility, the fundamental laws would need to include explicit rules for how to ameliorate any generated conflicts so that they accord with the principle that fundamental reality is consistent. Having noted this possibility, I have chosen, for the sake of a more readable presentation, to avoid continually raising the question of how my definitions would apply in models where events fix apparently conflicting events. To do so, I will tacitly assume discussion is restricted to models where the conflicts do not arise unless context dictates otherwise. This allows us to speak of *the* event c fixes for R, meaning the unique event $\overline{E}$ occupying R and fixed by c such that any other events that c fixes for R are subevents of $\overline{E}$. We can say of such an event that it is the *maximal* event fixed by c for R. Furthermore, conditional statements of the form, "If c fixes an event $\overline{E}$ for region R, . . ." are meant to pick out *the* unique maximal $\overline{E}$, not its subevents. (The assumption of uniqueness also applies to relations of probability-fixing, defined below.)

The fixing relation has nothing to do with the idea that as time passes, future events become fixed by becoming present or past. Indeed, for the purpose of understanding this chapter, readers will likely find it convenient to bracket any intuitions they may have concerning the metaphysical nature of the passage of time and attend to the sorts of resources available to static theories of time. An event c can fix a contextualized event $\overline{E}$ ten seconds in the future located in region R, and an event c' one second after c's occurrence can fix a different contextualized event $\overline{E}'$ for the same region R. It is even permissible for an event occurring temporally after R to fix (toward the past) yet another contextualized event occupying R. Which contextualized events are fixed for a given region depends on what events are doing the fixing, and many of these events occur at different times.

We can also extend the idea of fixing to apply to plain coarse-grained events.

> A fundamental event c *fixes a probability* p for a plain coarse-grained event E iff c nomologically suffices for a probability p for E.

To get a better picture of how *E*'s probability can be fixed, let us restrict discussion to some chosen *E*, all of whose members are full events of the same size and shape. Furthermore, let the region of the chosen *E* be R and consider a contextualized event, $\overline{E}$, that *c* fixes for R. If *c* does not fix an event occupying R, *c* does not fix a probability for *E*. Each member, e_i, of $\overline{E}$ will fall into exactly one of three classes. The first class includes every e_i that is not a member of *E* but is a subevent of some member of *E*. Intuitively, this occurs when e_i does not specify enough of the material contents in R to ensure that (the entirety of) *E* would occur. The second class includes every e_i that is a member of *E*. The third class includes every e_i that is not a member of *E* and is not a subevent of a member of *E*. If a measurable subset of $\overline{E}$'s members fall into the first class, *c* fails to fix a probability for *E*. Otherwise, the probability that *c* fixes for *E* is the proportion of $\overline{E}$'s members that fall into the second class.

An event does not fix a probability merely by implying specific values for quantities satisfying the axioms of probability theory. For example, under paradigmatic deterministic laws, a full event *c* at a single time can determine a full event *e* consisting of everything happening in some chosen region that includes exactly three coin flip outcomes: heads, tails, heads. Thus, *c* fixes an event—a trivial contextualization of *e*—that suffices for a frequency for tails of one-third. Because the frequencies implied by these three coin flips satisfy the axioms of probability, *c* nomologically suffices for a quantity that counts formally as a probability. As I am defining the terminology, this does not count as a scenario where *c* has fixed a probability of one-third for an occurrence of tails. It counts only as case where *c* has fixed a probability one for an occurrence of heads, tails, heads.

Fixing and probability-fixing should instead be conceived as metaphysical relations[14] whose existence depends on the character of the fundamental laws, and if there are any unresolved questions about whether a coarse-grained event has its probability fixed according to some candidate model of fundamental reality, that would be a matter to be settled simply by being more precise about what the fundamental laws dictate.

It is easy now to extend the definition of 'fixing' so that contextualized events can engage in fixing (and thus probability-fixing):

> A contextualized event $\overline{C}$ *fixes* a contextualized event $\overline{E}$ iff $\overline{C}$ nomologically suffices for $\overline{E}$.

The sufficing works here by using $\overline{C}$'s probability distribution to weight each of the contextualized events fixed by each member of $\overline{C}$ so that all of the resulting fundamental events are assigned an appropriate probability distribution (in terms of $\overline{C}$'s probability distribution and any chanciness in the dynamical laws). If any subset

[14] Because these relations in general hold between derivative events, they are derivative relations. In the special case where both events are trivially coarse-grained, the distinction between the relation being derivative rather than fundamental loses its significance.

of $\overline{C}$'s members (of non-zero measure) fix no contextualized event throughout $\overline{E}$'s region, $\overline{C}$ does not fix $\overline{E}$.

Finally, we can define a relation based on fixing that holds only between fundamental events.

> For any possible fundamental event *f* and any two of its subevents, *c* and *e*, *c termines* (is a *terminant* of) *e* iff *e* is a subevent of a member of a contextualized event that *c* fixes.

As in the case of determination, the events that *c* termines are specified in part by their location in the arena relative to *c*.

This definition implies a special case that is applicable to the actual world.

> An actual fundamental event *c termines* (is a *terminant* of) an actual fundamental event *e* iff *e* is a subevent of a member of a contextualized event that *c* fixes.

The essential idea behind a terminant is conveyed by imagining some possible arrangement of material contents in some portion of space-time. Within that arrangement, we can select any two events *c* and *e*. Let R be the region occupied by *e*. If *c* does not fix a contextualized event for R, *c* does not termine *e*. If *c* does fix a contextualized event for R, let us call it $\overline{E}$. If *e* is not a subevent of some member of $\overline{E}$, that means *e* instantiates more attributes than what *c* has fixed for R. In that case, *c* does not termine *e*. Otherwise, *c* has fixed an event, $\overline{E}$, that is instantiated by *e*. So, by definition, *c* termines *e*.

It is helpful to have a term for this generalization of determination. In order to avoid the unpleasant connotations of the word 'termination', I will refer to this relation as *terminance*. Terminance is by definition the fundamental causation-like relation that holds from a fundamental event *c* to a fundamental event *e* when *c* termines *e*.

What should we say if, for an event *c*, the laws specify what could happen in *e*'s region for two-thirds of the probability distribution, but fails to imply anything about what happens for the other one-third of the probability distribution? For the sake of simplifying some later proofs, it is convenient to insist that an event is a terminant of *e* only if it provides a complete probability distribution for what happens in *e*'s region. When *c* fixes part of a probability distribution for what happens at *e*, we can say that *c partermines e*. The concept of a parterminant is peripheral to the overall discussion of causation, but it will make a few fleeting appearances.

As an inspection of the definitions reveals, a determinant is always a terminant, but not vice versa. When an event *c* is a terminant of *e* but not a determinant of *e*, we say that *c* is an *indeterminant* of *e* and that a relation of *indetermination* holds from *c* to *e*. When *c* is a determinant of *e*, *c* makes *e* happen, but when *c* is an indeterminant of *e*, *c* probabilistically constrains what can happen at *e*'s location, but then leaves it up to chance which of the permitted events is instantiated.

Indeterminants reveal another discrepancy between our folk causal locutions and what is going on fundamentally. When *c* fixes a contextualized event $\overline{E}$ that happens to be instantiated by *e*, it fixes a probability for *e*, but it does not in general *make* *e* happen. In suitably simplified situations, if *c* fixes a very high probability for *e* and *e* occurs, then we might say *c* causes *e* because we tend to ignore the small possibility that *e* would fail to occur. But when the probability fixed for *e* is low enough, we tend not to think of *c* as causing *e*. In realistic theories, the contextualized events that are fixed by indeterminants usually include a continuous probability distribution over their members that assigns probability zero to each individual member.[15] In such cases, *c* will assign non-zero probability to some measurable sets of possible fundamental events but not to *e* itself. If we were to interpret the connection between terminance and causation too literally, we would find it strange that *c* can be the fundamental stuff that causes *e* when it makes *e* have a zero probability of occurring. Again, this is just another reminder of the mismatch between fundamental causation-like relations and what we ordinarily think of as causation. I will have much more to say about our practices of causal attribution later, but for now it is enough to note that the generalization of determination into stochastic theories widens the gap between our native conception of causation and what in fundamental reality is responsible for the utility of such talk.

Determination is a fundamental relation that somewhat resembles our ordinary notion of causation, at least by being able to play many of the conceptual roles that causation plays, but determinants are too restrictive for general use because they are unsuitable for fundamental theories that posit chanciness. In order to be more accommodating to the possibility of fundamental chanciness, we can just use terminant as the fundamental causation-like relation that plays the role of full cause. It turns out that much of what we want from determinants for understanding causation is provided just as well by the more general notion of terminance.

What justifies the importance of terminance is not anything in the fundamental theories themselves. One should not examine a fundamental theory by itself (ignoring its consequences for derivative metaphysics) to locate whatever plays the fundamental causation-like role. What makes terminance special is that relations of terminance among possible fundamental events are quality building blocks for a probabilistic notion of influence that will later prove to be optimal for representing all general causal relations. The only sort of argument I can provide for why terminance is an excellent concept to use as a basis for understanding causation is just to spell out its consequences for the derivative metaphysics of causation in the middle part of this volume.

---

[15] For readers unfamiliar with how events can happen even when they have zero probability, imagine an idealized arrow that is certain to hit a target of area *A* with a uniform probability distribution. Its chance of hitting a target patch of area *T* is simply T/A. Because a point has zero area, the probability of hitting any given point is zero. Yet, the probability of hitting somewhere among a continuum of points of finite area is non-zero.

At this point, the discussion warrants an important clarification about the operative conception of probability. There is a long tradition of attempts to construct probabilistic theories of causation along the lines of (Reichenbach 1956, Suppes 1970, Cartwright 1994). All such theories employ a different conception of probability from mine. On their views, probabilistic relations can hold between two (typically coarse-grained) happenings without there being any fundamental law connecting them. On my account, by contrast, the only probabilities that are ever invoked either arise from stochastic rules in the fundamental laws or are simply stipulated as part of one's (fundamentally arbitrary) choice of contextualized event. In chapter 5, I will discuss how the two kinds of probability postulated by my theory relate to statistical correlations, but for now, readers should note that my account never appeals to the kind of probabilistic relation employed by standard theories and so is not subject to most of the standard counterexamples to probabilistic theories of causation.

The difference in how my account models probability is neither a mere technicality nor a philosophical quibble. On my account, the kinds of events that can fix a probability for a given effect routinely turn out to be very big. Illustrated in Fig. 2.1 is the activation of an egg timer, *c*, followed three minutes later by *e*, the timer's buzzing. A standard probability-raising account of causation would claim for the imagined circumstances that *c* raises the probability of *e* or the probability of *E*, defined coarsely as the timer buzzing three minutes later. Or it might claim that if *c* occurs in an environment without anything to interfere with its operation, *E* will be highly likely to occur. On my account, no such probabilistic relations are invoked or required. Instead, the only kind of probabilistic relations at work in this example hold by virtue of much larger superevents of *c* that extend at least three light-minutes (about 900,000 km) in all spatial directions. Any smaller event will not have any probabilistic connection to *e* (except insofar as it is part of a large enough event). The superevent *c′* fixes a probability for *E*, which makes *c′* a terminant of *e*. This probabilistic connection does not require or allow any hedges like "if nothing interferes." Anything that could potentially interfere with the successful operation of the egg timer has already been included in the probability that *c′* fixes for *E*.

The fundamental laws might be such that as you get closer in time to *e*, the events that termine *e* become smaller and smaller. But even events that occur only

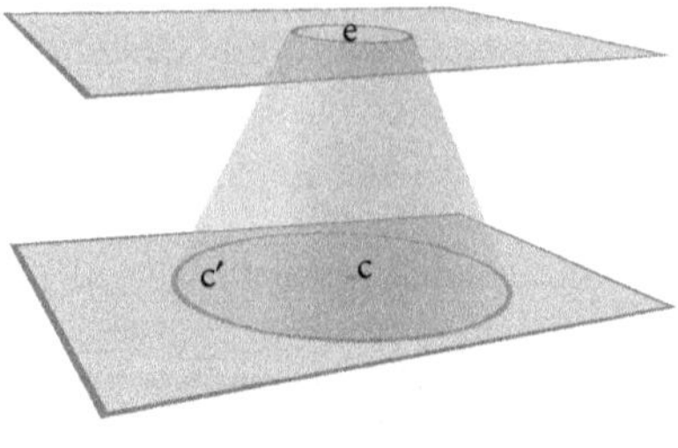

FIGURE 2.1 *Only events as vast as c′, which occupies the whole circle, fix a probability for some coarse-graining, E, of e. There is no probabilistic relation between the puny c, depicted as a dot, and E.*

a millionth of a second before *e* will still need to occupy a volume of well over a million cubic meters. So, we can say that for the kinds of causation we are usually concerned with, where the events are not microscopically small and where the times involved are longer than a microsecond, the fundamental events that play the role of full causes are much bigger than the events we ordinarily cite when providing causal explanations.

2.2.1 CAUSAL CONTRIBUTION

Terminants have been designated for the role of full cause in fundamental reality, and the role of partial cause will be played by a *causal contributor*, or just *contributor*.

> A fundamental event *c* *contributes* to (is a *contributor* to) a fundamental event *e* iff there exists a terminant c' of *e* such that c' with *c* excised from it is not a terminant of *e*.

To excise some event *c* from an event c' is to define a new event that is just like c' except without a specification of *c*'s material content in the region where *c* and c' overlap. When you excise, you do not substitute a vacuum or anything else; you in effect remain silent about whether the excised material content exists. For the sake of formality, we can define an event c'' as the *excision* of *c* from an event c' as the unique event whose initially defined region is the same as c' and which comprises every attribute in c' that does not exist at the corresponding location in *c*, taking care in the end to trim away any part of c'''s region where no attributes remain.

One can think of *e*'s contributors as all the fundamental events that play a role in the development of reality toward *e*. The formal definition captures the idea that a contributor is a partial cause in the following sense. If there is a terminant c' of *e* that stops being a terminant of *e* when you excise *c* from it, then *c* is necessary for c''s being a terminant of *e*. Because *c* is in this sense required for some "full cause" of *e*, it is (in one sense) a partial cause of *e*. We can also say that a contributor to *e* is any non-superfluous subevent of one of *e*'s terminants.[16]

Imagine some *c*, the presence of fuel in a certain rocket, that contributes to some *e*, the rocket's successful launch. Then, let c' be some event consisting of *c* conjoined with a totally unrelated event in the distant future far removed from *e*. It follows from the definition of 'contributor' that c' is a contributor to *e*. Because the definition does not require that all subevents of a contributor to *e* be contributors to *e*, the composite event c' counts as a contributor to *e* merely by virtue of the contribution of its proper subevent, *c*. Because it is handy to ignore contributors like

[16] The reason not to define a contributor as "part of a terminant" is that that would trivially make everything a contributor to *e*. The world event, *w* trivially determines any actual event *e*. So every part of *w* would be a contributor regardless of whether any fundamental laws link it to *e*.

c' that have superfluous components tacked on, we can introduce a more refined conception of contribution to exclude the extra baggage.

> A fundamental event c *purely contributes* to (and is a *pure contributor* to) a fundamental event e iff every subevent of c contributes to e.

A causal contributor bears a close resemblance to John Mackie's conception of a partial cause as an *inus condition* for some effect. An 'inus condition' is shorthand for "an *i*nsufficient and *n*on-redundant part of an *u*nnecessary but *s*ufficient condition." (Mackie 1973, p. 62) The most noteworthy differences between an inus condition[17] and my conception of a contributor are that (1) contributors can exist by virtue of fundamentally chancy relations whereas inus conditions require determination, (2) contributors exist only as fine-grained events characterized using fundamental attributes and so are not as general as Mackie's "conditions," which include absences and (arguably derivative) macroscopic happenings, and (3) contributors only contribute by virtue of fundamental laws whereas Mackie imposes no such restriction.

2.2.2 TRIVIAL TERMINANCE

Inspection of the definition of 'terminant' reveals that every event termines itself. This may seem like a commitment to a pervasive self-causation, but there is nothing metaphysically mysterious or dubious about self-terminance. Although a common maneuver in analyses of causation is to rule out by fiat (or at least to avoid pronouncing on) cases where an event causes itself, trying to incorporate the lack of self-causation into the concept of terminance does more harm than good by unnecessarily complicating the definitions. Instead of building into the definition of terminant a restriction to cases where the termining event is distinct from the termined event, it is better as a matter of conceptual design to identify classes of terminance that are somehow trivial and then to count trivial instances as being among the kinds of causation that we rightfully ignore when we talk of causation in the special sciences and in everyday language.

Here is one sufficient condition for triviality:

> A fundamental event c is a *trivial determinant* of a fundamental event e if the occurrence of c suffices (without any laws) for the occurrence of e.

Because every event is a determinant of itself and all of its subevents regardless of the laws, all self-determination is trivial.

---

[17] Judging by what Mackie seemed to have in mind, his inus condition is probably more accurately labeled an 'ns condition' because the 'insufficient' and 'unnecessary' in the acronym should not be considered strict requirements but merely reminders that the full cause *does not have to be* necessary for the effect, and the non-redundant partial cause *need not* by itself suffice for the effect.

Another variety of terminance that might count as trivial, depending on how one chooses to think of it, is determination by constitution. Imagine a hypothetical fundamental particle called a 'simplon'. Let the event c be the existence of a mass m located at point p. Let e be the event of a charge $+1$ existing at p. Suppose the only kind of particle that has mass m is a simplon and that simplons always have charge $+1$. Depending on how we construe fundamental reality, we might think of the uniform regular connection going from 'having mass m' to 'having charge $+1$' as a fundamental law. If so, c determines e. This is a case of *determination by constitution*. Let us say in general that c determines e by constitution iff c occupies at least all of e's region and every subevent e' of e is determined by some subevent of c that does not lie outside of e''s region. Determination by constitution is another example of a fundamental causation-like relation where we find it inappropriate to use causal terminology. We would not naturally say that having a mass m at point p *causes* the existence of a $+1$ charge at point p even when the fundamental laws imply that the mass at p determines the charge at p. We could complicate the definitions to exclude such relations from counting as genuine terminance, but I believe the definitions are better streamlined by allowing them and then explaining away our reluctance to think of them as causal by pointing out that we naturally tend to exclude as causal those cases where an event at p determines stuff at p. Also, note that even though determination by constitution is non-trivial in the sense that it requires a law of nature in order to hold, it does not have any deep consequences for derivative causation. If one can reliably create a mass of m at some point p, one can thereby make a charge of $+1$ exist there as well, but that holds true regardless of whether the connection between the mass and the charge exists by virtue of a fundamental law or merely by virtue of an accidental universal regularity.

Several other determination relations can be subsumed under determination by constitution and thus safely bracketed for the purposes of empirical analysis. One example comes from constraints on the kinds of quantities the fundamental laws allow. Suppose that fundamental reality includes a field quantity that is required by fundamental law to be a continuous real-valued quantity defined everywhere throughout the arena. Take any event e that includes a complete specification of all the fundamental attributes instantiated in a continuous region, and let that region include some point p in its interior. Let e' be identical to e except that it lacks a value for the field at the single point p. It follows merely from the law that the field must be continuous that e' determines e even though e' is a proper subevent of e. Such terminance might be considered trivial in the sense that it does not hold by virtue of any dynamical law but merely the fact that the field must be continuous.

Having noted this possibility, I want to withhold consideration of all such laws for the purpose of streamlining later definitions. The complication such laws of continuity pose is that they undermine the utility of any straightforward definition of a pure contributor. An unremarkable event could fail to have any pure contributors just because its terminants consist of attributes instantiated at points, which would fail to count individually as contributors because they are made redundant by any continuum of instantiated attributes surrounding them. This issue could be resolved by digging into the topological details, but that would constitute too large a digression for any philosophical payoff to the investigation of derivative causation. It will be best just to bracket this technicality with the observation that the definition of pure contributor can be modified to accommodate fundamental laws of continuity.

Sadly, I do not possess a comprehensive list of the forms of terminance that deserve to be labeled as trivial.

At this point, it may be helpful to take stock of the key conclusions reached so far. The most important laws for understanding non-trivial terminance and thus causation are dynamical laws, which govern how quantities develop from one time to another. The constraints imposed by dynamical laws can sometimes result in determination. In other cases, dynamical laws incorporate some stochasticity—that is, fundamental chanciness—into the relations among events. If either case obtains, we have an instance of terminance, which is the primary relation in the bottom conceptual layer of causation. A terminance relation exists anywhere a fundamental event determines a fundamental event or fixes a probability distribution over a set of possible fundamental events. An auxiliary concept is that of a contributor, which is a non-superfluous part of a terminant. A contributor is an extremely liberal and egalitarian conception of partial cause.

From this point until the end of the chapter, we will be wading through a lot of technical details. The primary reason for doing so is to identify and clarify several general principles that are obeyed by all four paradigm fundamental theories. These will be used in later chapters in order to prove several important claims about derivative causation, especially causal directness, which in turn bears on the asymmetry of causation. By demonstrating in this chapter that the principles and definitions hold for all (or nearly all) paradigm fundamental theories, I am attempting to inoculate my later arguments against the charge that they rely on an overly naïve model of fundamental physics. If the identified principles obtain in a wide variety of classical, relativistic, and quantum theories and are not in conflict with any known empirical phenomena, we have some reason to suspect that they hold true of the actual world.

A secondary reason for entering the technical swamp is to uncover some hidden (or at least underappreciated) aspects of fundamental causation. Of special importance is the signature relation between causation and time. Causation paradigmatically relates events at two different times, and shortly we will discover why time has a special relation to causation.

Experts who are familiar with the paradigm fundamental theories should be able to skim this material, but it will be beneficial to note my definitions of the main principles, which are not necessarily standard terminology. I believe that no prior knowledge of physics is strictly necessary to understand the rest of the chapter, but readers who are wholly unfamiliar with fundamental physics will probably need some patience and careful reading because I have had to impose more concision than is appropriate for an introductory textbook.

2.3 The Space-time Arena

Terminance is virtually always non-trivial when it relates events that occur in non-overlapping regions. In order to connect terminance to concepts of time and place, we need to explore how each paradigm fundamental theory structures its arena and laws to generate non-trivial terminance.

The most important fact about the paradigm fundamental theories is that they all posit arenas that distinguish some sort of temporal structure. That there is something in each theory distinguishing time is an important part of the story about why causation paradigmatically occurs between events occurring at different times. It will be convenient to constrain the discussion initially to fundamental theories that posit a space-time as their arena and only later to extend the conception to more general arenas. The goal in the next two sections is first to review standard conceptions of space-time in order to identify what in space-time corresponds to time, and second to use the notion of time implicit in the paradigm theories to make sense of what it means for an event to occur at a single time. The notion of a *state*—a full event that occurs at a single time—will help us to characterize non-trivial terminance relations and thus non-trivial causation-like relations among derivative entities.

A space-time can be represented in multiple ways. The most common representation employs a manifold, which is a mathematical object with enough structure to permit coherent distinctions concerning dimension, connectedness, differentiability of functions, and related concepts. It is doubtful that manifolds are optimal for representing space-times and arenas generally, but to avoid relying on them would likely make my discussion less accessible and for little gain. Only two space-times will be countenanced initially: Galilean space-time and Minkowski space-time.

Both of these space-times are four-dimensional and extend infinitely far in all straight directions without any looping around. Both are homogeneous, which means that the geometrical structure around any given point in space-time is qualitatively the same as around any other point. Both are connected, which means that there is always a space-time path between any two points.[18]

---

[18] In standard topology, there are multiple notions that attempt to represent connectedness among spaces. What I have in mind by 'connected' is closer to what is normally called 'path-connected', but the differences among these technical definitions do not matter for the sorts of space-times under consideration.

Both space-times distinguish among their four dimensions one that plays a privileged role with respect to time. This is known as the *time-like* dimension. The other three are *space-like*. Clarification of which formal structure makes the time-like dimension different from the other three will be postponed until §2.4.1 and §2.5.1. For now, it suffices to note that this distinction can be applied to paths in the space-time in the following sense. Straight line segments in space-time can be time-like or space-like or neither, depending on how they are situated with respect to the time-like dimension. The distinction extends to curved paths because they can be treated as the limiting case of a linked sequence of many short straight line segments, each of which is time-like or space-like or neither. When representing space-time as a manifold, a path can be time-like along some parts and space-like along others, though we will have no cause to consider such paths.

There are two commonly posited kinds of material contents countenanced in physics: corpuscles and fields. A *field*, as employed in this discussion of fundamental physics, is a quantity defined continuously over the arena. As mentioned in chapter 1, a *corpuscle* is defined as a point particle. Corpuscles are represented by differentiable paths in the arena, known as *world lines*. Often, the fundamental laws forbid corpuscles from popping into and out of existence, which we can model by stipulating that their world lines must stretch out as far as space-time allows. In ordinary and scientific language, one speaks of electrons and quarks as being fundamental particles without regard to whether a fundamental theory treats them as corpuscles or as some field quantity that behaves like a highly concentrated bit of stuff. I will follow the practice of using 'particle' to be ambiguous between fields that behave in a particulate way and corpuscles.

Temporal notions like duration and simultaneity are incorporated into the paradigm fundamental theories based on the distinction between time-like and space-like. One way this occurs is that when a fundamental theory includes corpuscles in its ontology, the theory typically restricts their world lines to be nowhere space-like. Paradigm theories always do so for ordinary particles like electrons and neutrinos, and there is currently no evidence of space-like particles.

A second way time appears in fundamental physics is that the paradigm theories include a *temporal metric*, so that there is a fact of the matter about how long a given segment of a time-like path is, say between points p and q, and this quantity is identified with the temporal duration of a corpuscle whose history stretches along that path from p to q. This temporal metric is not an idle element of the space-time structure but appears essentially in candidate fundamental laws like Newton's second law of motion and Bohm's equation.

A third way temporality is incorporated in paradigm theories is that non-trivial terminance relations obtain in the time-like direction but not in the space-like direction. For example, in relativistic electromagnetism, one can distinguish a *space-like surface*, defined as a three-dimensional connected region every point of which is space-like related to every other point in the region. A space-like surface that extends as far as possible is known as a *time slice*. Formally, we say a space-like surface is *inextendible* (and thus a time slice) iff it is not a proper subset of any space-like surface. When a full event's region is a time slice, that event counts as a *global state*. A global state corresponds to the idea of "everything happening at a single time." I will use '*state*' to refer to any full event occupying some subset of a space-like surface, so a state can inhabit a localized region. States are important because they are among the smallest events that are non-trivial terminants.

> A *state* is a full event whose region is a subregion of a space-like surface.
> A *global state* is a state whose region is a time slice, an inextendible space-like surface.

Occasionally, I will also refer to 'localized states' and 'localized events'. These are intended to be states or events that occupy subsets of a suitably small compact region. Because the discussion in this volume does not require a formal definition of local events or local causation, I will bracket the issue of how best to make these concepts precise.

In relativistic electromagnetism, some global states determine everything that happens throughout space-time, and there are also more localized states that determine everything within a more limited region. For contrast, consider the kind of surface one gets by taking a time slice and rotating it figuratively by a right angle so that the resulting surface is space-like in two of its dimensions but time-like in the other. The events on such surfaces do not determine events extending in the remaining space-like dimension and so do not maintain any noteworthy determination relations.

One way to encapsulate this feature of all paradigm theories is to note that their fundamental laws do not allow events to termine other events that are located space-like from them. We can make the notion of space-like terminance precise with a definition:

> An event *c* *space-like termines* an event *e* iff (1) *c* termines *e*, and (2) there exists a subevent c' of *c* and a subevent e' of *e* such that (A) every point of c''s region is space-like to every point of *e*'s region, and (B) the event c'' formed by excising c' from *c* (if it exists) does not termine e'.

This allows us to define a new principle, *non-spatiality*, that is obeyed by all four paradigm theories:

> The fundamental laws disallow space-like terminance.

The lack of space-like terminance is central to causation's distinctively temporal character. A catchy slogan for this idea is that time is the dimension of terminance.[19]

---

[19] Readers should explore some related conjectures by Craig Callender. I am not committing myself here to the strong claim that time exists because of terminance or even that terminance is primarily responsible for time having the characteristics it has.

There are some differences between Galilean and Minkowski space-time that bear on issues of causation. First, I will discuss Galilean space-time at length by spelling out its relation to classical gravitation. Second, I will discuss Minkowski space-time at length by spelling out its relation to relativistic electromagnetism. Then, I will return to discuss some common features of fundamental theories that employ either of these two space-times.

2.4 Classical Gravitation

Classical gravitation is the application of classical mechanics to situations where the dominant interaction is the inverse-squared gravitational force. In §1.6, I described the simple theory of classical mechanics, and the primary interpretation of classical gravitation to be discussed in this book—what I call 'the standard interpretation of classical gravitation'—is just a special case of the simple theory of classical mechanics. Remember that the simple theory of classical mechanics postulates a classical space-time, corpuscles, a mass property that adheres to the corpuscles, a distance relation between any two corpuscles at any given time, a relative speed relation between any two corpuscles at any given time, and a fundamental dynamical law governing how the arrangement of corpuscles evolve through time. It also includes charge properties that are responsible for corpuscles (nearly) colliding with one another, but we can ignore that part of the theory for present purposes. The dynamical law enforces, for every corpuscle, a functional relationship between its acceleration at the space-time point p in terms of the masses of all other corpuscles and their distances from p in the one and only time slice that includes p. I will also discuss a slightly more impoverished interpretation of classical gravitation that includes no fundamental speed relations. There exist other possible interpretations including one that posits a gravitational field throughout space-time and another that posits space-time curvature instead of a gravitational force. For brevity, I will not comment further on these alternative interpretations.

2.4.1 GALILEAN SPACE-TIME

Galilean space-time,[20] also known as Neo-Newtonian space-time, is a space-time perfectly adapted to classical gravitation. As illustrated in Fig. 2.2, it possesses structure that marks out a privileged *foliation*, a comprehensive partition of the space-time into non-overlapping time slices. Each time slice is intuitively what we call space (at some time t), and the totality of attributes instantiated in that time slice constitutes the global state of the world at t. There is an objective distance

[20] The name 'Galilean' is merely honorary, named so because the space-time makes useful the coordinate transformations named after Galileo; in Galileo's time, no one had formulated the concept of a space-time as we think of it today and some of Galileo's own physics does not accord with Galilean space-time. (Galileo 1590; see also McCloskey 1983).

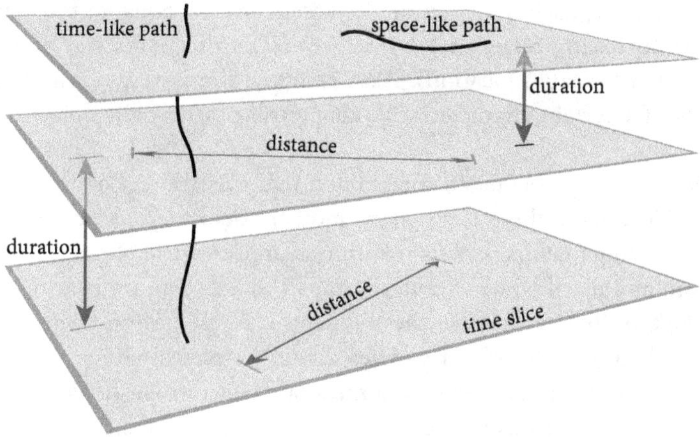

FIGURE 2.2 *Space-like paths remain within a single time slice. Time-like paths cut through time slices.*

function, called a *spatial metric*, defined on each time slice, which establishes objective distances and angular magnitudes in space. There is also an objective distance between any pair of time slices, called a *temporal metric*, corresponding to how far apart in time the slices are separated. Any corpuscle "experiences" the same duration of time passing regardless of the path it takes from one time slice to the other. Galilean space-time also has a special structure that grounds an objective distinction between motion that counts as accelerating (corresponding to a curved world line) and motion that counts as inertial (corresponding to a straight world line). The motivation for defining Galilean space-time as the space-time with these structures is that they make it ideally suited for classical mechanics, possessing minimal but sufficient structure for classical gravitation. For precision, we can now stipulate that the space-time posited by the standard interpretation of classical gravitation is Galilean space-time.

The structure of Galilean space-time allows us to categorize a line segment between points p and q. If p and q are the same point, they are *point-like* related. By definition, every point is point-like related to itself and not to any other point. If p and q are different points in the same time slice, then they are *space-like* related, and the unique straight line segment between them is space-like. If p and q are points in different time slices, then they are *time-like* related, and the unique straight line segment between them is time-like. Any well-behaved differentiable path can be cut up in pieces, each of which is space-like or time-like, depending on how it behaves infinitesimally.

2.4.2 TERMINANTS IN CLASSICAL GRAVITATION

In all models of classical physics, the motion of each corpuscle is connected lawfully to other fundamental attributes by way of its acceleration. In classical gravitation, the corpuscle's mass times its acceleration is equal to the sum of the gravitational forces impressed on it by every other corpuscle (by way of each one's mass and relative spatial location). As a result of the laws of classical gravitation, events of a certain characterizable shape will non-trivially determine other events.

For a long time, it was taken for granted that classical gravitation obeys determinism. Nowadays, there is a cottage industry dedicated to exploring special conditions proving that determinism fails to hold in classical physics. I will define determinism and discuss how counterexamples to *determinism* bear on relations of *determination* in §2.9, but for the moment, a single terminological clarification may be helpful to ward off worries about the plausibility of deterministic laws. To say that a fundamental dynamical law is *deterministic* does not imply that a theory or model incorporating that law obeys *determinism*. Determinism requires that absolutely every possible state compatible with the fundamental laws can be propagated lawfully in a unique manner throughout the entire history of the universe. We can declare a law to be deterministic, though, if *generic states* can be so propagated. That is, if we start with any state other than some specially contrived exceptions that can be set aside as unrealistic or fantastically unlikely, a deterministic dynamical law will propagate this state in a unique manner throughout the rest of the arena. In this sense, classical gravitation is deterministic.

For an illustration of how determination works in classical gravitation, let *c* be a generic global state, instantiating the masses and relative positions and speeds of all the corpuscles and a vacuum everywhere else. Because such states determine all the particles' accelerations, *c* determines the relative positions and speeds of the masses throughout the rest of the space-time. Because the corpuscles are the only fundamental entities in the standard interpretation of classical gravitation, *c* determines the world event.

Interestingly, any instantaneous event that is smaller than a global state termines nothing outside its region. The scope of a global state's determination is radically altered by excising any single point *p* from it. The state with *p* determines everything in the universe, and the state without *p* determines only itself. Metaphorically speaking, nature is not able to propagate the nearly global state through time without knowing whether there is a corpuscle at *p*.

One of the nice features of classical gravitation is the relatively simple structure of its fundamental causation-like relations, which we can characterize in terms of a minimal terminant:

> A fundamental event *c* is a *minimal terminant* of a fundamental event *e* iff *c* termines *e* and no proper subevent of *c* termines *e*.

The minimal terminants of classical gravitation include all nomologically possible global states, which are constituted by a full specification of the masses, relative positions, and relative speeds of all the corpuscles throughout a single time slice. And except for some special cases that can be quarantined,[21] all of the minimal terminants of classical gravitation are global states. Excising even a single point of empty space from any of these global states results in an event that termines nothing else whatsoever. Expanding any of these global states by adding a specification of where corpuscles are located at other times adds nothing new because the presence of these corpuscles is already determined. The only events in the standard interpretation of classical gravitation that instantiate matter and engage in non-trivial fundamental causation are superevents of global states. It is worth pausing to appreciate this remarkably simple causal structure of classical gravitation.

Moving on, we can also distinguish an ontologically sparse variant of the standard interpretation. The sparse interpretation differs from the standard interpretation merely by including no speeds in the fundamental ontology. This eliminates virtually every possible terminant that occupies just a single time slice.[22] A full specification of the masses and relative positions throughout a time slab of non-zero temporal thickness determines everything in the universe, so these fully specified time slabs serve as terminants. However, no matter how thin a time slab is, it can always be trimmed to a smaller time slab that determines the same events. Hence, there are no minimal terminants of significance in the sparse interpretation.

2.4.3 OVERDETERMINATION IN CLASSICAL GRAVITATION

One principle we rightly use to settle on which types of attributes are fundamental is whether they can be relegated to derivative reality without degrading our explanations of empirical phenomena. Fundamental theories that employ redundant determinants have a reason to be stripped down to a sparser ontology with the same empirical content and less redundancy. In philosophy, worries about redundant determinants go under the name *overdetermination*. What seems to be objectionable about overdetermination is that it reveals a pervasive redundancy in nature. From the point of view of formulating theories, having redundant

---

[21] There do exist non-trivial determinants that do not require a fully specified time slice. Imagine a bowl-shaped region of empty space-time with the open end pointed toward the future. The fact that, by law, corpuscles never spontaneously spring into existence and that they always exist along time-like paths ensures that the inside of the bowl will be devoid of corpuscles. So the bowl determines that its interior will be empty. Such terminance has only minor consequences for causation so far as I can tell. It demonstrates the curiosity that some less-than-global terminants exist, and it provides a limited counterexample to the principle of content independence, which I will introduce in §2.6. Also, one could modify a global state by taking a portion without any corpuscles and warping it a small amount in a time-like direction in order to form an empty bowl-shaped region, a dimple. Some of these dimpled states nomologically suffice for the state from which they were derived and thus serve as minimal termiants as well.

[22] If one adopts some arguably innocuous auxiliary assumptions, a completely empty time slice would still determine an empty universe.

determination is often a sign that there is an opportunity for ontological elimination, to be remedied by whittling down the fundamental ontology or by reformulating an entirely new ontology without the overdetermination. However, some versions of overdetermination are worse than others from the standpoint of a scientific theory of fundamental reality. What is problematic about the worst forms of overdetermination is that their redundancy has a conspiratorial character. I will very briefly discuss four versions of fundamental overdetermination in order of increasing objectionableness.

First, overdetermination in its most general sense occurs when two different events determine the same effect *e*. Such overdetermination is commonplace in deterministic theories. In generic models of classical gravitation, for example, every event is determined by every global state. This overdetermination is uncontroversially viewed as unproblematic because the determinants occur at different times.

A second version of overdetermination that is philosophically irritating, but only mildly so, obtains when co-instantiated events consisting of distinct properties determine the same effect and do so on a regular basis. A good example of this is the standard interpretation of classical gravitation with respect to its relative speed relations, as mentioned at the end of the previous subsection. A generic full event *c* that specifies the masses and relative distances and speeds throughout a time slab of non-zero temporal thickness determines the world event. The subevent formed by stripping all relative speeds from *c* also determines the world event. What is irritating about the standard interpretation is that these speed relations can be stripped out of fundamental reality (resulting in the sparse interpretation) with no cost to the predictive and explanatory content of the theory. The irritation is only mild because the relative speed relations are already implied by the relative distance relations, which ensures that the redundancy does not imply anything beyond what is already implied by the corresponding sparse interpretation. What's more, treating relative speeds as fundamental provides some convenient components, especially minimal terminants, so that stripping velocities out of fundamental reality may have some (presumably minor) practical costs.

A third sort of overdetermination that is a bit more irritating involves augmenting the standard interpretation with some additional fundamental ontology that uses some empirically inaccessible properties to add redundancy to the determination relations. An example of this occurs when one adds a gravitational field to the material content. The gravitational field is a smoothly varying quantity defined virtually everywhere in space-time. Its value on any single time slice is determined by the masses and relative positions of the totality of corpuscles on that slice plus a fundamentally arbitrary parameter, which usually takes the form of a boundary condition at spatial infinity. Just like thermal energy, the magnitude of the gravitational field at any given location is not made determinate by the fundamental attributes of the standard interpretation alone. If we were to add the gravitational field to the fundamental ontology by attributing a determinate strength to it at every location, we could reproduce the structure of the standard interpretation as follows. Any event on a time slice that comprises every fundamental attribute except the gravitational field would (in conjunction with the boundary condition) determine the complete gravitational field on that time slice. The complete gravitational field on that time slice, together with all the other fundamental attributes on the time slice would then determine the corpuscle motions at other times and thus the world event. This posited redundancy is a bit more irritating than the case of the fundamental relative speeds because it adds another attribute to fundamental reality that could have been treated as a fundamentally arbitrary parameter instead. There might be good reasons to add the gravitational field to the fundamental metaphysics despite this ontological bloat, but ceteris paribus, converting a fundamental quantity into a fundamentally arbitrary parameter makes the ontology trimmer and thus preferable.

Fourth, there exists an important kind of overdetermination that is uncontroversially objectionable. This occurs when two co-located determinants of *e* are composed of property types not linked to each other by any law or other metaphysical constraint. What's worrisome about such situations is that it violates the expectation that independent fundamental properties should not exhibit conspiratorial coincidences. If two different kinds of fields independently determine the evolution of things and no law relates them, then varying one field while holding the other constant will—under virtually any interesting theory—make them disagree about what will happen elsewhere. Thus, the only way to ensure harmony in the actual world is to suppose that somehow the universe's boundary conditions (including initial conditions) are set up to avoid inconsistency. In any realistic theory, this consistency requires an extremely coincidental arrangement, which is a hallmark of bad theoretical design. I will provide an example of such overdetermination in §2.5.3.

2.4.4 INSTANTANEOUS CAUSATION

As a further illustration of terminance, we can take up the question of whether classical gravitation possesses instantaneous causation. I will discuss a form of instantaneous influence among *derivative* entities later in §4.12, which applies to classical gravitation among other theories, but for this chapter, attention needs to be restricted to fundamental relations. In doing so, I think it is fair to construe 'contribution' as one good way of rendering 'influence' precise as a fundamental relation.

We can identify several questions specifically about classical gravitation. Does classical gravitation involve instantaneous contribution? Does it involve instantaneous terminance? Should the fundamental classical gravitational interaction count as arbitrarily fast rather than instantaneous? Answering these questions will not tell us anything new about classical gravitation, which we already know operates across arbitrarily large distances without any kind of limiting speed, but our investigation of these questions does suggest how to regiment the causation-like concepts of the fundamental layer.

We can define *arbitrarily fast contribution* to be the lack of any limit as to how fast (in space) contribution can go. Because (1) there are generic global states at time t that determine a spatially localized event e at time $t + \epsilon$ (with $\epsilon > 0$) no matter how small ϵ is, and because nothing less than the entire global state will suffice to determine e, e will have (time-like) pure contributors stretching out arbitrarily far in space for any fixed temporal duration ϵ. Thus, there is arbitrarily fast contribution according to the theory of classical gravitation.

We can also characterize conditions under which contribution occurs without any time passing whatsoever:

> An event c is an *instantaneous contributor* (or a *space-like pure contributor*) to an event e iff c purely contributes to e and every point of c's region is space-like to every point of e's region.

The standard interpretation of classical gravitation forbids instantaneous contribution because all non-trivial terminants of gravitation existing within a single time slice must occupy the entire time slice.

So, instantaneous events in the standard interpretation of classical gravitation exert their fundamental gravitational influence with unbounded speed but not instantaneously. The illegality of instantaneous contribution implies that the standard interpretation of classical gravitation must also forbid space-like terminance, defined in §2.3. As mentioned previously, all the other paradigm fundamental theories rule out space-like terminance as well.

It is notable that if we were to augment the standard interpretation of classical gravitation by construing a corpuscle's acceleration as an additional fundamental property, as discussed by Larry Sklar (1977), then all the masses contributing to that acceleration would be space-like contributors. Because the acceleration at p also depends on the mass at p, however, not all of the contributors would be space-like to p. The terminance issuing from the global state (not including the acceleration at p) would then space-like termine the acceleration at p by virtue of its instantaneous contributors.

Nevertheless, because generic global states determine all the positions and velocities of the corpuscles at other times, the added fundamental accelerations would be superfluous in the sense that they must—by virtue of the laws and the meaning of 'acceleration'—agree with all the facts concerning how particle velocities change. Because having the more ontologically loaded conception of accelerations does not augment the story about terminance (and hence causation) in any fruitful way, one may leave acceleration attributes out of the material content of fundamental reality. Accelerations undoubtedly play a crucial role in the Newtonian laws of motion, but it is not obligatory (and is redundant given the standard or sparse interpretations) to include them as part of what needs to be specified for a full event.

2.5 Relativistic Electromagnetism

Relativistic electromagnetism is the relativistic, non-quantum-mechanical theory of how electrically charged particles interact. The ontology of its standard interpretation includes a space-time for its arena and two types of material content: corpuscles and an electromagnetic field. The electromagnetic field is defined over all of space-time, with its value at each point being a special multi-dimensional vector-like quantity with six independent magnitudes. The corpuscles have a constant mass property, a constant charge property, and spatio-temporal relations among all the parts of all their world lines.

My discussion of classical electromagnetism in this chapter will ignore that it treats charged particles inconsistently, as explained in §1.8. Despite my previous insistence on the importance of treating theories of fundamental reality according to the STRICT standards of adequacy that permit no genuine conflicts, I will continue to include it as one of the paradigm theories of fundamental physics in order to encourage suitable generality in my theory of causation. Its inconsistency will not hurt because the general principles to be extracted from relativistic electromagnetism will themselves be consistent.

2.5.1 MINKOWSKI SPACE-TIME

Just as Galilean space-time is perfectly adapted to classical gravitation, Minkowski space-time has structures ideally suited to relativistic electromagnetism. Minkowski space-time is similar to Galilean space-time, remember, by being four-dimensional, homogeneous, and infinitely extended without any holes or edges. The geometrical structure of Minkowski space-time, as usually conceived, is different from Galilean space-time mainly by having a single *spatio-temporal metric*, which defines objective distance and angular magnitudes in *space-time*, rather than having two separate metrics, spatial and temporal.

One important feature of the Minkowski metric is that it picks out one of the space-time dimensions as the special time-like one by defining a distance function such that paths going primarily along that one dimension have positive length-squared. The paths going primarily along the three other dimensions have negative values for the square of their lengths, and these quantities can be thought of as spatial distances, though there are no known laws that require reference to them. This motivates thinking of spatial distance as a derivative relation in theories where Minkowski space-time is the arena.

For every space-time point p, one can distinguish four separate classes of points. The first class is just the point p itself. As before, we say p is *point-like* related to itself and not to any other point. The other three classes are standardly identified by the length-squared of the unique straight line segment going from that point to p. The points located at a distance-squared of zero from p, using the

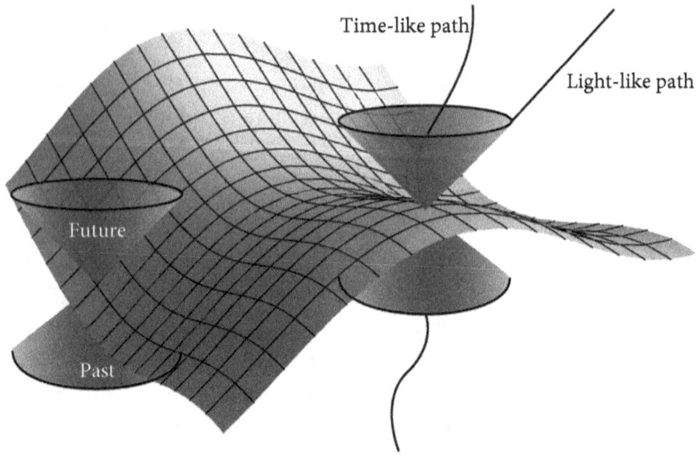

FIGURE 2.3 *A pair of light cones, each composed of a future and past light cone.*

special Minkowski metric for measuring distances in space-time, constitute the null cone or light cone of p. They are all said to be *light-like* related to p.

In diagrams like Fig. 2.3, we omit one spatial dimension, which makes the light cone look like two cones that touch at p, a future light cone and a past light cone. In reality, the light cone consists of a two-dimensional surface of a sphere (embedded in three dimensional space) that, as time goes by, collapses on p at the constant and invariant vacuum speed of light and then expands away from p. We usually speak of p's future and past light cones together as a single light cone.

The points that are located at a negative distance-squared from p are said to be *space-like* related to p and occupy the connected region outside p's light cone. They form the third class. The points that are a positive distance-squared from p are said to be *time-like* related to p and form two regions disconnected from each other: the interior of p's future light cone and the interior of its past light cone. They form the fourth and final class. Note that every point of Minkowski space-time is qualitatively just like any other; every point has its own numerically distinct light cone structure, but every light cone is qualitatively the same, and they are all connected uniformly to each other without any twisting of the light cone structure from here to there.

The light cone plays a significant role in causation through its constraint on both corpuscles and fields. Unlike in classical mechanics, the intrinsic mass property of a corpuscle in relativistic electromagnetism comes in two varieties: zero and positive. If a corpuscle has positive mass, its world line is always time-like. If its mass is zero, its world line is always light-like. One physical interpretation of the light cone is that the surface of the future light cone comprises all the points that a massless corpuscle at p, like a photon, can possibly reach by flying in a straight line. The surface of the past light cone comprises all the points that a massless corpuscle at p could have come from by flying in a straight line. Similarly, the interior of the future light cone comprises all the points that a massive particle starting at p can possibly reach, and the interior of the past light cone are all the points that the massive particle at p could have come from. We can talk about the light cone of an event in general by stipulating that its surface and interior is the union of all the surfaces and interiors of the light cones of the points in the event's region. Then, the surface of this region is (the surface of) the event's light cone.

One major role of the light cone in the behavior of the electromagnetic field is to enforce the rule that an event c can be a pure contributor only to an event whose region intersects the surface or interior of c's light cone. It follows from Maxwell's laws that no portion of the electromagnetic field purely contributes to events situated wholly space-like to it. Furthermore, all known particles, when construed as corpuscles in a relativistic theory, always remain in the interior or on the surface of the light cones of the points they pass through. This suggests the consequence that all forms of influence and causation in relativistic electromagnetism are forbidden from extending any faster than light can travel. This turns out to be not quite right, but my suggested amendments to this statement must be postponed to §4.12.

An important difference between Galilean and Minkowski space-time is that every point in Galilean space-time sits inside a single time slice, but in Minkowski space-time, each point p sits inside a continuum of time slices. Any way of carving an inextendible space-like surface through p constitutes a legitimate time slice. Each corresponding global state counts as one legitimate way of cashing out "what is happening at the time when p occurs." Minkowski space-time has no preferred foliation.

2.5.2 TERMINANTS IN RELATIVISTIC ELECTROMAGNETISM

The laws of nature governing relativistic electromagnetism differ from those of classical mechanics in a few important respects. First, in classical gravitation theory, the law governing the gravitational field has no interesting dynamical development and can even be omitted from the fundamental ontology as it was from the standard interpretation. The electromagnetic field, by contrast, is not easily eliminated from the universe's dynamical development. It obeys a dynamical law that holds even where no corpuscles are present.

Second, the determination in relativistic electromagnetism is more localized than in classical gravitation. The properties that constitute a full event in relativistic electromagnetism are (1) the value of F throughout the region and (2) the mass, charge, and world line of every corpuscle in the region. Some global states determine everything throughout the future and past. However, there are also global and non-global events that serve as non-trivial determinants of more limited regions. To get an idea of what some of the more localized determinants look like, one can start with any event e whose region does not stretch out infinitely far. As illustrated in Fig. 2.4, any full event that completely cuts across the entire width

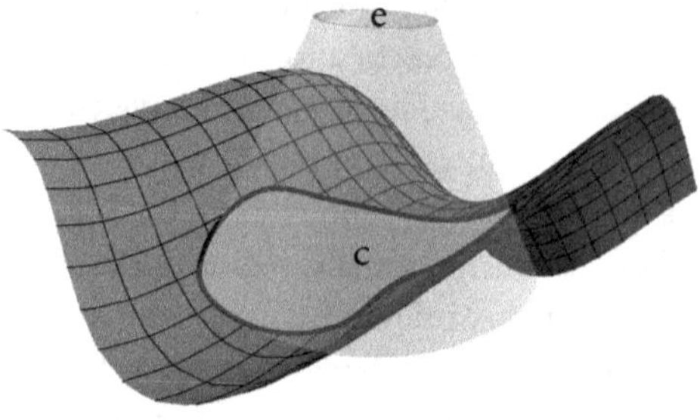

FIGURE 2.4 *Every full event spanning e's light cone determines e.*

of *e*'s light cone counts as a determinant of *e*. There also exist global states that are tucked entirely within the past or future light cone of a point. In such cases, subregions of the light cone are determined, but not the world event.

2.5.3 CLASSICAL UNIFIED FIELD THEORY

We currently know of two other kinds of fields that are similar to the classical electromagnetic field: the weak field and the strong field. The weak and strong fields can be described classically using more or less the same kinds of mathematical representations used in relativistic electromagnetism. As it turns out, though, the classical description of these interaction fields does not approximate the actual behavior of particles because at the scales at which these fields would be significant, quantum-mechanical effects drown them out. Their classical description is only used in practice as a heuristic for formulating quantum field theories.

For the purpose of studying how the fundamental causation-like relations generalize, though, it is useful to think about a theory just like relativistic electromagnetism except that it includes two new fields, *W* and *S*, corresponding to the weak and strong interactions respectively. *W* and *S* are represented mathematically by a multi-vectorial field quantity very similar to *F*. To maintain the analogy with electromagnetism we need to add four new fundamental properties that corpuscles can bear, a weak charge, and three strong charges known as red, blue, and green. There has been comparatively little philosophical exploration of how to interpret theories with classical strong and weak fields where one maintains that particles are corpuscular and obey a generalization of the Lorentz force law. So, I hereby christen this new theory *classical unified field theory*.

Classical unified field theory is just like relativistic electromagnetism as described above except that it includes strong and weak charges and fields. Because this is just a toy theory, I will make no attempt here to specify precisely how these new fields are implemented except to say that the issues are discussed in theories of classical gauge fields. Corpuscles in the classical unified field theory are meant to correspond to known fundamental particles like photons, electrons, gluons, and quarks by having the appropriate fundamental masses and the appropriate amounts of the five different kinds of fundamental charges. To keep the model simple, I will assume that the three ontologically independent field quantities do not interact with each other directly. For example, the weak field can only affect and be affected by corpuscles that carry the weak charge, and it can thus indirectly interact with the electromagnetic field by way of corpuscles that carry both weak and electromagnetic charge, but it does not interact with the electromagnetic field except through the mediation of appropriately charged corpuscles. The laws of classical unified field theory include three sets of field equations—Maxwell's laws and similar formulas for the weak and strong fields—as well as a single force law analogous to the Lorentz force law. The standard Lorentz force law for a corpuscle at p sets the corpuscle's mass times its acceleration at p equal to its electromagnetic charge times the (inner) product of F at p with the tangent of the corpuscle's world line at p (normalized as a unit vector). The extended Lorentz force law for classical unified field theory includes similar terms for the W and S fields.

One notable feature of the four paradigm fundamental theories is that they all obey the principle of *content completeness*. The intuitive idea behind content completeness is that in order for an event *c* to termine everything that happens at the location of *e*, it must include a complete specification of all the fundamental attributes in the fundamental ontology. More formally, we can define content completeness as follows:

> *Content Completeness*: For any terminant *c* of any possible full event *e*, there is a full subevent of *c* that termines *e*.

For example, a spatial sphere *e* extends one light minute in radius and specifies the values of F, W, and S defined throughout the volume and all the corpuscle world lines and their fundamental charges and masses. Suppose *c* is a full event occupying a spatial sphere of radius two light minutes and that it just barely spans *e*'s past light cone. Then, *c* is a determinant of *e*. Now, let the event c_{-} be any non-full event that occupies the same region as *c*, for example by not specifying the value of the W field in some subregion R of *c*'s region. According to content completeness, none of c_{-}'s full subevents can specify any material contents for R. Thus, c_{-} cannot span the past light cone of *e*, which results in its being unable to termine *e*.

Although the paradigm theories all obey content completeness, one might wonder what grounds we have for thinking that the actual world obeys content completeness. Here are two suggestive arguments. The four well-established types of physical interaction—electromagnetic, weak, strong, and gravitational—are linked with one another so that events formed from anything less than a full specification of each type do not termine anything. The first reason to think interactions are linked is that we know from experiments in high-energy physics that particle interactions of any one type are affected by possible interactions involving the other types. This can be seen in Feynman diagrams of particle interactions where the diagrams corresponding to higher-order corrections to various measured parameters involve particles of other types being virtually created and destroyed. Without having to make any detailed assumptions about how quantum field theory works, the fact that measured quantities depend on a system where all three kinds of interactions mix together gives us some reason to believe that fundamentally, the interactions are not working wholly independently. The second reason involves the universality of gravity. The three non-gravitational interactions make a difference in how various corpuscles and fields are distributed in space-time and all these material contents have an associated stress-energy, which in turn plays a role in gravitational interaction, which affects everything regardless of whether it is electromagnetic, weak, or strong. So, even if *F* does not affect *W* directly or indirectly through particles with both electromagnetic and weak charge, *F* at least affects *W* by affecting the gravitational field (or the curvature of space-time), which in turn affects *W*.

I have engaged the topic of multiple attribute groupings and corresponding principles of content completeness in order to explore the possibility that the fundamental laws allow multiple routes of terminance, situations where a full event *c* termines the existence of an electron (which carries both weak and electromagnetic charge) by virtue of one law governing electromagnetic interactions and independently via some other law governing weak interactions. If an electromagnetic law implied that the value of *F* throughout *c* (with some generic attributes like mass) determines the position of the electron and a separate weak law implied that the value of *W* throughout *c* (also with some generic attributes) determines the position of the electron, then it would be very likely that the actual initial conditions for *W* and *F* would always need to be coordinated so that if *F* were different, *W* would also be different in exactly the right way to ensure that they always determine the same events. This example of the bad kind of overdetermination mentioned in §2.4.3 is forbidden by content completeness.

Multiple routes of terminance are also a potential worry for some arguments to come later in §6.2 because they might permit multiple routes of influence that could give rise to non-redundant backtracking influence.

2.6 Content Independence

A remarkable feature of paradigm fundamental theories is that the arena structure by itself is extremely informative of causal structure. Often, the question of which full events termine which other events can be answered merely by examining their relative locations in the arena. One form of this principle, *content independence*, holds whenever it is the case that for any full event *c* and any region R, whether *c* termines (or fixes) an event occupying R is independent of *c*'s material content. We have seen several examples of content independence already. A generic global state in classical gravitation determines everything else in the universe regardless of where the corpuscles are located, what their masses are, and what their relative speeds are.[23] In relativistic electromagnetism, determination relations between full events are fixed entirely by the light cone structure. For example, any full event that spans *e*'s light cone will determine *e* no matter what electromagnetic field values or corpuscles it instantiates. In any theoretical framework where content independence holds, we can expect there to be facts about causation that hold regardless of what new particles or fields are added to the theory, as was done when extending relativistic electromagnetism into classical unified field theory. When content independence holds, causal relations are insulated from many esoteric details of the fundamental ontology.

Whether content independence holds for the actual world is questionable because it is violated by the standard interpretation of general relativity, where the structure of the arena depends on how material contents are situated. That the actual world exhibits quite a bit of evidence confirming general relativity ought to count as at least some evidence that content independence does not actually hold. However there are non-standard interpretations of relativity that posit a fixed arena structure, and it is not far fetched to speculate that some richer theory with a fixed arena will be able to explain the signature evidence that confirms general relativity. It would also not be too surprising for a fundamental theory to blur the distinction between matter and space-time, as many contemporary theories do. If such a theory used some non-spatio-temporal arena and explained why space-time and its material contents are correlated as derivative entities, then perhaps content independence would gain credibility. Content independence is worth discussing despite serious doubts about its truth because it is pedagogically useful to hone several definitions for the simplified environment where content independence holds and then to examine how they extend to more general theories where it does not hold.

In full generality, we can isolate three kinds of regions that are interesting for issues concerning terminance. The following definitions refer to the geometry of the arena. They are easier to picture as applied to space-time, but the definitions are suitable for more general arenas.

> An event *c*'s *domain of terminance* is the union of the regions occupied by all the events *c* termines.
>
> An event *c*'s *domain of influence* is the union of the regions occupied by events all of whose subevents *c* contributes to.
>
> An event *c*'s *domain of contribution* is the union of the regions occupied by *c*'s pure contributors.

---

[23] Content independence does not hold for classical gravitation because of the bowl-shaped regions and dimpled states mentioned in §2.4.2, but this shortcoming has little bearing on the study of causation so far as I can tell.

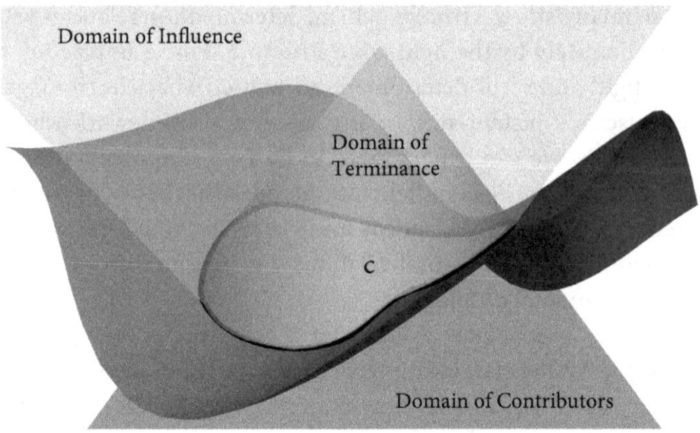

FIGURE 2.5 *Relativistic electromagnetism with no past-directed determination.*

The domain of influence for any event *c* is the region where *c* plays a fundamental role in producing what happens. An event *c*'s domain of contribution is the region occupied by everything that plays a fundamental causal role in producing *c*. In Fig. 2.5, the laws of nature are imagined to be just like those of electromagnetism except with a fundamental temporal asymmetry where events determine nothing toward the past.

We can now relate these terminance-based definitions to space-time structure. Two helpful concepts for expressing the relations are a c-path and c-connection. A c-path is what Earman (1986) calls a 'causal curve', and although the 'c-' is inspired by the word 'causal', it facilitates discussion to avoid overusing that word.

> A *c-path* is an everywhere differentiable path whose tangents are nowhere space-like and are well-defined and non-space-like in any mathematical limits along the path.[24]
>
> *Points* p and q are *c-connected* iff $p = q$ or a c-path exists between p and q. (The c-path in question has endpoints p and q.)
>
> Two *regions* are *c-connected* iff some point in one region is c-connected to some point in the other.
>
> Two *events* are *c-connected* iff their regions are c-connected.

In Minkowski space-time, a point is c-connected to itself and every point in and on its future and past light cones. In Galilean space-time, a point is c-connected to itself and every other point not simultaneous with it.

[24] The restriction on how the path behaves in limits exists to reduce a number of problematic issues in classical mechanics discussed in §2.11.1.

Let us now focus attention on two paradigm fundamental theories that include no fundamental asymmetries in their terminance relations and obey content independence. Specifically, let us examine the structures that are at least common to Galilean and Minkowski space-time by examining relativistic electromagnetism, which obeys content independence, and by restricting ourselves to the portion of classical gravitation that obeys content independence by excluding events in classical mechanics that have no global state as a subevent. For such theories, we can link the terminance-based concepts listed above with corresponding arena structures using a few simple rules.

Two preliminaries are needed. Let us say that a c-path is *inextendible* iff it is not a proper subset of an everywhere differentiable path. Intuitively, an inextendible path stretches as far as the arena allows. Then, let us adapt some terminology from Earman (1986, p. 35) by defining a full event *e*'s *domain of dependence* to be the set of all points *p* such that every inextendible c-path intersecting *p* also intersects *e*'s region. Intuitively, in relativistic theories, these are the points whose future or past light cone is completely spanned by *e*'s region.

Now we can state the connection between terminance and the arena structure for any full event *e* in relativistic electromagnetism and for any superevent or subevent *e* of a global state in classical gravitation.

- A point *p* is in *e*'s domain of terminance iff *p* is in *e*'s domain of dependence.
- A point *p* is in *e*'s domain of influence iff *p* is c-connected to *e*.
- A point *p* is in *e*'s domain of contribution iff *p* is c-connected to *e*.

Note that the domain of influence and domain of contribution end up being identical because of the temporal symmetry in the kinds of locations over which an event exerts determination. If the dynamical laws incorporate a temporal asymmetry where events fix a probability distribution over future happenings but never over past happenings, then the domain of influence and domain of contribution come apart as shown in Fig. 2.5. Such asymmetric terminance occurs in some possible fundamental theories, although realistic fundamental theories with chancy dynamical laws typically involve some subtleties regarding past-directed terminance that will be explored in §2.10.2.

The principle of content independence does not hold in the standard interpretation of general relativity (GR). However, the definition of c-connection nevertheless remains applicable because GR possesses an initial value formulation that is well posed. As discussed by Robert Wald (1984, p. 244), this means GR can be formulated in terms of a specification of variables representing the fundamental attributes of the space-time geometry and any material contents on a subset of a time slice such that those variables can be propagated toward the future in a well-defined way. Furthermore, the variables at point *p* do not contribute to what happens outside *p*'s light cone. This ensures that a state *s* can be propagated toward the future through an event *e* only if *s* spans the past light cone of *e*. The result is that the three principles listed in the previous paragraph hold for GR as well, even though it does not obey content independence.

2.7 Continuity and Shielding

A notable feature of the paradigm fundamental theories is that their relations of terminance never hop over portions of the arena. This idea involves two related principles: continuity and shielding. In order to define continuity and shielding efficiently, it helps to establish some preliminary definitions.

> A region R is *intermediate* between *c* and *e* iff (1) every point of R is c-connected between some point of *c*'s region and some point of *e*'s region, and (2) there is a connected space-like subregion Q of R such that every c-path from a point in *c*'s region to a point in *e*'s region intersects Q.

We can also say an *event* is intermediate between *c* and *e* iff its region is. A nuance worth noting is that because of the way 'c-connection' was defined, the intermediacy relation is inclusive of the endpoints. So, *c* and *e* can count as intermediate between *c* and *e*.

Let us say that a contextualized event $\overline{I}$ is a *fixed intermediate* on the way from *c* to *e* iff the region R occupied by $\overline{I}$ is intermediate between *c* and *e* and $\overline{I}$ is the unique maximal contextualized event fixed by *c* for R and $\overline{I}$ fixes an event that includes a superevent of *e* as a member. For any possible fundamental event *f* that includes subevents *c*, *e*, and *i*, where *c* termines *e* and *i* is a member of a fixed intermediate on the way from *c* to *e* that also termines *e*, we can say that *i* is an *intermediate terminant* on the way from *c* to *e*.

One final auxiliary definition that will be needed shortly is that of an *e-wards domain of influence* for a fixed intermediate (or intermediate terminant), which is intuitively its future domain of influence when *e* is entirely toward its future (and to the future of *c*) and its past domain of influence when *e* is entirely toward its past (and to the past of *c*). More formally,

> For any fixed intermediate $\overline{I}$ (or any intermediate terminant *i*), occupying region R on the way from *c* to *e*, a point *p* is in $\overline{I}$'s (or *i*'s) *e-wards domain of influence* iff *p* is not in R and there is a c-path from *p* to *c* that intersects a c-path from *e* to *c* and passes through any of the previously designated subregions Q. (These Q, recall, are subregions of R such that every c-path from a point in *c*'s region to a point in *e*'s region intersects Q.)

First, let us consider continuity. The key idea behind continuity is that the fundamental interaction between events at separate locations is always mediated by what happens in between. When an event *c* determines *e*, it also determines any events occupying the regions intermediate between them, and all those events also determine *e*. When an event *c* fixes a probability for *E* (instantiated as *e*), it also fixes contextualized events in the regions intermediate between them, and all of these fixed intermediates fix that same probability for *E*.

We can formalize one version of continuity applying to fundamental reality as follows:

> *Continuity of Terminance*: For any possible fundamental event *f* (occupying region F) with any subevent *e* and any subevent *c* that termines *e* and any subregion R of F intermediate between *c* and *e*, there exists an intermediate terminant on the way from *c* to *e* occupying R.

Note that intermediate terminants are only required to be as continuous as the arena itself, so this continuity principle is compatible with the possibility that the arena is a discrete lattice rather than a continuum.

Second, all paradigm fundamental theories obey the principle that if *i* is an intermediate terminant on the way from *c* to its termined *e*, then augmenting *i* with stuff termined by *c* on the side of *i* opposite from *e* will not result in any change to the probabilities fixed by *i* for what happens at *e*'s location. We can say of such a situation that *i shields c*'s terminance of *e*.[25]

This property of a fundamental theory can be formalized as follows:

> *Shielding of Terminance*: For any possible fundamental event *f* with any subevent *e* and any subevent *c* that termines *e*, the probability any intermediate terminant *i* (on the way from *c* to *e*) fixes for any coarse-graining *E* of *e* is equal to the probability fixed for *E* by any superevents of *i* that are termined by *c* and do not intersect *i*'s *e*-ward domain of influence.

As a corollary, *i* fixes the same probability for *E* that is fixed by superevents of *i* that are termined by *c* and *i* and cannot be c-connected to *e* without intersecting *i*.

Fig. 2.6 depicts shielding in a relativistic setting. The event *c* fixes a unique maximal contextualized event $\bar{I}$ for the region where the time slice intersects the past light cone of *e*. The event *i*—which is the member of $\bar{I}$ that is actually instantiated—fixes a probability for any coarse-graining *E* of *e* that is the same as the probability fixed by $c \cup i$. Furthermore, for any other event *j* that is termined by *c* and lies on the opposite side of *i* from *e*, the probability fixed for *E* by $j \cup i$ is the same as that fixed by *i* alone. The upshot of shielding is that—in the evolution of nature from *c* to *i* to *e*—once we already take into account the existence of *i*, everything else that happens as a result of *c* on the way to *i* (or even previous to *c*)

---

[25] The principle I am labeling 'shielding' is sometimes identified by saying that a theory is Markovian. I have chosen not to use 'Markovian' because it is often defined in a temporally asymmetric way, often defined too imprecisely in terms of times rather than arena regions, often defined using conditional probabilities rather than unions of events, and often not defined for non-full events. One should also avoid conflating shielding with claims about Markov processes or the screening of probabilistic correlations because these two ideas are standardly defined using an alternative conception of probabilistic relations among events.

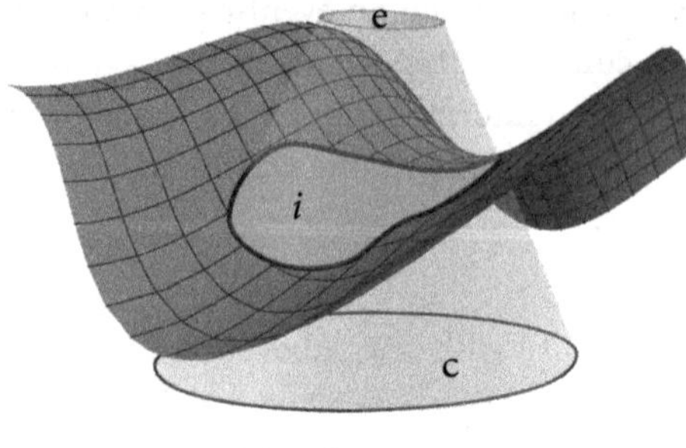

FIGURE 2.6 *The event c termines e with an intermediate terminant i. The probability that i fixes for E is the same as the probability fixed by any of i's superevents that are termined by c and do not inhabit i's future domain of influence.*

is irrelevant to the probability of E. (In order to make shielding as weak a condition as possible, I have deliberately not excluded the possibility that there exists an event k prior to i and not termined by c such that $i \cup k$ does not fix the same probability for E that i alone fixes. Such a possibility could occur if there are multiple routes of terminance, as mentioned in §2.10.2.)

It is easy to imagine possible worlds where continuity and shielding do not hold. Suppose there are fundamental laws of magic such that waving a wand with the proper incantation determines that all rabbits will vanish after precisely one day. The state of the universe in the meantime is exactly the same as it would have been without the magical spell having taken place. In such a world, the spell does not have its effect on rabbits by way of something fundamental in the intervening times but does play a role in what happens one day later. More realistically, in general relativity, there are possible worlds with space-time wormholes where none of the usual intermediate terminants exist on the most direct route from c to e because c and e can be c-connected circuitously through a distant wormhole. This possibility is noteworthy because it suggests that space-time wormholes might circumvent the principles that prevent exploitable forms of past-directed influence.

2.8 Transitivity

Determination is trivially transitive, but indeterminants are transitive as well. Applied to terminance, transitivity can be defined in the usual way:

> *Transitivity of Terminance*: For any e_1, e_2, and e_3, if e_1 termines e_2 and e_2 termines e_3, then e_1 termines e_3.

Note that the probability e_1 fixes for some coarse-graining E_3 of e_3 is not in general the same probability that e_2 fixes for E_3. This is intuitive when one considers that as things evolve forward in time from e_1, something chancy might occur such that by the time e_2 occurs the chance of E_3 has gone up or down.

Readers who are uninterested in further technical details related to fundamental theories of physics are cordially invited to skip ahead to §2.14 for a review of all the principles relevant to the discussion of causation among derivative entities. The important material in the rest of the book can be understood using only what I have covered so far.

2.9 Determinism

Some subtleties in the concept of determinism are worth highlighting at this stage of the discussion. So far, 'determination' has been employed in the sense of a fundamental event *c* nomologically sufficing for a fundamental event at a chosen location relative to *c*. The much-referenced concept of determinism can be glossed in terms of determination as follows. Determinism is roughly the thesis that any nomologically possible global state determines a complete history, a world event. There are several ways to make this definition more precise.

First, it can easily be weakened to allow for temporally directed versions of determinism.

Second, the definition speaks of global states, and one might want to characterize a spatio-temporally localized form of determinism to make it more useful for discussing relativistic electromagnetism, which countenances localized determinants.

Third, a shortcoming in this definition of determinism is that 'global state' and 'time slice' were earlier defined as space-like surfaces that stretch out as far as possible, but this stretching is allowed in relativistic models to lie entirely in the past light cone of some point *p*. Such global states do not termine what happens outside *p*'s past light cone. We could attempt to restrict the class of global states to so-called Cauchy surfaces as in (Earman 1986, p. 59), but such a choice would be overly restrictive when applied to general relativity. For rhetorical convenience, I will speak informally of *global states of suitable shape* to designate the states that occupy space-like surfaces that stretch out across as much space as possible rather than hiding inside a light cone, but below I will characterize determinism in a way that does not depend on the shape of the events that engage in determination.

A fourth deficiency in the rough definition of determinism is that it does not accommodate the possibility of non-maximal space-times. Intuitively, a space-time is non-maximal if it has an edge or a hole or a boundary even though nothing in the laws and material content forces the space-time to come to an end there. Formally, an arena *a* of some model *w* is non-maximal iff the laws permit a model w' with arena a' where *w* is a submodel of w' and *a* is a proper substructure of a'. For example, a space-time that is just like the actual space-time but lasts only a minute is a non-maximal space-time. If the laws permit such arenas, unproblematic exemplars of determinism might fail to satisfy the definition of determinism because no global state would suffice either for the existence or for the non-existence of a world event that lasts more than a minute.

More important, failing to address this threat will undermine later appeals to counterfactual dependence. Although we have not yet delved into how to think about the relationship between terminance relations and contrary-to-fact possibilities, it ought to be easy enough to see that if we model counterfactual possibilities in terms of what the fundamental laws imply about hypothesized events, then claims about what would happen counterfactually virtually always turn out trivially false or undefined unless an assumption is made about how far the counterfactual arena must extend. Suppose I wonder, "What would have happened if some global state *c* had been the case?" If the laws allow a possible world with a space-time barely big enough to instantiate *c* and nothing else, nothing of interest would follow from the existence of *c*. Left unmodified, counterfactual claims could not be understood in terms of what the laws imply about hypothesized events. One possible modification is to presume that the laws only permit *maximal* arenas, meaning arenas that cannot be extended any further. Another way is to construe the fundamental laws as having a hedge clause. Claims of the form "*c* determines *e*," would be understood as implicitly of the form, "If there is enough of the arena stretching from *c* to reach the entire location of the possible *e*, then *c* nomologically suffices for *e*." Below, I will adopt a version of the first strategy by maintaining that the laws suffice for the existence of the arena wherever an event would have a domain of terminance.

Assuming we have an adequate grasp of what counts as a global state of suitable shape, we can distinguish two different conceptions of determinism. The first conception is *unique-propagation determinism*. which holds just in case every nomologically possible global state of suitable shape has a unique lawful extension throughout its arena. A popular alternative way to frame the thesis of determinism is in terms of nomologically possible worlds (Earman 1986). According to this conception, *possible-worlds determinism* holds just in case for any nomologically possible global state of suitable shape, there exists exactly one nomologically possible world event that instantiates that state.

The approach I prefer is to adopt unique-propagation determinism as the primary notion of determinism for two reasons.

First, if the fundamental laws permit space-times with components that are disconnected from one another, there could be multiple worlds consistent with a given global state without any deterministic laws being violated. Under any theory resembling the paradigmatic deterministic theories, a global state of suitable shape in one component of the universe would only suffice for what happens in its own component of the universe. So, if the actual world has just a single component arena by happenstance and not by law of nature, it could obey unique-propagation determinism without obeying possible-worlds determinism because a disconnected component would be nomologically possible and could contain stuff undetermined by the material contents of our component. Possible-worlds determinism, then, turns out to be an inferior conception because we are unlikely ever to be able to make a reasonable assessment of whether the arena is a single connected space, unless we make it true by stipulation.

Second, unique-propagation determinism avoids the kinds of confusion that have given rise to the famous hole argument. According to the hole argument, if a relativistic space-time is conceived as a kind of substance that serves as a container for all matter, then its states will fail to determine a unique world event because the laws do not determine which space-time regions will be inhabited by which bits of matter. The advocate of the hole argument notes that this failure of determinism occurs "cheaply" in the sense that it arises by virtue of degrees of freedom in the theory that have no predictive significance, not because of some chanciness or randomness in the laws. The advocate of the hole argument then attempts to leverage the objectionability of having determinism fail regardless of the dynamical laws to argue that space-time should not be conceived as a substantial container. In order to avoid the force of the hole argument, various proposals have been offered, but if the unique-propagation conception of determinism is adopted, the threatened failure of determinism can never occur. Because determination relations going from *c* to *e* have the relative location of *e* automatically built in—defined only in terms of physical degrees of freedom without any additional gauge degrees of freedom—there is no problematic ambiguity in how later stuff is laid out relative to later locations in the arena. Or, if there is a problematic ambiguity for some chosen model of fundamental reality, that is a problem with the model itself, not a problem with interpreting its arena as a substantial container.

Building on the unique-propagation conception of determinism, we can now eschew reference to "global states of suitable shape." As a preliminary, let us say that a *maximal domain of dependence* for a fundamental event *c* is a region R produced by applying the fundamental laws to *c* (extending the arena if necessary) until R includes all and only those points that are definitive of *c*'s domain of dependence: every point *p* such that every inextendible c-path intersecting *p* also intersects *c*'s region. Determinism can then be defined to hold iff the following obtains:

> *Determinism*: For any nomologically possible full event *c*, *c* determines a unique full event throughout its maximal domain of dependence.

This definition of determinism refers to a maximal domain of dependence in order to force the (possible) arena which contains the possible fundamental event *c* to expand automatically to whatever size is appropriate, using whatever fundamental laws establish (or propagate) the geometrical structure needed to classify the relevant c-paths.

This choice of conceptual design has two minor disadvantages. First, it can be trivially satisfied by states if their maximal domains of dependence do not extend beyond the state itself. For example, it classifies completely lawless worlds as satisfying determinism. I do not think this is a significant shortcoming because in realistic theories and presumably in the actual world, there are laws governing domains of dependence to ensure that many suitable events have non-trivial maximal domains of dependence. We should simply demand that theories incorporate adequate laws for domains of dependence to avoid trivial determination.

Second, this definition would be inappropriate if the actual world happens to have a non-maximal arena. We currently have no good scientific reason to believe that the actual arena is non-maximal, so it is not clear to me how serious a deficiency this is. Even so, one could argue that this potential shortcoming should be trumped by the need to ensure that contrary-to-fact conditionals are not rendered trivially false or undefined. When thinking about what follows from a (possibly contrary-to-fact) occurrence of event c, one should propagate the arena structure and its material contents out as far as the laws will allow. This method is especially appropriate for counterfactual claims governed by fundamental theories like general relativity where the structure of the space-time arena depends on how its material content is arranged.

2.10 Stochastic Indeterminism

A stochastic fundamental theory includes laws allowing fundamental events to suffice nomologically for a probability distribution over a set of possible events without determining which particular member of the set will be instantiated. Two versions are worth discussing by virtue of their application to important questions about causation. The first kind of theory postulates only a stochastic rule with no trace of non-trivial determination. The second kind proclaims the existence of a default deterministic rule plus rules about how and when the default is overridden by a chancy jolt.

2.10.1 STOCHASTIC LATTICES

Consider a toy theory where the arena is a three-dimensional space-time made of an infinite cubical lattice of points with a constant distance that separates adjacent points. Also, suppose there is an objective foliation dividing the space-time into distinct two-dimensional time slices. The only property in the universe is a single scalar field, represented by a real number for each space-time point. The fundamental dynamical law relates the state s at one time to the next state in the future direction by way of a stochastic rule. The state s at some space-time location (x, y, t) is equal by law to the average of the value of s at those nine space-time points plus a random real number chosen from a uniform distribution between -3 and 3.

For illustration, let e be a full event located at some point $p = (x, y, t)$ in the time slice at time t. As depicted in Fig. 2.7, the non-trivial terminant of e that precedes it by one unit of time is the full event comprising the values of s at the points

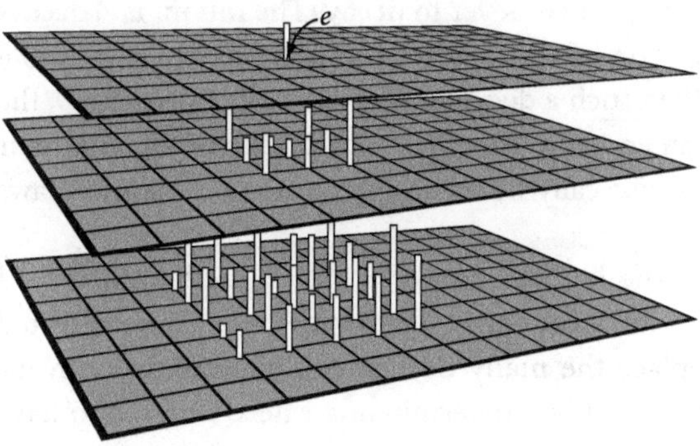

FIGURE 2.7 *A lattice space-time with purely stochastic dynamical laws.*

enclosed by $(x \pm 1, y \pm 1, t - 1)$. The non-trivial terminant of e that precedes it by two units of time is the full event comprising the values of s at the points enclosed by $(x \pm 2, y \pm 2, t - 2)$. In general, the non-trivial terminants of e are full events that occupy squares centered on e's spatial location.

The notable feature of this purely stochastic model is simply that terminance relations are temporally directed. There are no terminance relations that are past-directed. In such models, there is a simple explanation for the asymmetry of causation, although whether it is the best available explanation is a question that will be addressed in chapter 7.

2.10.2 A TOY THEORY OF PARTICLE DECAY

This section investigates a different kind of stochasticity, one that occurs in the form of a default deterministic rule that is occasionally overridden with a chancy jolt. This kind of stochasticity is worthy of consideration because it occurs in some versions of quantum mechanics, to be discussed in §2.13.3, and has consequences for the asymmetry of causation. By discussing a toy model of particle decay based on the classical unified field theory of §2.5.3, we will be able to identify one sort of past-directed parterminance[26] that also appears in realistic stochastic versions of quantum mechanics, but without needing to appeal to any of the technicalities specific to quantum mechanics.

The toy theory of particle decay is constructed by modifying the classical unified field theory so that corpuscles of any given type have new attributes representing a rate of decay for some number of decay modes as well as rules that specify the products produced by each decay mode. The rate for any given decay mode is represented by a constant probability per unit of time. We can imagine that these rules match what we know about empirically observed particle decays. Electrons

[26] Parterminance was defined toward the end of §2.2.

in this toy theory, for example, are posited never to decay. The muon, μ^+, decays in several modes, the most common of which is for the corpuscle to split into a positron and two neutrinos. When such a decay occurs at space-time point p, the world line of the μ^+ comes to an end and three new world lines begin with their initial velocities determined stochastically in accordance with conservation laws and symmetry considerations.

Adding these attributes and the laws of decay to the classical unified field theory has several consequences for terminance. With regard to future-directed terminance, the decay rules replace the many determination relations with indetermination relations, so we have only indeterminants where we used to have non-trivial determinants.

It is unclear, as stated so far, how the laws apply to the past-directed dynamical development of states. If we have a state consisting of a single μ^+ at rest, we can unproblematically propagate that state forward in time to get a probability distribution over some set of future outcomes. But to propagate that state back in time, we would need to specify more about the fundamental laws of the toy theory. Especially, we would need to identify rules for how corpuscles—picturing them as if time were flowing backwards—collide in ways consistent with their having been formed (in the ordinary future-directed sense) by a μ^+ decay. One option is that there are no laws for how to propagate a given state toward the past. A second option is that the same default deterministic rule applies toward the past together with a stochastic jolt and some rule establishing a probability distribution over some set of possibilities for the past. Although this second option is coherent, its rules for propagating a state toward the past will almost certainly serve as an extremely poor guide to the behavior of the actual world, for reasons having to do with temporal asymmetries in the macroscopic character of matter discussed in §5.4. The temporally asymmetric character of our local portion of the universe would effectively mask the expected empirical consequences of past-directed indetermination relations, so that experimentation provides little advice for how to choose among candidate dynamical laws that fix probabilities for past events.

The notable feature of the toy theory of particle decay is that even if there are no explicit rules for past-directed terminance relations, there still exists a remnant of past-directed terminance by virtue of the default deterministic rule. If the rules for propagating a generic state s fix a positive probability per unit time that no stochastic events occur (and hence that the default deterministic rule will apply), then there exists a positive probability that the past-directed development of s throughout some earlier region R will satisfy the default deterministic rule. Let s_R be the unique event occupying R that one gets by developing s toward the past using the deterministic dynamical law. Because s fixes a positive probability for s_R and no probabilities for any events that instantiate a chancy jolt, s partermines s_R. This parterminance constitutes a limited form of past-directed influence.

2.11 Non-stochastic Indeterminism

There are many ways for laws to be indeterministic without incorporating chanciness. In the extreme case where there are no fundamental laws, the only states that termine a chosen event e are ones that include e as a subevent; only trivial terminance exists in such worlds. If it turns out that there are plentiful sources of non-stochastic indeterminism, that would be a major obstacle for my account of causation because then there would not be any terminance to ground the utility of our talk of cause and effect. Depending on the precise nature of the indeterminism, there are some maneuvers that can be made on behalf of my account in order to rescue it from the lack of terminance. But, my theory would lose much of its appeal if it had to abandon the simple explanation that what fundamentally connects causes with their effects is terminance. Although there are no good ways of generalizing over all possible forms of non-stochastic indeterminism, we can examine a number of recognized sources of non-stochastic indeterminism in the paradigm fundamental theories. What I intend to demonstrate in this section is that their existence would not undermine the utility of my metaphysics of causation.

2.11.1 NEWTONIAN INDETERMINISM

Although classical mechanics in general and classical gravitation in particular are often thought to be exemplars of determinism, there is a cottage industry dedicated to the exploration of several ways classical physics can defy determinism. I will mention three examples.

Because there are no bounds on corpuscle speeds, it is apparently consistent with the laws of classical physics, either by way of corpuscle collision or perhaps by gravitational interaction, that a particle can accelerate unboundedly over a finite amount of time, resulting in a corpuscle being able to fly off to spatial infinity, disappearing off the edge of space-time during a finite span of time (Mather and McGehee 1975, Earman 1986). By the time-reversal invariance of classical mechanics, it would then be possible for corpuscles to fly into existence from the boundary at spatial infinity if the existing particles are arranged correctly. These *space invaders* are counterexamples to determinism because the states previous to the corpuscle's entry into space-time do not determine whether a space invader will appear from spatial infinity.

Another source of indeterminism, noted in (Laraudogoitia 1996) for example, can appear by virtue of the true point-like nature of classical corpuscles. If an infinite sequence of qualitatively identical elastic corpuscles $\{p_1, p_2, \ldots\}$ is lined up right to left next to a meter stick so that p_1 is located at the one meter mark, p_2 is located at the half-meter mark, and p_i is located at $1/i$ meters. If another particle p_0 were to fly in from the right and strike p_1, replacing p_1 and sending p_1 to the left to strike p_2 and so on, then in a finite amount of time, there would be a sequence $\{p_0, p_1, p_2, \ldots\}$ where p_i is located at $1/p_{i+1}$, which is qualitatively identical to the initial configuration. By time reversal invariance, if there are particles in this configuration, a particle could "enter" space-time by the zero mark on the meter stick, knock successive corpuscles to the right, eventually expelling the outermost particle, all without there being anything in the previous states to determine the appearance of the corpuscle motion. When corpuscles begin to move spontaneously by virtue of a localized arrangement of an infinite number of corpuscles, let us say we have a *gremlin*.

The simplest example, because it does not require multiple particles or special boundary conditions, is Norton's (2008) dome, where a particle sits perfectly on top of a dome under the force of gravity at rest for an undetermined duration and then spontaneously moves in some undetermined direction.

These and other examples violate determinism, but a failure of determinism does not imply a failure of ubiquitous determination. The truth of determinism depends on what the laws dictate about *all* nomologically possible events, but determination holds merely by virtue of the nomological consequences for *actual* events. We can define *ubiquitous determination* to hold just in case the following condition obtains:

> *Ubiquitous determination*: For any actual full event c, c determines a unique full event throughout its maximal domain of dependence.

The previous comments from §2.9 apply: ubiquitous determination holds non-trivially only if the rules governing domains of dependence are adequate.

Determinism is a much more stringent condition than ubiquitous determination because counterexamples to determinism can be constructed using highly idealized special conditions. It is much harder to identify plausibly actual events that would violate ubiquitous determination, and the difficulty is not merely due to our ignorance of the actual laws. All three counterexamples to determinism in classical physics require highly tuned initial conditions or special kinds of matter and so do not suggest a failure of ubiquitous determination in cases where initial conditions are not conspiratorial. The cited failures of determinism would not constitute a problem for a theory of causation based on determination unless they hold in more generic situations, which has not yet been demonstrated.

The fact that ubiquitous relations of determination can exist without determinism should help to overcome the kind of skepticism expressed by Norton (2003, 2007) about the plausibility of a "general principle of causality" in fundamental physics.[27]

[27] I wholeheartedly agree with Norton's thesis that what he calls "causal fundamentalism" is an implausible hypothesis.

2.11.2 CONTRIBUTION EXTENDED

Terminants are a generalization of determinants used for handling stochastic theories of physics. The reason I introduced the concept of terminance was to show that my metaphysics of causation is flexible enough to accommodate fundamental theories that incorporate chanciness. It is possible, though, that actual fundamental laws have enough structure to explain the empirical phenomena that motivate our having causal concepts but not enough structure to fix probabilities. So at this stage, it is worth noting that the concept of contribution can be extended to accommodate fundamental theories whose indeterminism is more restricted than "anything goes" but less restricted than fixing a determinate probability distribution over possible fundamental events.

One possibility is that the fundamental laws could impose constraints on what worlds are nomologically possible, laws describable with fairly simple mathematical formulas but where there is no way to specify a state such that the laws generate what happens later. In the physics literature, such models are said to fail to solve the initial value problem. In such a case, we ought to say that if *c* plays a role in constraining what happens elsewhere, it should count as a contributor.

Another possibility is that fundamental reality unfolds according to a dynamical law, but the law is of an indeterministic and non-stochastic variety. Perhaps the world evolves according to laws that are fundamentally mentalistic, meaning that nature at bottom behaves according to principles that take into account the purely volitional act of a Cartesian spirit, which is neither determined by previous states nor chancy nor governed entirely by mathematical constraints nor completely ungoverned. One could also postulate fundamentally theological substances or behaviors so that the volitions of gods constitute part of the development of the world. Because I know of no plausible theories that specify how such laws would work in combination with the laws of physics, terminance is not defined to accommodate them, but if there are such rules, we can define a causal contributor more broadly as any fundamental event that plays a role in the development of fundamental reality according to whatever fundamental laws exist. This allows the concept of causal contribution to be a more inclusive notion of partial cause within the context of fundamental metaphysics without imposing undue complications for my terminance-based model of fundamental causation.

2.12 General Relativity

The theory of general relativity (GR) is Einstein's theory of gravitation. It proves especially helpful for clarifying determination and contribution because it is significantly different from the other paradigm theories in several respects. For brevity, I will only discuss the standard interpretation of GR, which postulates a space-time that is similar to Minkowski space-time on an infinitesimal scale, but differs by permitting spatio-temporal curvature, including the relative tilting of the light cone structure at different points.

The signature feature of general relativity is that its space-time structure is dynamically responsive to the distribution of a world's material contents. For example, space-time is more curved where there is a stronger concentration of corpuscles or fields. Although GR thereby fails to obey the principle of content independence, this does no great harm to our attempt to accommodate GR as a paradigm fundamental theory because generic states in GR can be propagated through time, generating non-trivial determination relations. Just as relativistic electromagnetism provides laws where a full state determines what happens at other times, general relativity provides laws whereby full states defined to include a specification of its spatio-temporal geometry determine what happens at other times, barring the presence of pathological geometrical structure.

One issue raised by allowing the geometry of an event's region to vary systematically with the event's material content is the need to be more subtle about using coarse-grained events whose members all have the same size and shape. Strictly speaking, there are very few coarse-grained events that can differ in material content and have exactly the same shape. The best we can do, which is good enough, involves using only very large fundamental events of the same size and shape that match each other near their edges and differ from one another only in the interior. Then, any other coarse-grained events that are defined relative to this initial event—later effects, for example—can have the relative location of their members adjudicated by referring back to the edge of the initial event using whatever arena relations are needed.

2.12.1 SPATIO-TEMPORAL INDETERMINISM

Although in general relativity the default rule for the propagation of a state is a deterministic rule, indeterminism can appear from a number of sources. One concern is the possibility of *naked singularities*. A singularity is an edge or hole in the space-time manifold.[28] A singularity is naked if it generates indeterminism toward the future that is unconstrained by probability distributions. An example of a singularity that is clothed is the center of a black hole. There, space-time is singular because the curvature tensors that represent the black hole's effect on space-time become infinitely large. Nevertheless, the boundary conditions on the black hole singularity do not pose any problem for determination toward the future, because matter existing well within the black hole's event horizon will never escape owing to the black hole's intense gravity. An example of a possible naked singularity is the time reverse of a black hole, known as a white hole. The singularity at the center of a white hole is quantitatively similar to that of a black hole, but because particles

[28] Singularities are rather difficult to define using the resources of standard topology, as is evident from (Earman 1995).

are expelled from a white hole rather than drawn in, space invaders entering the world from the singularity can have observable effects.

An even simpler example of a naked singularity is a simple point-like perforation of the Minkowski space-time. It does not violate any laws of nature to have particles going into and out of that perforation, and because GR imposes insufficient constraints on the data associated with such boundary conditions, the resulting indeterminism is not cloaked under a probability distribution.

Although naked singularities violate determinism, they do not prevent the existence of non-trivial determinants. For a world with a single naked singularity, all events wholly to its past exhibit the same determination relations to each other as they would without the singularity being present. The same holds for determination relations wholly to the future of the singularity. The previously existing determination fails, though, when the naked singularity is temporally in between *c* and *e*, and the normal regularities cannot be expected to hold in such cases. It is possible that if naked singularities exist, some useful laws govern what comes out of them, but in general one can only make causal claims conditional on what springs out of the naked singularity. One can get unconditional non-trivial determination relations around a naked singularity only in terms of events that specify what happens around the boundary of the naked singularity. Such events shield the problematic source of indeterminism, but because they are presumably not pertinent to the behavior of matter in our actual environment, I will set aside further discussion of them.

2.12.2 CLOSED TIME-LIKE CURVES

Because GR only places local constraints on the geometry of the space-time manifold, it is possible according to the standard interpretation to have space-time topologies with *closed time-like curves* (CTCs) or, more generally, c-paths that are loops. Because massive corpuscles are presumed to be represented by inextendible time-like paths, they can exist in a loop of finite temporal length. A simple example of this can occur if space-time includes an appropriately oriented wormhole. Such possibilities conflict with determinism because global states that occur before or after the wormhole do not determine whether a corpuscle appears on the loop. So far as relations of determination are concerned, though, options that are available in the case of naked singularities are also available for indeterminism resulting from CTCs.

The additional issue that CTCs pose for understanding causation is that wormholes permit physical interactions to go from *c* to *e* by skipping over events that appear to lie in between *c* and *e*. For this reason, the consequences of shielding for backtracking influence and past-directed causation, to be discussed in chapter 6, will not be operative for worlds with certain kinds of CTCs. As a result, my discussion of causal asymmetry in chapter 7 will be compatible with the observation that GR permits causal loops.

2.13 Quantum Mechanics

A fully adequate discussion of causation in the context of quantum mechanics requires too much space to be included in this book. Instead, I will just set up enough of the terminance structure of non-relativistic quantum mechanics to make clear that my central arguments regarding how derivative causation depends on fundamental causation are also applicable to quantum mechanics. My task in this section is just to counter any suspicion that my later arguments are not general enough. I will only discuss non-relativistic quantum mechanics because the foundations of relativistic field theory are too unsettled to serve as a firm foundation for understanding causation.

2.13.1 THE QUANTUM ARENA AND ITS CONTENTS

To discuss terminants in quantum mechanics in detail requires an explicit statement of the (fundamental) ontology of quantum mechanics, which is famously controversial. For manageability, I will restrict discussion to the subset of interpretations I find conceptually clear enough to be instructive and leave the extension of my arguments to other interpretations as an exercise for the reader.

In the formulation I will use, there are two related mathematical spaces for representing fundamental reality's material contents. One space is just the mathematical space we use to represent the arena of Galilean space-time, discussed in §2.4.1. The other is the so-called quantum configuration space. A *configuration space* is a high-dimensional mathematical space for representing the global state of the universe at any given time. For illustration, let N be the number of particles in a given quantum-mechanical world. In non-relativistic quantum mechanics, setting aside the possibility of particle decay, fundamental particles never pop into or out of existence, so N remains constant. The appropriate configuration space is modeled mathematically as N copies of a Euclidean R^3, with three dimensions for each existing particle. Fig. 2.8 depicts these spaces for a pair of particles inhabiting a two-dimensional Galilean space-time with one space-like and one time-like dimension. The space-time on the left has time slices depicted as horizontal lines. The history of configuration space on the right has one temporal dimension and two spatial dimensions. (2 particles $\times$ 1 dimension of physical space = 2 dimensions of configuration space.) The time slice, A, in space-time is associated mathematically with a corresponding time slice in the history of configuration space.

The primary fundamental attribute in quantum mechanics is a holistic state called the *quantum state* of the universe, and it is represented using a complex-valued[29] mathematical function, ψ, called the *wave function*, which is defined

[29] To model the intrinsic spin of some fundamental particles, this complex-valued quantity is multi-dimensional, modeled as a spinor or multi-vector.

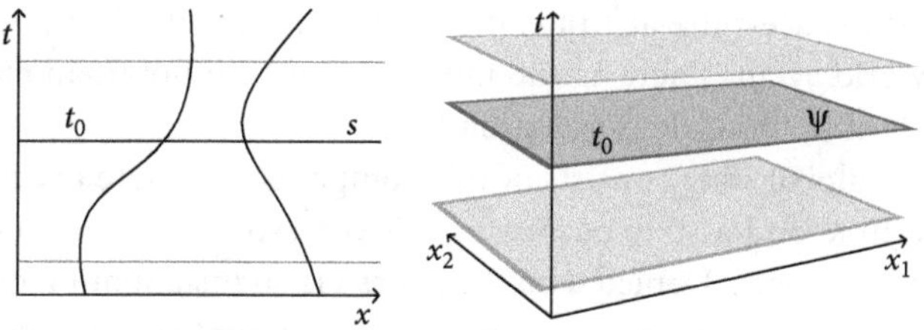

FIGURE 2.8 *A Bohmian world with two corpuscles in a one-dimensional space.*

throughout the configuration space. There are some nomological constraints on the magnitude of ψ so that the quantum state will be well-defined at all times, but such niceties can be ignored for current purposes. The global state of the quantum-mechanical universe at any given time t consists of the quantum state at t plus any other fundamental attributes instantiated in the spatio-temporal time slice at t.

The least revisionary construal of the arena in quantum mechanics is just to understand it as Galilean space-time. However, one might venture an interpretation of quantum mechanics where the history of configuration space is adopted as a separate component of the complete arena of fundamental reality. This might be motivated on the grounds that ψ is not representable as a function on a time slice of Galilean space-time, only as a function on a time slice of configuration space. Furthermore, ψ evolves in configuration space much like a physical wave evolves in ordinary physical space. I will not consider such an interpretation any further, but everything I say about causation is compatible with the possibility that configuration space is treated as a separate component of the arena so long as suitable identifications are made between the time slices in space-time and their corresponding time slices in configuration space.

2.13.2 BOHMIAN MECHANICS

In Bohmian mechanics, (Goldstein (2013), named after David Bohm, the quantum state evolves at all times according to a deterministic rule that incorporates information about every particle's fundamental properties. In addition to the quantum state, Bohmian mechanics also postulates that all particles are corpuscles whose histories are time-like paths in Galilean space-time. The corpuscles obey a deterministic equation called Bohm's equation. The evolution of the quantum state, represented by ψ, is deterministic in a way that depends only on the value of ψ at other times and not on the positions or velocities of the corpuscles. The velocity of any given corpuscle at time t depends in general on the positions of all the other corpuscles at t as well as the quantum state at t. So, the determinants in Bohmian mechanics come in two varieties. First, any event instantiating the quantum state throughout a time slice at time t is a minimal determinant of the quantum state at all other times. Second, any event instantiating all corpuscle positions on a time slice of Galilean space-time at t plus the quantum state at t is a minimal determinant of a world event. There are no indeterminants in Bohmian mechanics; its models exhibit ubiquitous determination.[30]

In Fig. 2.8, a global state consists of two components, a state occupying a time slice in space-time and a state occupying the corresponding time slice in quantum configuration space. Depicted on the left as s is an instantiation of exactly two corpuscle positions at some time t_0. Depicted on the right as ψ is a quantum state throughout all configuration space at time t_0. (In some interpretations of quantum mechanics, one locates ψ holistically in space-time instead of in quantum configuration space.) That global state composed of s and ψ determines the world event. The quantum state alone determines the world's full history of quantum states but nothing about where the corpuscles are located in space-time.

2.13.3 SPONTANEOUS COLLAPSE INTERPRETATIONS

In traditional spontaneous collapse interpretations, such as the (1986) Ghirardi-Rimini-Weber interpretation,[31] the quantum state evolves according to the same deterministic rule as in Bohmian mechanics, except for special violations that affect randomly selected particles at randomly selected instants. When a violation of the default deterministic rule kicks in, we say that the particle has been hit, and at that time, t_c, the quantum state makes a discontinuous, fundamentally chancy transition to a new quantum state. As a result of this instantaneous collapse process, ψ becomes squished along the three dimensions of configuration space corresponding to the spatial position of the hit particle. The resulting quantum state is characterized by the difference between the wave function in the limit just before the collapse, $\psi^-(t_c) \equiv \lim_{t \to t_c^-} \psi(t)$, and its value in the limit as time approaches t_c from the future, $\psi^+(t_c) \equiv \lim_{t \to t_c^+} \psi(t)$. The post-collapse state $\psi^+(t_c)$ is derived by first determining a density function from $\psi^-(t_c)$, then selecting a random location x probabilistically weighted by the density function, and finally multiplying $\psi^-(t_c)$ by some theory-specified function highly focused around the centerpoint x, such as a Gaussian function. Such spontaneous localizations are responsible for the fundamentally wavy quantum state behaving enough like a particle for us to make sense of our ordinary particle talk.

Indeterminism exists in traditional spontaneous collapse interpretations through two routes. First, a new fundamental constant sets the temporal rate at which collapses occur. Second, when a particle is hit, the standard quantum-mechanical probability distribution constrains where the collapsing quantum state will become concentrated. Together, they ensure that any global state, c,

---

30 For the ubiquitous determination to hold, certain conditions on the wave function must hold. See Wüthrich (2011).

31 See Lewis (2006) for a helpful review of its variants.

which instantiates the quantum state at any time t as well as any material contents in space-time at time t, is an indeterminant of any later event e. Regarding c's relationship to past events, the fundamental collapse rate fixes, for any duration Δt, some positive probability for the possibility that state c is not preceded by any collapses within the most recent span of time, Δt. However, there is no probability distribution for what states preceded a collapse, even when taking into account which particle was selected for collapse and the post-collapse state $\psi^+(t_c)$. Thus, in traditional spontaneous collapse interpretations of quantum mechanics, global states partermine past events.[32] This constitutes evidence that the kind of indeterminism postulated in quantum mechanics exhibits the same kind of past-directed influence described in §2.10.2 for the toy theory of particle decay.

In some versions of the spontaneous collapse interpretation, the occurrence of every collapse in configuration space is matched by the existence of matter in space-time. In particular, *flash interpretations* of quantum mechanics postulate fundamental entities known as *flashes* that each occupy a single point of space-time. In a traditional spontaneous collapse interpretation with the flash ontology, whenever a collapse squishes ψ in configuration space, there is a corresponding flash in space-time at the centerpoint of the collapse. This results in each quantum state being an indeterminant of later events in space-time that specify the existence and non-existence of flashes.

In a recent modification (Allori et al. 2008) to the traditional spontaneous collapse interpretation using the flash ontology, the quantum state is stipulated to evolve entirely deterministically. It is able to reproduce the same predictions, though, by specifying conditional probabilities for the appearance of flashes in space-time. The event c, consisting of the quantum state at some initial time t_0 and a complete specification of all flashes in space-time from t_0 until some time t, fixes a probability for whether there will be a flash in any chosen space-time region at t, and that probability matches the probability dictated by the traditional spontaneous collapse interpretation. This raises the possibility that this modified interpretation of flashy quantum mechanics can be construed in ways that deny the principles of shielding and continuity. Unfortunately, it would take too much of a digression to address this issue adequately. The best brief response I can offer is that the interpretations that circumvent shielding and continuity always postulate an initial condition for the quantum state. This initial state ensures that all causal relations hold by virtue of what probabilities are fixed by the initial state in conjunction with later events, including the occurrence of flashes. This kind of causation will be accommodated in §4.12 so that this new-fangled flashy interpretation of quantum mechanics will not constitute a counterexample to my account of causation.

---

[32] Recall the definition of parterminance from §2.2.

2.13.4 OTHER INTERPRETATIONS OF QUANTUM MECHANICS

Almost every interpretation of quantum mechanics either resembles (1) Bohmian mechanics, by relying entirely on relations of determination in the fundamental dynamical development, or (2) the spontaneous collapse interpretations, by supplementing the deterministic dynamical laws with some fundamentally chancy violations of the default deterministic rule. For all such theories, the structure of terminance will be largely the same as in these two interpretations. Some interpretations remain cagey about the precise characterization of quantum chanciness (and more generally about how quantum mechanics is to be understood as a complete and consistent theory of fundamental reality), but to the extent that an interpretation of quantum mechanics does specify a rich enough theory of fundamental reality, it should fit within the framework laid out in this section.

2.14 Summary

This chapter covered a lot of topics and introduced quite a bit of jargon, so it is useful at this stage to emphasize the conclusions that are most important for understanding how causation can be understood in terms of fundamental physics.

My most important claim is that causation among fundamental events consists of terminance. One might want to formulate this principle in an overly simplistic slogan by claiming, "What really causes an event are its terminants." The 'really' in the slogan emphasizes that components of fundamental reality like terminants and contributors have a privileged ontological status and that other kinds of causes exist only derivatively. The slogan tells us that terminants are entities that *cause*, but a more accurate way to portray their status is to say that terminants are the only entities that engage in those metaphysical relations that ultimately vindicate talk of causation. The real work that terminants do comes in the form of determining or fixing. These relations do not necessarily have the logical properties we ordinarily associate with causation, like irreflexivity and asymmetry. Nevertheless, terminance and contribution constitute the relations of singular causation that fundamentally bind various parts of the universe together.

The terminants that play a role in causation among macroscopic events are almost always big. For a cause-effect pair whose events occur more than a few nanoseconds apart, the terminant serving as the cause subsumes a vast swath of space, encompassing far more partial causes than we ordinarily cite as causes of *e*. For relativistic theories, the only prior states that termine a full event *e* span at least the entirety of *e*'s light cone. For paradigm non-relativistic theories, the only prior states that termine *e* span all of space.

Furthermore, terminance does not come in degrees. If some fundamental event *c* does not include a complete enough specification of its material content, which is typically a full specification, the fundamental laws imply absolutely nothing about what happens at *e*'s location. If *c* includes a complete enough specification, further specification of the material contents in its region does not "take away" its terminance of *e*.

Another remarkable observation about the paradigm theories is that many facts about terminance depend only on the arena structure. This allows many of the conclusions we draw about causation from current physics to be insulated from changes we expect to occur to our fundamental theories by virtue of discovering new types of particles and fields. Some kinds of discoveries in physics could motivate changes in our assessments of what termines what. For example, if we believe that generally any event's domain of contribution is its light cone, and then we discover a new space-like particle interaction, that should motivate us to reject the thesis that *e*'s minimal terminants are exactly the states that just barely span its light cone. However, if we were to discover a particle that is just like an electron but has charge +5, that would not force us to revise which full events termine which others. This insulation allows us to draw powerful inferences about causation just from our knowledge about the arena and some very basic facts about how the dynamical laws depend on the arena. The most important such generalization that holds for all paradigm fundamental theories is that every point occupied by *e*'s pure contributors are connectable to *e* by way of an everywhere differentiable path that is nowhere space-like.

There is an important related pair of conditions that hold in all paradigm theories: continuity and shielding. The essential idea behind them is that the physical state of the world evolves continuously through time. Nomic connections never skip across time; they always operate through intervening states. (If time itself comes in discrete steps, continuity should be understood as the claim that causal connections never skip over these steps.) Shielding can be thought of as the claim that as nature evolves dynamically in an ordered sequence or continuum of suitably large states toward the future, each of these states incorporates all the relevant information from its past for anything it fixes toward the future. Similarly, if we think of nature as evolving dynamically toward the past, shielding implies that states incorporate all the relevant information about the future for anything they fix toward the past.

Nowhere in the account of fundamental reality is a temporal asymmetry presupposed. It is possible for a fundamental theory to incorporate a fundamental temporal asymmetry, as was seen in theories with a fundamentally asymmetric stochastic rule, but causation in general and the asymmetry of causation in particular can be adequately explained without any fundamental temporal asymmetry. My account of causation permits events to influence the past and to bring about past events. Although such consequences sound counterintuitive, they are consistent with everything we know empirically about the temporal asymmetry of causation. Until we take up the issue of causal asymmetry in chapter 7, keep in mind that terminance and contribution can be future-directed or past-directed.

Here is a summary list of the principles that will play a role in later arguments, rephrased in non-technical terms to communicate the basic ideas informally.

- *Non-Spatiality*: The fundamental laws disallow events from fixing any events that occur elsewhere at the same time.
- *Transitivity*: If e_1 termines e_2 and e_2 termines e_3, then e_1 termines e_3.
- *Continuity*: The fundamental laws disallow fixing relations that skip over intermediate regions.
- *Shielding*: Events incorporate all the relevant information about the past for the future events they fix.

My general approach to fundamental aspects of causation draws inspiration from John Stuart Mill (1858), who emphasized that the real causes of events include far more than the events we ordinarily cite when called upon to give causal explanations:

> ... [M]ankind are accustomed with acknowledged propriety so far as the ordinances of language are concerned, to give the name of cause to almost any one of the conditions of a phenomenon, or any portion of the whole number, arbitrarily selected, without excepting even those conditions which are purely negative, and in themselves incapable of causing anything; it will probably be admitted without longer discussion, that no one of the conditions has more claim to that title than another, and that *the real cause of the phenomenon is the assemblage of all its conditions*. (Mill, Book III, Ch. 5, §3; Ed. 1, Vol. 1, p. 403, emphasis added.)

While I do not agree with Mill's precise claim here, he appears to hint at the crucial insight. Pure contributors have equal standing under the law. The benefits of treating all causation and influence as existing only insofar as it proceeds through "real causes"—what I have identified as terminants—can be adequately appreciated only after we have examined how it implies causal directness. To grasp that argument, though, we first need to investigate difference-making, the subject to which our discussion now turns.

{ PART II }

The Middle Conceptual Layer of Causation

{3}

Counterfactuals and Difference-making

The concept of a causal contributor captures one sense in which an event can make a difference to an effect. When *c* is a contributor to *e*, *e* must by definition have a terminant *c′* that would fail to termine *e* if *c* were absent and unreplaced. Every pure contributor *c* makes a difference to *e* in the sense that *c′* needs *c* to termine *e*. Correspondingly, every non-contributor makes no difference as to whether some *c′* termines *e*. It should be no surprise that this sort of difference-making is far too coarse to be of much use in the special sciences because it counts every contributor equally as a difference-maker. It cannot ground the distinction we apparently need to draw between stuff that is causally important for making effects happen and stuff that plays an insignificantly small role in the background.

It is now time to address this issue by developing a formal conception of difference-making to underwrite the difference between causally important foreground contributors and negligible background contributors. According to the method of empirical analysis, the target notion of difference-making is supposed to be engineered for explaining the empirical phenomena relevant to the metaphysics of causation. At the beginning of chapter 5, I will specify an experimental schema intended to encapsulate the relevant phenomena, but until then it will suffice for readers to keep in mind that in my account, difference-making is intended to serve the needs of a *metaphysics* of *general* causation.

3.1 General Causation

In §1.4, I drew a distinction between general causation and two kinds of singular causation. 'General causation' is our label for the kind of lawful relationship expressed by sentences like, "Whale breaches cause splashes." 'Singular causation' is our label for some kind of causal relationship between particular occurrences. The two varieties of singular causes in my account are (1) fundamental singular causes, exemplified by terminants and causal contributors, and (2) derivative singular causes, exemplified by culpable causes.

Existing theories of causation vary in their accounts of the precise connection between general and singular causation, but many (if not most) of them understand claims of general causation to hold by virtue of patterns of culpable causation (Lewis 1973b, Carroll 1991, Mellor 1995). Nancy Cartwright (1994, p. 95) states explicitly that a "generic claim, such as 'Aspirins relieve headaches', is best interpreted in terms of singular claims: 'An aspirin can relieve a headache'; and the surest sign that an aspirin can do so is just that sometimes one does do so." Even authors who do not explicitly endorse the priority of culpable causes often posit it implicitly. A common presumption, for example, is that the temporal asymmetry of causation consists in the fact that effects generally occur after their (culpable) causes.

By contrast, in my approach, general causation is understood in terms of possible *fundamental* singular causes—specifically, terminance—not in terms of culpable causes. This way of modeling general causation, I believe, turns out to be superior to standard approaches because it allows us to bypass the long-standing obstacle of providing a principled, demonstrably consistent (STRICT) account of culpable causation and to focus instead on what makes some contributors generally more effective than other contributors at bringing about certain effects. An adequate account of why, for example, whale breaches are effective at generating splashes does not require facts about (or rules concerning) what makes a localized (whale-sized) event effective in a single case. Even when there is an uncontroversial instance of what looks like a whale breach causing a splash, there does not need to be any STRICT account of culpability that declares the breach (rather than, say, the flight of some distant bird) as one member of a privileged set of events that were effective at bringing about that one splash. As I will discuss much later in chapter 9, the principles concerning general causation help to make sense of why our intuitions about culpable causation serve as a decent practical guide for understanding the world, but the key point is that, in my approach, we do not need facts about culpability to be susceptible to a rigorous, consistent systematization.

To summarize, most analyses of causation attempt to provide a STRICT account of culpable causes, which supports a STRICT account of general causation; my metaphysics of causation provides a STRICT account of terminants, which supports a STRICT account of general causation, which in turn plays a significant role in a RELAXED account of culpable causes.

My strategy will be to build an account of general causation by first investigating the idea of influence and then by interpreting claims of general causation as a specially prescribed kind of influence. 'Influence', as I use the term, refers to a family of concepts including our intuitive notions of influence as well as technical precisifications based on the idea that *c* influences *e* whenever *e* depends on *c* by virtue of fundamental laws. I will use 'influence' as an informal term compatible with many different varieties of such dependence.

There are a number of ways to make sense of dependence. I noted at the beginning of this chapter how causal contribution captures one sense of dependence. My task in this chapter is to engineer a new notion of dependence, *prob-dependence*, which will be linked in the next chapter to its corresponding notion of influence. Prob-dependence is short for 'probabilistic dependence holding by virtue of the fundamental laws'. What makes prob-dependence especially useful for metaphysics is that it abstreduces to fundamental reality. It is defined in terms of fundamentally arbitrary parameters such that once these parameters are filled in, the fundamental laws suffice for a specific value for the degree of prob-dependence of one event on another. Thus, it is a derivative relation that can be linked to fundamental reality in STRICT fashion. (Recall the discussion of §1.7 for further details concerning abstreduction.) The other benefit of prob-dependence is that it is a fully general relation, applicable to causal generalities in organic chemistry, ornithology, and economics. (I am not claiming here that practicing scientists will be able to gain an advantage by trying to express their causal regularities in terms of prob-dependence.)

General causation will turn out to be a special case of the kind of influence associated with prob-dependence. Specifically, a statement of the form, "*C*'s cause *E*'s," will eventually be rendered as a claim that the probability of *E* is higher with the chosen contextualized event representing the occurrence of *C* than with the chosen alternative contextualized event representing the non-occurrence of *C*. There is a long tradition of understanding influence and causation in terms of such difference-making, which is traditionally known as *counterfactual dependence*. My account is a part of this broad tradition in that it attempts to cash out the idea that *C* is causally important to *E* in terms of what would have happened had *C* not occurred. However, my treatment of counterfactual dependence is different from standard counterfactual accounts of causation in that (1) my notion is not closely tied to the logic and semantics of ordinary language and (2) my notion is designed to apply only to general dependence among event types, not single-case dependence among token events.

The purpose of this chapter is to scavenge historically prominent accounts for any materials that can be appropriated for a notion of difference-making optimized for a comprehensive model of causal generalities. Along the way, I will attempt to motivate the hypothesis that prob-dependence performs better than traditional versions of counterfactual dependence, but because there is no way I could ever argue that my specific formulation of difference-making is superior to every possible alternative, the justification for my choices of terminology cannot be absolutely decisive. My constructed concepts can only be judged in light of how well they perform in an empirical analysis of causation, which will not be evident until later chapters.

3.2 Counterfactuals

The term '*counterfactual*' is traditional shorthand for '*counterfactual conditional*' or '*contrary-to-fact conditional*' which is a statement expressing what would happen if certain non-actual conditions had obtained. A *conditional* in general is a statement of the form "If *A*, then *C*," including variants like, "If *A* is true, then *C* is true," "If *A* were true, then *C* would be true," and "If *A* had been true, *C* would have been true." The variables standardly represent propositions—the one in the 'if' slot being the *antecedent* and the one in the 'then' slot being the *consequent*.

It is standard practice to allow counterfactuals to have true antecedents even though that goes against the idea that counterfactuals are supposed to be about situations that are counter to fact. For convenience, I will use 'counterfactual' inclusively to permit true antecedents and use 'contrary-to-fact' conditional when the falsity of the antecedent is being presupposed.

It is also standard to allow counterfactuals to have antecedents that are not literally propositions about what would happen if certain non-actual conditions had obtained. In the account of counterfactuals I will be constructing in this chapter, though, it often proves convenient to restrict antecedents to the special case where they assert the occurrence of some contextualized event, as defined in §2.2.

Because the purpose of my exploration of counterfactual conditionals is to gather scientifically useful notions of counterfactual dependence, it is important to avoid hobbling the investigation by requiring it to obey the logical structures appropriate for understanding natural language counterfactuals. Even the most central features of ordinary language conditionals—that they obey modus ponens, that they ignore fantastically unlikely possibilities, that they have a kind of modal character—are all potentially deleterious to an adequate account of influence. If the structures used to explicate influence do not come close enough to matching the syntax of ordinary counterfactual conditionals, then their subsumption under the label 'counterfactual dependence' might be faulted as misleading. I think it is fair to say my particular notion, prob-dependence, is in the same conceptual neighborhood as those posited by other counterfactual theories of causation because it evaluates causal claims largely in terms of what would have happened had certain (typically non-actual) possibilities occurred. But as we proceed through the discussion, it should become clear that my model of the counterfactual that underlies counterfactual dependence departs so much from standard language-based accounts that it counts as a counterfactual conditional only under a liberal construal of 'conditional'. Nevertheless we should avoid getting bogged down in terminological quibbles about what constitutes a genuine conditional or genuine counterfactual dependence and just look for the best regimentation of difference-making we can find.

3.3 Goodman's Account of Counterfactuals

Nelson Goodman was an early advocate of connecting counterfactual conditionals with reasoning about how the world operates, following ideas that have been around since at least John Stuart Mill's (1858) *System of Logic*. Goodman's (1947) account is aimed primarily at interpreting counterfactuals of scientific interest, including causal and evidential claims. For the purpose of gleaning insights from Goodman that are useful for understanding causation, we can harmlessly focus on the special case where counterfactuals take the form, "If C had occurred, then E would have occurred," with C and E being mundane events. I will use Goodman's (1954) symbolization, $C > E$, where the logical connective is called the '*corner conditional*', to represent counterfactuals insofar as they are interpreted according to Goodman's theory. Each capitalized variable here represents a token event coarse-grained as some event type, and it also stands for the proposition that that event occurred.

Goodman evaluated counterfactuals using truth conditions. The skeletal rule he provided was that $C > E$ is true iff there is an appropriate set, S, of truths *cotenable* with C, such that C and S and the laws of nature together entail E. A proposition is cotenable with a set of propositions only if their conjunction is logically consistent, but there are other conditions Goodman wanted to specify in order to designate a special class of appropriate propositions. He seemed to conceive of his project as identifying a general rule for isolating a "correct" set of propositions that result in the "correct" truth value for the counterfactual conditional.

A commendable feature of Goodman's account is that it makes solid contact with a scientific account of the world and causation. When applied to counterfactual conditionals with antecedents that contradict logic or the actual laws of nature or involve a denial of important ingredients of reality, it does not seem to work at all or tell us anything interesting. But if we restrict Goodman's analysis to antecedents that proclaim the existence or non-existence of some event, we get something approximating a reasonable story about counterfactual dependence. I will now sketch how Goodman's theory of counterfactual conditionals can generate a version of counterfactual dependence. The resulting account of dependence will be a useful reference point for my own account of dependence. My own account is not based on cotenability but is inspired by Goodman's more general idea that counterfactual conditionals are to be evaluated by examining what the laws dictate about some hypothetical (and typically non-actual) situation.

Imagine a universe with a single deterministic dynamical law where there is a person standing in an ordinary room with an ordinary box of matches. She takes one of the matches, strikes it on the box's surface, and the match lights in the ordinary way. Let M be the striking of the match and let E be the match lighting. To analyze this situation, consider a state s, which is chosen to be the actual state of the universe at some time t during the striking. Now, let s_{-} be constructed by excising the striking of the match from s in some determinate region, and then let the state $s_{\neg M}$ be constructed by filling in the excised region from s_{-} with some contextually appropriate (and maximally determinate) non-striking of the match. The actual state, s, leads deterministically to the match lighting, and the counterfactual state $s_{\neg M}$ leads deterministically to the match remaining unlit.

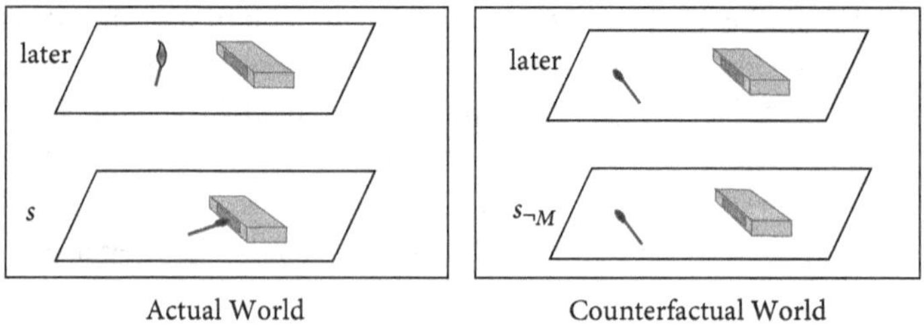

FIGURE 3.1 *Goodman-inspired counterfactual dependence.*

It is possible that some contextually relevant ways of instantiating the non-striking of the match evolve into the match lighting, but such states are of a kind very rarely found in the actual world. To keep things simple for the time being, let us set aside such states as fantastically improbable, and we will return later in §5.2 to investigate the complications involved in accommodating such possibilities.

One reason that $M > E$ is true is that there is a true proposition, ⟨that s occurred⟩, which is seemingly cotenable with M and together with the laws entails E. This helps to make sense of our willingness to agree with, "If the match had been struck, it would have lit." Similarly, $\neg M > \neg E$ is true because there is a proposition seemingly cotenable with $\neg M$, namely ⟨that s_{-} occurred⟩, such that together they entail $\neg E$. This makes sense of our willingness to agree with, "If the match had not been struck, it would not have lit."

It is now possible to characterize a notion of counterfactual dependence corresponding to these two counterfactual truths. The lighting of the match E counterfactually depends on the striking of the match M in the sense that s determines E whereas $s_{\neg M}$ determines the non-occurrence of E. Because s and $s_{\neg M}$ only differ with regard to what is going on locally with the striking of the match, this form of counterfactual dependence accords with the intuitive idea we are trying to make precise—that the striking of the match in particular makes a crucial difference to the lighting of the match.

In greater generality, Goodman's account of counterfactuals appears to support the following definition: E counterfactually depends on C iff $(C > E)$ & $(\neg C > \neg E)$. If I were going to use this version of counterfactual dependence myself, I would clarify how it applies to a comprehensive range of cases, but because I will be providing an improved definition of my own, I want to focus on a few salutary components that can be salvaged.

For an example of counterfactual independence, consider the true proposition F, that there is a fly on the wall of the room. Goodman's account judges $F > E$ as true because s determines both F and E. This corresponds with our judgment that if there had been a fly on the wall (which there was), the match would have lit (which it did). Then, consider an alteration to s that replaces the chunk of physics instantiating the fly on the wall with a chunk of physics that instantiates no fly being on the wall. Assuming the lighting of the match is suitably stable under microscopic perturbations to the striking of the match, just about any such counterfactual state will determine the lighting of the match. This vindicates the intuitive appeal of $\neg F > E$, "If the fly had not been on the wall, the match would (still) have lit." Because $(F > E) \& (\neg F > \neg E)$ is false, E does not counterfactually depend on F, which accords with our judgment that the presence of the fly made no difference to the lighting of the match.

We can repeat the reasoning used in the two previous examples to vindicate the claim, "If there had been no oxygen in the room, the match would not have lit," and thus that the lighting of the match counterfactually depended on the presence of oxygen.

Already at this stage, we can see something in Goodman's corner conditional that relates to what we are looking for when we seek a distinction between causally important and causally insignificant events. The striking of the match and the presence of oxygen count as important parts of the determinant s because E counterfactually depends on them. Had either one not been present, there would have been no flame. The fly on the wall does not count as an important contributor because E does not counterfactually depend on it. Had it not been on the wall, there would still have been a flame. Because this kind of counterfactual dependence does not ground a difference between the striking and the oxygen, it does not ground the intuitive distinction between being a foreground cause and being merely a background (enabling) cause, but some significant progress has been made in distinguishing important causal factors from unimportant background contributors.

Goodman thought that one serious deficiency of his account was the lack of a satisfactory, principled rule for identifying the relevant set of truths to serve as background conditions. Goodman's discussion of counterfactuals does not provide enough guidance, in my opinion, as to what would constitute a satisfactory rule. Perhaps the underlying idea is that we have a prior grip on the truth values of some uncontroversial counterfactuals, and our rules for identifying the relevant condition are satisfactory if they reproduce what we already believe about the uncontroversial counterfactuals.

In any case, this problem of relevant conditions that Goodman fretted about arises even in the best case scenario where there are laws of nature that provide the kind of determination relations among macroscopic conditions that would make certain uncontroversial counterfactuals true. Suppose it is correct (in circumstances where the match was not struck) that had the match been struck, it definitely would have lit. In the best case scenario, Goodman's account reproduces this truth because there are enough cotenable propositions—that oxygen was present, that the match was dry, that there is not a strong wind, etc.—such that when conjoined with the laws of nature and the proposition that the match was struck, it follows that there would be a burning match. But this very same scenario also seems to result in the truth of, "If the match had been [struck], it would not have been dry," because there are enough cotenable propositions—that oxygen was present, that the match did not light, that there is not a strong wind, etc.—such that when conjoined with the laws of nature and the proposition that the match was struck, it follows (according to Goodman) that the match would not have been dry. The problem of relevant conditions, as Goodman saw it, was that there needs to be a rule that selects a set of background conditions like the former but not the latter, and there seems to be no plausible tractable candidate rule that operates with sufficient generality.

My own take on Goodman's project is that he was wrong to think of his relevant conditions problem as a serious problem. The identification of the appropriate background conditions to import into our evaluation of counterfactuals is something we have to deal with in ways other than coming up with a general rule that can be specified in advance. My solution will be just to permit anyone to use whatever background conditions she wishes, and then to take the chosen background conditions into account when judging the correctness of the resulting counterfactual.

3.4 The Nomic Conditional

Several deficiencies with Goodman's overall account make its application to counterfactual dependence unacceptable. Although I applied one notion of counterfactual dependence together with Goodman's account of the corner conditional in the special case where the universe is governed by a deterministic dynamical law, significant modification is needed to make it compatible with a suitably large range of fundamental theories and to make it useful for the explanation of effective strategies. Fortunately, the changes are not difficult to implement, but cumulatively they will take us further away from the account Goodman explicitly provided. In this section, I will construct a new conditional to be known as the *nomic conditional*. Its purpose is to serve as a scientifically improved version of Goodman's counterfactual conditional. Let us first focus on contrary-to-fact conditionals, ignoring how to handle counterfactuals whose antecedents are true.

A major problem with Goodman's account is that in order to avoid virtually every mundane[33] counterfactual claim coming out false, it requires laws of a kind that are extremely implausible. His examples involving matches, for example, are supposed to render true such claims as, "If that match had been scratched, it would have lighted," in situations where the match was not scratched but where the background conditions were favorable. Goodman writes that "[t]he principle that permits inference of 'That match lights' from 'That match is scratched. That match is dry enough. Enough oxygen is present. Etc.' is not a law of logic

[33] A mundane counterfactual may be thought of as a counterfactual whose antecedent and consequent in effect assert the existence or non-existence of some mundane event.

but what we call a natural or physical or causal law" (Goodman 1947, p. 116). Although Goodman may be correct that we sometimes call such inferential principles 'laws', his model requires entailment from the *proposition* that the match is scratched (together with other laws and facts) in order for the counterfactual to be counted as true. But there are no such laws. On the one hand, there may well be deterministic laws, but all remotely realistic versions of such laws entail nothing whatsoever from a proposition like ⟨that match is scratched⟩ even if it is combined with other propositions concerning actual background facts. Realistic deterministic laws require a full specification of every last microscopic detail of how the match is being scratched. On the other hand, one might want to count many of the rules of thumb present in the special sciences as full-fledged laws, but such laws are almost certainly not deterministic. Owing to some discoveries in the branch of physics known as statistical mechanics, we have good reason to believe that for just about any coarse-grained event *C*, there are possible ways *C* can be instantiated so that things behave perfectly ordinarily for a reasonable span of time, but then evolve later into extremely abnormal behavior. It is physically possible, for example, for a rock to spontaneously spring into the air and for an unremarkable tree to quickly reassemble into a wooden replica of Michelangelo's *David*. These possibilities are sometimes labeled 'anti-thermodynamic behavior' because historically they are associated with violations of reliable thermodynamic regularities, but because they violate virtually every special science regularity, I will refer to them as *bizarre developments* and *bizarre evolutions* and any possible world that contains such an evolution a *bizarre world*. There exists a slight difference between 'bizarre development' and 'bizarre evolution' because I use 'evolution' to designate how the *actual* world (or postulated fragment of a possible world) changes as we figuratively move through time. I use 'development' to designate how a possible state changes as we use the fundamental dynamical laws to propagate it through time. The relevance of bizarre worlds for Goodman's theory is that so long as a false antecedent (together with as many cotenable propositions as you like) is compatible with a bizarre development that fails to secure the truth of the consequent, the counterfactual is false. For virtually any counterfactual dealing with mundane affairs, it is extremely plausible that some bizarre world exists where the antecedent is true and consequent is false, which makes the counterfactual false. On Goodman's account, virtually every non-trivial contrary-to-fact conditional dealing with causal happenings is false.

The possibility of these bizarre evolutions, I can note in passing, are not problems for Goodman alone. Any theory of causation that requires determination or strict regularities among macroscopic states or properties is going to face the problem that realistic theories of fundamental physics support the non-existence of such relations.

In order to draw out any consequences from the fundamental laws for ordinary counterfactual antecedents, the antecedent must be fleshed out in full fundamental detail. One way to achieve the necessary precisification of the antecedent is to stipulate a sufficiently large state for hypothetical consideration. Often, the state will be one that is just like an actual state except modified (if necessary) to make the antecedent true. The state $s_{\neg M}$ from §3.3 was just such a state. We arrived at it by altering a localized patch of one of the states when the match being struck so that it would instantiate a precise way for the match not to be struck. Then, we looked at what the fundamental laws implied about that state to figure out what to think about the counterfactual. Because $s_{\neg M}$ determines the non-occurrence of E, we concluded that $\neg M > \neg E$ is true.

The problem with precisifying an antecedent (and background conditions) as a maximally fine-grained event, though, is that it mischaracterizes the content of the counterfactual claim. When we are considering what would have happened if this match had not been struck, we are not considering one microscopically precise way of instantiating its not being struck but rather a range of relevant possible ways the match could remain unstruck. Modeling the counterfactual with a fundamental event fails to capture the spirit of the antecedent.

If we try to use a plain coarse-grained event to capture the imprecision of the antecedent, we run into the problem that the laws fail in general to deliver a univocal answer as to what counterfactually happens. A coarse-grained non-striking of the match itself is too spatially small to determine anything interesting, so the relevant coarse-grained event needs to include all the background conditions stretching out far enough into space to constitute terminants of E. But even if all the members of the coarse-grained event are big enough to determine whether the match lights within the relevant amount of time, there is no guarantee that they will all agree on whether E is determined. Indeed, the prevalence of bizarre worlds suggests strongly that typical coarse-grained events will be unable to pronounce on whether E is determined, for some members will determine E and others will determine $\neg E$. A common solution is to let the value of the counterfactual be true if and only if *all* the relevant worlds with the antecedent true also have a true consequent. This maneuver, however, results in the bizarre worlds making false virtually every mundane counterfactual, which makes it unsuitable for understanding influence.

The problem can be remedied by rendering the antecedent as a contextualized event. Remember that a contextualized event was defined in §2.2 as a coarse-grained event with a probability distribution over its members, and recall as well that useful contextualized events are almost always full and spatially expansive, specifying all the detailed background conditions extending far out into space to fix a probability for whatever event E we are interested in examining. Fortunately, owing to the contextualized event's built-in probability distribution, we are guaranteed to get a univocal answer as to what it implies for E. For illustration, let $\overline{\neg M}$ be a contextualized event representing the non-striking of the match. Every one of its members, let's say, is a full event occupying at least a sphere of radius ten light-seconds. Each member agrees with s_{-} concerning what is instantiated over the entire region s_{-} covers, but each disagrees on how the s_{-} state is augmented to instantiate a match not being struck. Every member instantiates the non-striking of the match but does so with a different microscopic arrangement of particles.

Let us further assume that $\overline{\neg M}$ is a contextualized event that corresponds reasonably well to what we are thinking of when we consider, "What if the match had not been struck?" To get an initial grip on an appropriate representation of the match being struck at time t, imagine a state three seconds before t, s_{t-3}, and let it develop according to fundamental laws that have enough microscopically localized stochasticity to affect macroscopic happenings over a relatively short period of time. The contextualized event fixed by this procedure should then be stripped of any members that are inconsistent with what is intended by "the match is not struck." After renormalizing the probability distribution, its full subevent that occurs exactly at t is a reasonable choice for $\overline{\neg M}$. The vast majority of its members are plausible ways the world could have evolved to result in the match remaining unstruck. For example, most of them will instantiate in the person's brain no intention to strike the match, and the vast majority of them will not instantiate a wet match, and the vast majority of them will instantiate more or less the same density of air around the match, etc. Unlike the maximally fine-grained $s_{\neg M}$, the contextualized event more accurately matches the intended content of the antecedent.

My proposal is to model the counterfactual, "If the match had not been struck, it would not have lit," with the probability that the contextualized event $\overline{\neg M}$ fixes for E. This *probability*, rather than some truth, is what underlies the correctness of the counterfactual. If one likes, one can think of the probabilities fixed by contextualized events as substitutes for truth conditions. For counterfactuals based on nomic relationships among physical things, talk of their truth is a crude substitute for probabilistic relations.[34]

The move to model counterfactual dependence probabilistically using contextualized events marks a significant improvement. A major problem with Goodman's account is that it requires relations of determination in order to avoid wrongly evaluating a vast class of contrary-to-fact conditionals as false. If there is any source of fundamental chanciness, located temporally between an event hypothesized in a false antecedent A and an event specified in a false consequent C, such that it is possible for C not to obtain given A, Goodman's account will render $A > C$ false. This makes counterfactual relationships that are nearly certain count as very different from relationships that are absolutely certain, which is

[34] In (Kutach 2002), I called this semantic device 'objective assertibility' because it resembles the subjective assertibility that had been used to model reasoning concerning indicative conditionals by Ernest Adams (1975). The subjective assertibility of an indicative conditional $A \rightarrow B$ is roughly an agent's credence that A conditional on B. The objective assertibility of a counterfactual conditional $A \,\Box\!\!\Rightarrow B$ is the credence an agent ought to assign to B given the situation described by A if that agent knew all the laws and all the material contents of the universe and could infer everything that the laws imply about hypothetical situations. I will not employ this terminology in order to avoid confusing issues of influence with issues of semantics.

a bad design feature if we are interested in accounting for effective strategies. In thinking about patterns of influence that humans can learn and exploit, it matters little whether an event *E* follows with certainty or with probability fantastically close to one. Because Goodman's corner conditional is based on truth values in a way that exaggerates that fine difference by making the former true and latter false, it is suboptimal for representing difference-making.

By contrast, incorporating stochasticity into my account is trivial. The probabilities that a contextualized event fixes do not require determination but only terminance. One only needs to formulate one's counterfactuals in terms of contextualized events all of whose members are large enough and filled out enough to count as terminants of *E*'s instance. Every such contextualized event will fix a well-defined probability for *E*.

Another notable benefit of contextualized events is that they allow us to model the seeming chanciness of macroscopic phenomena even when the fundamental laws are deterministic. This allows us to accommodate chancy causation in the special sciences and in everyday life in a way that is insulated from whether there is any fundamental chanciness in the laws of nature. Using Goodman's theory, bizarre worlds will render virtually any mundane counterfactual false. On my account, any reasonable probability distribution over the members of $\overline{\neg M}$ will include states with bizarre future developments but will assign them extremely low probability.[35]

One potential worry about modeling counterfactual situations in terms of very expansive contextualized events is that it might make it difficult to characterize localized events adequately. I will explain in the next section how to make sense of localized counterfactual alterations and how localized events can counterfactually depend on other localized events.

Here is a good place to gather together the partial insights we have gained so far. I will abbreviate statements like, "If $\overline{C}$ were to happen, then *E* would happen," as $\overline{C} \; \Box\!\!\Rightarrow E$ and call this construction the *nomic conditional*. I will not discuss how the nomic conditional can be incorporated into a broader logic but will merely offer clarification for the special case where $\overline{C}$ is a proposition expressing the occurrence of a contextualized event. In any such case, the value of the nomic conditional is defined as $p_{\overline{C}}(E)$, the probability $\overline{C}$ fixes for *E* (using the fundamental laws). Note that this probability is not the same thing as the conditional probability of *E* given $\overline{C}$.[36] The probability distribution included in $\overline{C}$ is stipulated as part of what it is to

---

[35] I take up this topic in more detail in §5.2.

[36] There exist prominent theories that evaluate counterfactuals in terms of some combination of what the laws indicate follows from a certain situation and what we should believe if we were somehow to adopt hypothetically the antecedent, e.g. (Adams 1975, 1976; Skyrms 1981; Edgington 2004). The probabilities invoked by these accounts resemble the probabilities I employ but they are far from equivalent. The basic idea behind these theories is that an important magnitude counterfactual conditionals often express is one's subjective grasp of the objective chance (at some time shortly before the time pertaining to the antecedent) of the consequent conditional on the obtaining of the antecedent. On my

be a contextualized event; the probability distribution is not generated through a process of conditionalization on some prior probability distribution.

Ordinary language counterfactuals are only given a semantic value by my account if the antecedent can be translated into a contextualized event. For example, "If the match had not been struck, the match would not have lit," can be translated into $\overline{\neg M} \mathrel{\Box\!\!\Rightarrow} \neg E$, where $\overline{\neg M}$ is a contextualized event that (1) extends suitably far out in space to be able to fix a probability for E, (2) has all of its members instantiating the match not being struck, and (3) captures in its members and probability distribution all the contextually relevant aspects of what we intend when we hypothesize about the match not being struck. When we consider what would have happened if the match had not been struck, we ordinarily do not consider global time slices that are completely devoid of matter, because context dictates that we intend to consider situations somewhat like the actual world only without the match striking and with whatever alterations are appropriate in light of the absence of the striking.

Modeling counterfactual relationships merely in terms of what is fixed by contextualized events provides several advantages over modeling them with propositions. For one, a contextualized event is able to represent that when we envisage a counterfactual possibility occurring, we typically imagine it could happen in different ways, some of which are more likely than others. For example, when we wonder, "What if the weather had been bad today?" we may think of a range of different ways the rain could have been bad, consistent with the kinds of weather typical for the environment and season, but probabilistically weighted toward the more likely kinds of bad weather. Propositions alone do not have enough structure to weight some possibilities more than others. We certainly do not always consciously conceive of counterfactual possibilities as weighted probabilistically, but when we think about what might happen if the weather were bad, we implicitly ignore possibilities that we know are possible but extremely unlikely. Contextualized events let us incorporate any probabilistic weighting we choose, while proposition-based accounts of counterfactuals disallow non-trivial probabilistic weighting.[37]

The nomic conditional represents the nomic implications of a hypothetical contextualized event for the consequent. It is not intended to match how we humans instinctively reason about counterfactual possibilities. Its ultimate purpose is to provide a structure to help explain the empirical phenomena associated with effective strategies, broadly construed, and thereby provide conceptual

account, (1) subjective probability is irrelevant, (2) one does not consult actual chances, and (3) one does not conditionalize on the antecedent but instead assumes the chosen antecedent contextualized event.

[37] Additional probabilistic machinery could be added to Stalnaker's (1968) proposition-based account of counterfactuals to perform the same function, though such an approach would require a significant alteration in how the modified semantics fits within a broader proposition-based semantics, and so should probably not be thought of as merely a tweak of Stalnaker's theory.

regimentations of causation and influence suitable for making sense of how causal behavior studied in the special sciences can hold by virtue of a fundamental reality that resembles models of fundamental physics. Because much of our counterfactual reasoning concerns causal behavior, to some degree the way we think about what follows from hypothetical physical situations should resemble the nomic inference I am elucidating, but there are certainly discrepancies owing to the many psychological shortcuts we use when reasoning about counterfactuals and to our use of counterfactuals to express relations that are epistemic or logical.

3.5 Comparison to Ordinary Language Conditionals

An appropriate activity for this volume would be for me to conduct a comprehensive comparison of my nomic conditional with popular alternative models of counterfactuals in order to justify my suggestion that the nomic conditional is better suited to a scientific understanding of the empirical phenomena associated with effective strategies than existing accounts of counterfactual conditionals. Unfortunately, space limitations prevent me from conducting even the more limited activity of demonstrating that natural language counterfactuals are suboptimal for understanding influence and causation. So instead, I have made my arguments on this topic publicly available elsewhere, and as a poor substitute I will briefly summarize my conclusions regarding the benefits of my nomic conditional. I apologize to readers who rightly expect me to defend the claim that my nomic conditional outperforms alternatives, but enough readers have indicated to me that they find it antecedently extremely implausible that the logic of natural language counterfactuals provides a promising theoretical structure for modeling influence and causation. Because such a discussion does not help to elucidate anything later in this volume, I ask interested readers to examine the supplementary material.

To summarize quickly, there are three main differences between my nomic conditional and natural language counterfactuals. First, the nomic conditional maintains a clean division between the content that is stipulated as part of our free choice of which hypothetical situation we want to consider and the content that is supplied by the objective structure of fundamental reality. With the natural language counterfactual, the aspects that are stipulated and the aspects that are objective are mingled in a complicated way. The stipulated part comes partially through our specification of the antecedent and partially through context-dependent parameters (such as a choice about the relevant conception of similarity to employ) in order to arrive at the relevant counterfactual worlds. The objective part comes partially through the laws of nature but also from accidents concerning how the material content of the actual world is arranged, which unhelpfully mixes patterns we can expect to be reliably repeated with patterns we can expect not to be repeated.

Second, the nomic conditional is especially handy for science because its semantic value is unadulterated by contingencies from the actual world that have no bearing on the general effectiveness of strategies. Specifically, it does not take into account the actual future outcomes of fundamentally chancy processes. Natural language counterfactuals, by contrast, possess several channels through which accidental facts about the layout of matter, such as fundamentally chancy outcomes, affect the semantic values of counterfactuals and make them suboptimal for measuring the strength of causal regularities.

Third, the nomic conditional is optimized for representing *general* relations of influence: what *kinds* of events are connected to each other in terms of probability-fixing relations. Unlike natural language counterfactuals, the nomic conditional avoids making any claims about what events depended on what other events on some particular occasion. See my previous discussion in §2.1.1 for a reminder of how coarse-grained events, including contextualized events, play the role of event types.

3.6 Prob-dependence

We can now apply the nomic conditional to establish a firmer grip on counterfactual dependence. The version of counterfactual dependence constructed in terms of the nomic conditional is prob-dependence. Recall that 'prob-dependence' stands for "probabilistic dependence holding by virtue of the fundamental laws" and that its purpose is to hone the idea of counterfactual dependence in order to improve our understanding of general causation.

Prob-dependence arises from a comparison of what is fixed by two contextualized events, conceived as possible occupants of a single region of an arena. To get a sense of how it works, imagine an empty arena and insert $\overline{C_1}$ anywhere you like. Next, let the fundamental laws go to work on $\overline{C_1}$ to specify its nomologically consequences for other regions. Then, reconsider the arena in a second pass by replacing $\overline{C_1}$ with $\overline{C_2}$ and emptying the rest of the arena. Then, let the laws dictate what follows from $\overline{C_2}$. Any coarse-grained event E wholly located in the domain of terminance of both $\overline{C_1}$ and $\overline{C_2}$ will have two probabilities assigned to it: $p_{\overline{C_1}}(E)$ and $p_{\overline{C_2}}(E)$. The difference between these two numbers is the amount that E prob-depends on the difference between $\overline{C_1}$ and $\overline{C_2}$. If E does not have probabilities fixed by both $\overline{C_1}$ and $\overline{C_2}$, the prob-dependence is undefined.

The degree to which E *prob-depends* on $\overline{C_1}$ relative to $\overline{C_2}$ is $p_{\overline{C_1}}(E) - p_{\overline{C_2}}(E)$.

The degree of prob-dependence thus ranges from -1 to 1. Positive values signify that $\overline{C_1}$ makes E more likely than $\overline{C_2}$ makes it, and negative values signify that $\overline{C_1}$ lowers the probability of E relative to $\overline{C_2}$. A zero value signifies that E is just as likely regardless of which of these two events occurs.

For illustration, let us attend to the simplified account of Goodman-inspired counterfactual dependence from §3.3. Remember that in the actual world, the match is struck under normal conditions and it lights. We are trying to represent that the striking of the match is causally important to the match's lighting by constructing a suitable notion of counterfactual dependence such that the lighting counterfactually depends on the match strike. The limitation when we first discussed this example was that we only used fine-grained events and assumed determinism. We can now drop both assumptions. Let the contextualized event $\overline{M}$ include some fuzzing of the actual state of the universe at the time when the match was struck. Formally, $\overline{M}$ is defined to be a set of possible states that are very similar to the actual state macroscopically but have a range of microscopic differences. Let $\overline{\neg M}$ be a contextualized event that agrees with $\overline{M}$ everywhere except that the striking of the match is replaced by a contextually relevant non-striking of the match. The amount that the lighting of the match, E, prob-depends on the match's being struck is $p_{\overline{M}}(E) - p_{\overline{\neg M}}(E)$. That is, the lighting prob-depends on the striking to the extent that the probability of the lighting is higher with the strike than without. In this particular case, $p_{\overline{M}}(E)$ is a reasonably high number depending on the quality of the match and the competence of the striker, while $p_{\overline{\neg M}}(E)$ is virtually zero. The significantly large positive number represents the fact that the lighting prob-depends significantly (and positively) on the striking of the match. One can repeat this procedure to calculate the prob-dependence of the lighting on the presence of oxygen as very nearly equal to one, and the prob-dependence of the lighting on the presence of the fly as very nearly equal to zero. These examples serve as evidence that the magnitude of prob-dependence matches our intuitive grasp of probabilistic influence.

3.7 Contrastive Events

Because prob-dependence is going to be a crucial component for later discussion, it will be convenient to introduce a fourth conception of events in order to make it easier to express relations of prob-dependence. Remember that I have so far used one kind of fundamental (fine-grained) event and two kinds of derivative (coarse-grained) events, plain coarse-grained events and contextualized events.

Now, we can define another kind of derivative coarse-grained event:

A *contrastive event* is an ordered pair of contextualized events.

Let's call the first contextualized event in the ordered pair the '*protrast*' and the second contextualized event the '*contrast*'. I will always use a tilde over a capital letter to signify a contrastive event. The tilde is not a function or an operator; it is just part of the label of the contrastive event, just as the bar over a letter is (by convention) a part of the label for a contextualized event. As with any other coarse-grained event, it does no harm to restrict our consideration mostly to contrastive events that have a determinate size and shape in the arena, which is accomplished formally by having two contextualized events that occupy the same kind of region, and informally by thinking of the contrastive event as two different possible inhabitants for a given location. (In full generality, one should specify the location of the contrast in terms of the location of the protrast or the other way around.)

We can easily extend the idea of fixing to contrastive events.

> A contrastive event $(\overline{C_1}, \overline{C_2})$ *fixes* a contrastive event $(\overline{E_1}, \overline{E_2})$ iff $\overline{C_1}$ fixes $\overline{E_1}$ and $\overline{C_2}$ fixes $\overline{E_2}$.

Just as there are cases of trivial determination, there are also cases of *trivial fixing*. Trivially, every contrastive event fixes itself and all of its subevents.

We can also define, for any chosen contrastive event $\tilde{C}$, a unique maximal contrastive event that it fixes.

> The *maximal contrastive event* fixed by $\tilde{C}$ is the contrastive event consisting of whatever $\tilde{C}$ fixes for the entire arena trimmed to exclude any region where $\tilde{C}$ does not fix a contrastive event.

We can similarly speak of the contrastive event that $\tilde{C}$ fixes for a chosen region R.

The purpose of a contrastive event is to simplify claims about prob-dependence. Instead of saying E prob-depends on $\overline{C_1}$ relative to $\overline{C_2}$, we can communicate the same thing just by saying E prob-depends on $\tilde{C}$, where $\tilde{C}$ is defined to be $(\overline{C_1}, \overline{C_2})$.

> The degree to which E *prob-depends* on $(\overline{C_1}, \overline{C_2})$ is $p_{\overline{C_1}}(E) - p_{\overline{C_2}}(E)$.

Contrastive events also provide a richer representation of a mundane event in its background environment. Instead of representing the striking of the match merely as a localized coarse-grained event, the discussion in the previous section in effect represented the striking as a contrastive event. Just think about a region that could be occupied by either $\overline{M}$ or $\overline{\neg M}$. Both of these contextualized events can be (and were above) chosen to agree everywhere except in the region where the match was located. There, $\overline{M}$ specifies a striking of the match, and $\overline{\neg M}$ does not. So, we can think of the striking of the match as the contrastive event $\tilde{M}$, defined as $(\overline{M}, \overline{\neg M})$. That allows us to satisfy the demands of fundamental physics by having $\tilde{M}$ stretch out far enough into space to fix probabilities for any potential effects of interest, but it also satisfies our desire to focus on the more localized cause, the striking of the match. We can clarify this formally by distinguishing between two regions of $\tilde{M}$. The *background* of $\tilde{M}$ is the region where both of $\tilde{M}$'s contextualized events agree about what is happening. The *foreground* of $\tilde{M}$ is the region where its two contextualized events disagree about what is happening. In ordinary language, we can usually be a bit sloppy and say that the probability of some event E depends on a localized event (in the foreground) even though it can only prob-depend on the whole contrastive event. For example, we say that the probability of the match's being lit depends on whether it is struck, but the proper observation to make is that the probability of the match's being lit depends on the existence of a vast background state together with a localized striking of the match rather than a non-striking of that match. The background is essential for prob-dependence because realistic fundamental laws dictate almost nothing about what is likely to occur by virtue of the localized events alone.

Note that nothing in the definition of 'foreground' and 'background' implies any restrictions on the events being represented. In particular, a contrastive event can be "all foreground" by having no region where the two contextualized events agree. This is important because one of the main advantages of the concepts I am constructing in this chapter is their general applicability to all nomologically possible situations, fuzzed to whatever degree you like.

3.8 Summary

In this chapter, I formulated a new kind of counterfactual conditional, the nomic conditional. Statements of the form, "If C were to occur, then E would occur," are to be coaxed into the form $\overline{C} \mathrel{\Box\!\!\Rightarrow} E$, where $\overline{C}$ is a contextualized event and E is a coarse-grained event. The value of $\overline{C} \mathrel{\Box\!\!\Rightarrow} E$ is $p_{\overline{C}}(E)$—the probability assigned by $\overline{C}$ through the fundamental laws to E—also known as the probability $\overline{C}$ fixes for E.

The conception of counterfactual dependence I formulated using the nomic conditional is prob-dependence. Prob-dependence can be modeled in terms of a pair of nomic conditionals. A coarse-grained event E prob-depends on the occurrence of $\overline{C_1}$ rather than $\overline{C_2}$ to the extent that the value of $\overline{C_1} \mathrel{\Box\!\!\Rightarrow} E$ exceeds the value of $\overline{C_2} \mathrel{\Box\!\!\Rightarrow} E$, which is $p_{\overline{C_1}}(E) - p_{\overline{C_2}}(E)$. An equivalent formulation of prob-dependence uses a contrastive event $\tilde{C}$ defined as the ordered pair $(\overline{C_1}, \overline{C_2})$. The coarse-grained event E prob-depends on $\tilde{C}$ to the extent $p_{\overline{C_1}}(E) - p_{\overline{C_2}}(E)$.

To illustrate, if $\overline{C_1}$ instantiates a person striking a match and $\overline{C_2}$ instantiates the very same event as $\overline{C_1}$ except altered so that the person is not striking the match, then the contrastive event $\tilde{C}$ represents the striking (rather than the non-striking) of the match, embedded in a background environment.

Prob-dependence is a measure of difference-making among derivative events, quantifying how much difference *in general* one contrastive event makes to the probability of a plain coarse-grained event. It does not pronounce on whether, in some particular historical instance, one chosen event depended on another probabilistically or causally. It only measures the degree to which the probability of events of type E are made larger by $\overline{C_1}$-events than by $\overline{C_2}$-events by virtue of the fundamental laws.

Finally, I have abstreduced prob-dependence to fundamental reality by providing two fundamentally arbitrary parameters, a choice of one contrastive event to play the role of a cause and a choice of one coarse-grained event to play the role of the effect (including information locking down their relative position in the arena). These parameters together with the (presumably adequately rich) fundamental laws suffice for a determinate magnitude of prob-dependence between any two appropriately located and appropriately specified events. When conjoined with my forthcoming story about how the contrastive events can be chosen in order to extract some reasonable level of utility from relations of prob-dependence, my precisification of counterfactual dependence will have been reduced to fundamental reality in the sense captured by 'abstreduction' from §1.7.

{4}

Derivative Causation

In the previous chapter, I constructed a measure of difference-making called 'prob-dependence'. In doing so, I provided an abstreduction of prob-dependence to fundamental reality by specifying a complete set of fundamentally arbitrary parameters such that once they are filled in with determinate values to represent how to abstract away from fundamental reality, fundamental reality with no further supplementation suffices for relations of prob-dependence with determinate magnitudes. So far, though, nothing definitive has been claimed about how prob-dependence is linked with influence and causation. Now the stage has properly been set for the main attraction: an account of general causation. The layout of this chapter involves first examining influence and how prob-dependence can be equated with one specific conception of influence, dubbed 'prob-influence'. General causal claims will then be interpreted in terms of prob-influence. After demonstrating how this formalism makes sense of some commonly cited features of causation, I will describe a third conception of influence that will figure in later chapters.

4.1 Influence

As noted in §3.1, I use the word 'influence' throughout this volume informally to range over a vague collection of concepts including folk conceptions of influence as well as any appropriate technical precisifications. I will now identify what I think is one under-appreciated aspect of our intuitive grasp of influence: its linkage with fundamental laws. The observations in this section are not meant as part of a justification for my eventual definition of 'prob-influence' but are meant to serve as rhetorical grease. I suspect readers will be more likely to accept my three regimentations of 'influence' as precisifications of influence if I first point out the intuitive appeal of using fundamental laws to guide more rigorous formulations of influence.

One excellent reason for believing in influence is that we antecedently believe we can manipulate or control some sorts of events. Reality includes a vast pattern of paradigmatic manipulation: people deciding to place a book on the shelf rather than the desk followed by the book being on the shelf, and so on. We need not inquire too deeply into what manipulation or control amounts to in order to recognize that influence is a broader, more permissive notion than manipulation and control. We can say of all events under our control that we influence them, but our influence extends to other events over which we exert such a weak degree of control that they can hardly be said to be under our control at all.

It is uncontroversial that the moon influences us. Its position and brightness affect people who look at it, and it also generates oceanic tides. In some ancient civilizations, though, it was a prominent opinion that we do not influence the moon. This makes sense in light of the total absence of any detectable manipulation of the moon. No one had observed anyone shoving the moon to a desired location or altering its appearance at will. Yet, upon coming to believe in a universal law of gravitation that operates among all massive bodies, we humans correspondingly find it natural to change our opinion. "Of course we influence the moon. How could we not? If the moon is overhead and we climb on a ladder, the inverse-square law of attraction implies a difference in the tidal forces we exert on the moon." Such an effect is far too weak to be noticed or useful for redirecting the path of the moon, but our beliefs about the fundamental laws dictate that we have some influence. Throughout history, as our conception of fundamental physics has changed, so has our attitude toward what we influence. When the theory of relativity was adopted, people were motivated to disbelieve in faster-than-light influence. When quantum non-locality was confirmed through violations of Bell's inequalities, that suggested we should reconsider the extent of our influence. Perhaps quantum mechanics does imply that influence extends beyond the light cone. The lesson is simply this: the range of what we think we influence is expanded and contracted by what we think the fundamental laws dictate.

Another instructive example of influence is the effect we have on the detailed happenings of the distant future. Owing to chaotic interactions, what I do now arguably influences which humans exist thousands of years from now. Because there is no good way to know how my behavior today will affect the specific genetic makeup and activities of any person in the distant future, I cannot manipulate or control the details of their behavior, yet this seems to be best interpreted as the result of their activities being extremely sensitive to the influence exerted by my precise present actions rather than not being influenced by me.

These two features of our ordinary notion of influence suggest that toward the future, what we are able to influence includes all the events we purely contribute to. This is why, in §2.6, I labeled this region the 'domain of influence'. Although I think our ordinary notion of influence is not determinate enough to equate it with future-directed pure contribution, they are nearly the same because our ordinary notion of influence applies paradigmatically to events we are able to influence usefully and extends throughout the future by way of the fundamental laws to other events regardless of whether that influence is weak or uncontrollable or unobservable.

4.2 Prob-influence

Contribution is too indiscriminate a notion of influence to be used as a foundation for a metaphysics of causation. It never comes in degrees; for any given *e*, every event is either a contributor to it or not. What's more, events almost always have a vast multitude of contributors, so that contribution does not select some partial causes as more important than others. We need an alternative notion of influence that is sophisticated enough to capture the kind of difference-making needed to make sense of causal generalities.

I offer now for your consideration a precisification of influence that expresses the same thing as prob-dependence.

> A contrastive event $\tilde{C}$ *prob-influences* a coarse-grained event *E* to the degree that *E* prob-depends on $\tilde{C}$.

Prob-influence is the kind of influence that is by definition equivalent to the prob-dependence of a coarse-grained event on a contrastive event. If the degree of prob-influence is positive, $\tilde{C}$ is said to *promote E*. If negative, $\tilde{C}$ is said to *inhibit E*. We can also say $\tilde{C}$ *exerts* prob-influence on *E* when there is a well-defined degree to which $\tilde{C}$ prob-influences *E*.

Prob-influence is the most important conception of influence for understanding the metaphysics of causation. The fact that prob-influence is equivalent to prob-dependence raises several issues demanding clarification.

First, as I have defined it, prob-dependence incorporates no temporal asymmetry. Events can prob-depend on future events, and thus events can prob-influence the past. People may think that contemporary events can counterfactually depend on the future in a probabilistic way (or the past on the present), but this dependence is usually interpreted as a by-product of future-directed influence. On my account, when *E* prob-depends on a future event, that implies the future event is prob-influencing it. Past-directed influence might strike one or two readers as counterintuitive, but I will continue to postpone clarification of this feature until chapter 7 when it can be given a fair hearing. At this point, I will only note that nothing I say about prob-influence forbids a person from imposing a fundamental direction of influence by fiat. In that case, future-directed prob-influence is set equal to prob-dependence, and past-directed prob-influence is stipulated out of existence.

Second, in ordinary language, some kinds of counterfactual dependence apparently have an epistemic and non-causal character. If the thermometer in the isolated room were presently three degrees lower, the local air would be about three degrees cooler a few milliseconds later. We do not normally interpret this true assertion to imply that the thermometer is influencing the air in a way that makes it warmer. The specific kind of counterfactual dependence inherent in prob-dependence, though, is entirely causal in the sense that when *E* prob-depends on $\tilde{C}$ that always implies that $\tilde{C}$ prob-influences *E* even if the probabilistic connection between the events would be normally interpreted by people as largely or entirely epistemic. This revisionary interpretation of counterfactual dependence demands more attention, but for now I just want to flag its existence. In chapter 6, I will briefly discuss the relation between causal and evidential dependence in a way that is compatible with the equivalence of prob-dependence and prob-influence.

Third, prob-dependence and influence generally should not be associated in any essential way with agency or actions. In prototypical instances of agency, such as when rational human beings are informed and alert and are acting on the basis of information and reasons, we say that such agents influence what happens later, and we think of their intentional behavior as actions. For some philosophical purposes, it helps to distinguish some cases as genuine actions and other cases as not. For the purpose of understanding causation, though, no such distinction should be made because the goal is to understand influence in its application to all things, including among physical objects that are remote from all agents. Likewise, constraints stemming from concerns about free will ought to be set aside. Some incompatibilists might think that if our world exhibited ubiquitous determination, agents would not exert any genuine influence. However, that reasoning would invoke an overly narrow conception of influence that is irrelevant to the topic of causation simpliciter.

Fourth, one could quibble that prob-influence is artificially precise, that it models our influence over events as if it always has a precise magnitude when arguably nothing in our evidence concerning causation dictates a specific quantity to us. The situations we ordinarily entertain are not microscopically precise and so there are a range of different contextualized events that represent more or less what we have in mind when we are thinking about counterfactual possibilities, and these contextualized events can exert different degrees of prob-influence over any given E. Yet, it usually turns out that most contextualized events exert approximately the same degree of prob-influence, and little depends on the precise value. What matters is whether the value is nearly zero or significantly negative or significantly positive or very nearly one or very nearly negative one. The artificial precision in the model is employed in order to give the fundamental laws something they can process, but it can usually be discarded when we use prob-influence to interpret causal regularities.

Note that one could define 'prob-dependence' and 'prob-influence' in other ways to make it scale differently. I have chosen to keep things simple by just defining it as a difference between two probabilities.

Fifth, in my initial discussion of events way back in §2.1.1 and in my introduction of contrastive events from §3.7, I pointed out that fine-grained (fundamental) events play the role of singular (or token) events, whereas coarse-grained events, including contextualized events and contrastive events, are abstractions that are intended to help model type-level claims about causation and influence. When some $\tilde{C} \equiv (\overline{C_1}, \overline{C_2})$ prob-influences E, that is a relation that holds by virtue of what

$\tilde{C}$ is defined to be, what E is defined to be, and what the fundamental laws are. To say that $\tilde{C}$ prob-influences E to degree x is to say the fundamental laws of the actual world are such that the probability $\overline{C_1}$ fixes for E is x more than the probability $\overline{C_2}$ fixes for E. It does not depend on what events are instantiated in the actual world, a.k.a. fundamental reality. Nor is the prob-influence relation itself a part of fundamental reality; it is in general a *derivative* metaphysical relation.

4.3 General Causation

We say, "Moisture causes iron to rust," and what makes it reasonable for us to do so is that there is a causation-like relation holding between the potential presence of moisture on iron and the likelihood of rust forming later. The cause can be represented as a contrastive event $\tilde{C}$, using a contextualized event $\overline{C}$, which comprises possible instances of moisture on iron (with oxygen nearby as part of the environment) and a contextualized event $\overline{\neg C}$, each member of which is just like a corresponding member of $\overline{C}$ except that its moisture is replaced with a lack of moisture. The presence of rust on the iron is represented as a coarse-grained event E, comprising all the contextually relevant possible happenings following $\tilde{C}$ (within an appropriate amount of time) that instantiate the piece of iron being rustier than it was at the time of $\tilde{C}$. The goal, now, is to characterize such relations in general terms.

The main thesis defended in this chapter is the following:

> The important content of general causal claims is adequately represented by relations of the form, $\tilde{C}$ promotes E.

What makes promotion count as the important content is that it explains (in one important sense) the empirical phenomena corresponding to a canonical test of how much E is made more likely by $\overline{C}$ than by $\overline{\neg C}$. In the next chapter, I will explain how claims of promotion can be tested, which will help to substantiate my claim that promotion is a concept optimized for understanding the empirical phenomena behind our belief that causes are means for bringing about certain distinctive events. This chapter will focus on what the promotion relation is and how to interpret its relation to causal claims.

I will first explore two examples to illustrate how general causal claims can be rendered in terms of the structures I have introduced in previous chapters.

For the first example, imagine a global state, s, depicted in Fig. 4.1, with a light switch in the on position in region C and a distant pristine beach in region E. From this state we can construct contextualized events conceived as possible occupants of this single global time slice. Let $\overline{C_1}$ be a reasonable contextualization whose members very closely resemble s everywhere except in the cubic region C, where they instantiate different ways the microscopic structure of the switch can be arranged so that the switch is on. Let $\overline{C_2}$ be a contextualized event that matches

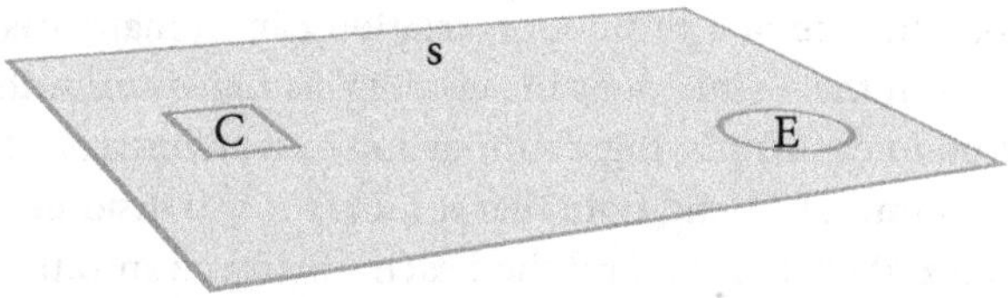

FIGURE 4.1 *A contrastive event whose foreground is region C will typically not (in the near term) significantly prob-influence events in the beach's region, E.*

$\overline{C}_1$ everywhere except in the cubic region where it instantiates the switch's being off in microscopically different ways. Then, $\tilde{C}$ is stipulated to be $(\overline{C}_1, \overline{C}_2)$, which represents the switch's being on rather than off in environments that are very similar to the actual state *s*. Assuming the switch is connected to a functioning light in a prototypical manner, $\tilde{C}$ promotes the light's being on rather than off. By contrast, $\tilde{C}$ does not significantly promote anything having to do with the beach in the near term future. The beach is hundreds of miles away and having the switch on rather than off makes virtually no difference to the probabilities of what may soon happen at the beach.

When expressing $\tilde{C}$'s promotion in ordinary causal language, care is required to interpret the contrastive event in a fair manner. You are always within your rights, when considering prob-influence, to pick any contrastive event that interests you. Yet, whichever contrastive event you choose must be interpreted correctly so that all the consequences of your choices are properly assigned to the foreground or background of $\tilde{C}$.

One potentially misleading way to think about what the switch setting promotes is to reason as follows. I am permitted to pick any contrast I want because it is a fundamentally arbitrary parameter. So for fun I will pick a contrast $\overline{C}_3$ that is just like $\overline{C}_2$ except that in the spherical region, E, its members all instantiate trash on the beach. Then, I will represent the switch's being on as $\tilde{C}' \equiv (\overline{C}_1, \overline{C}_3)$ which I am free to do because it is a fundamentally arbitrary parameter. When I examine the nomological consequences of $\tilde{C}'$, I note that the switch's being on promotes a clean beach. Thus, there is a spurious causal connection between the switch and the beach.

This reasoning is incorrect because it mischaracterizes the $\tilde{C}'$ event as 'the switch's being on'. It is fine to choose $\tilde{C}'$ as a contrastive event; you can choose any contrastive event you want. And $\tilde{C}'$ does indeed promote the cleanliness of the beach, but the correct interpretation of $\tilde{C}'$'s promotion in terms of ordinary language goes as follows. If the switch were on and the beach were pristine, then the beach would be more likely to be clean than if the switch were off and the beach were filled with trash. What that shows is that $\tilde{C}'$ trivially promotes a clean beach.

The lesson is that caution is needed when thinking of promotion as a relation that arises from a local event that is then relativized to a background condition and contrast. The significance of a promotion relation can in many cases reasonably be glossed as a claim that "what does the causing" is the event in the foreground with the background events playing a supporting or enabling role. So one should look at the foreground of $\tilde{C}'$ and note that it includes two disjoint subevents, the switch's being on rather than off and the beach's being clean rather than trashy. So what is "doing the causing" in $\tilde{C}'$ are two spatially separate subevents neither of which significantly prob-influence what happens in the neighborhood of the other.

A more accurate way to conceive of promotion is that it always holds by virtue of the entire contrastive event. A good rule of thumb is that it is fairly safe to interpret prob-influence in terms of its foreground and background when the contrastive event occupies a single time slice or a time slab of short duration. Such a decomposition can be especially misleading, though, when the contrastive event consists of multiple subevents that are temporally disconnected, like when it is composed of everything that happens at noon together with something that happens at midnight. These more complicated cases will be discussed in §4.12.

A second example should help to convey how the probabilistic character of prob-influence comes entirely from the fundamental laws and a stipulation of one's chosen contextualized event. Prob-influence (and hence causation) should not be conceived in terms of probabilistic correlations between fairly localized events. For example, in drought conditions, there can be correlations between forest fires and thunder that we think of as a non-causal probabilistic connection holding by virtue of a common cause: lightning. Let us now evaluate several prob-influence relations that hold in this scenario. Fig. 4.2 depicts on the left a state at $t = 0$ just as a prototypical stroke of lightning is beginning to interact with the tree. At $t = 1$ sec, there will very probably be thunder spreading away from the location of the strike, depicted as a gray annulus, plus there will be hot wood where the lightning hit, depicted as a flame. An hour later, the fire will likely have spread, and the horse that happened to be wandering nearby will very likely have bolted after hearing the thunder.

Promotion only exists by virtue of possible terminance relations. Given that we suspect the fundamental laws obey content independence and that influence spreads at least at the speed of light, it is fairly safe to conclude that the only possible terminants for a full event at $t = 1$ sec are full events that span at least one light-second in radius at $t = 0$. Thus, the lightning can only promote later fire and thunder by way of the possible terminance exerted by spatially vast full events. To construct a contrastive event for assessing the lightning's probability-raising, one can in general choose any pair of contextualized events one wishes, so long as they differ only by having one of them instantiating lightning and the other not. An appropriate selection in order to isolate what the lightning itself promotes is to choose a contextualized event representing the lightning occurring in its full environment, and then form a corresponding contrast contextualized event by altering the microphysics to eliminate the lightning and nothing else. Then, one compares

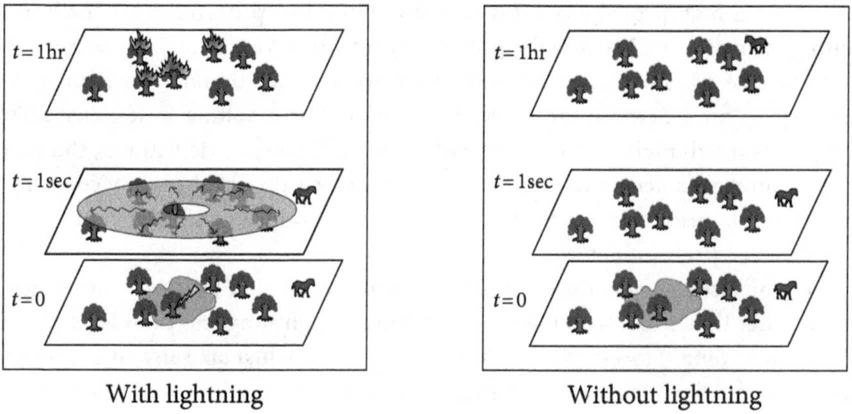

FIGURE 4.2 *Lightning promotes forest fires and frightened horses.*

both situations based on what the fundamental laws imply about such events. On the right hand side of Fig. 4.2, removing the lightning strike from the area around the tree creates a contextualized event that is just a normal forest with nothing special going on; thus, it most likely develops toward the forest remaining without fire and with the horse continuing to wander around.

For comparison, one can evaluate what *thunder* prob-influences by considering a prototypical situation in which thunder occurs as shown in Fig. 4.3. The contextualized event at $t = 1$ sec pictured on the left sets a high probability for the horse bolting in fear shortly after $t = 1$ sec and for a sizable forest fire at $t = 1$ hour. To figure out what the thunder is promoting, one strips the thunder from the contextualized event at, say, $t = 1$ sec, generating the contextualized event pictured on the right hand side. It has a very hot tree but with no concentric spheres of alternatingly compressed and rarified air centered on the tree. This contextualized

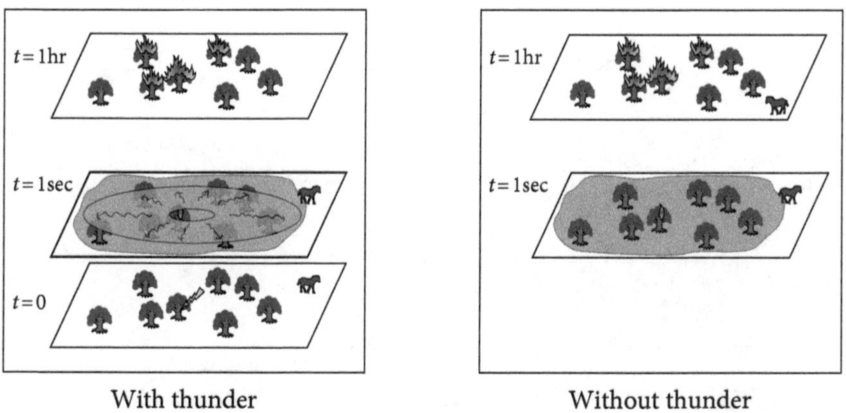

FIGURE 4.3 *Thunder does not promote forest fires, but does promote frightened horses.*

event fixes a high probability for a later fire, but a low probability for the horse bolting immediately. Because the probability for fire is very nearly the same high value for the $t = 1\,\text{sec}$ states on both sides, we say that thunder does not promote forest fires. Because the probability for the horse bolting in fear just after $t = 1\,\text{sec}$ is much higher on the side with thunder than the side without, thunder does promote the fleeing of horses. (Figs. 4.3 and 4.4 do not show any contrary-to-fact events prior to $t = 1\,\text{sec}$ in order not to introduce confusion concerning past-directed prob-influence.)

One can also examine the promotional effects of thunder by investigating situations other than those with the usual sequence of lightning and thunder. Fig. 4.4 depicts an ordinary forest where nothing is happening. Just take any large enough state, like the $t = 1\,\text{sec}$ state, contextualized however you like, and that will fix events on the left hand side. For the right hand side, fiddle with that same contextualized event to incorporate thunder, and let the fundamental laws go to work on that event to fill out the fragment of history shown on the right hand side. Because the probability for fire is very nearly the same low value for the $t = 1\,\text{sec}$ states on both sides, we say that thunder does not promote forest fires. We get the same result as in the previous comparison, which helps to illustrate how thunder fails to promote fires in a wide variety of environments.

There are certainly some environments where thunder does promote fire. In a forest where some guy has been hired to start a fire if he hears thunder, thunder can promote fire. The general statement, "Thunder does not promote fire," is worthy of being declared true because the contextualized events that are useful for drawing inferences about the actual world do not put a high probabilistic weight on environments with guys being hired to start fires upon hearing thunder. This is defensible on the grounds that such situations are rare in the actual world, and it is usually in our interest to ignore rare background conditions. One is free to consider the nomological consequences of whatever contextualized event

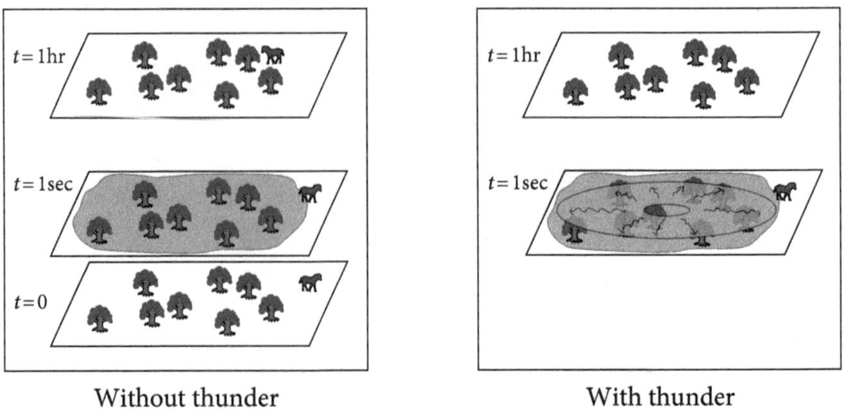

FIGURE 4.4 *It does not matter whether the contextualized events are actual.*

one wants, including weighting the probability of hired fire-starters heavily. Such events promote fires, but it is misleading to say they demonstrate that thunder promotes fires. It is more perspicuous to say they demonstrate that thunder promote fires in environments with suitable fire-starters.

Theories of causation as probability-raising often seek to establish rules for which background conditions are relevant for judging causation, and one might think my account should say something about which contrastive events are appropriate for judging causal claims. However, the primary virtue of my account is that it does not restrict the background or foreground conditions that can be evaluated. One is free to use any contrastive event one likes, and I do not offer much in the way of advice about which contrastive events will prove more useful than others. How to select contrastive events judiciously is largely a matter for the special sciences themselves to uncover. This parallels my observation that providing the fundamentally arbitrary parameters needed for a determinate amount of thermal energy does not require providing a general rule to indicate which corpuscle groupings and which choices of rest frame are more useful for the practicing scientist. Such judgments are better made through context-appropriate evaluations based on practical experience. In the next chapter, I will discuss how to relate contrastive events with the empirical data we gather when testing causal claims. For the present, though, I only want to warn readers that it is unwise to think there is a definitive right and wrong way to choose background conditions and also unwise to think of the probability distributions built into contrastive events (and thus fixing relations) as either objective or subjective. A better way to think about the probabilities is that they can include fundamental (and objective) chances as well as probabilities that have been simply stipulated, which makes them non-objective but also not subjective in the usual sense. Whether the stipulated probability distributions prove useful in practice is often decided by a non-subjective reality. There are often objectively better and worse ways to make one's fundamentally arbitrary choices of probability distributions.

It may be helpful at this stage to explore the task of finding tractable descriptions of promoters and inhibitors, which should illustrate how my account distributes the scientific labor of investigating causation between the metaphysics of causation and the special sciences themselves. Uncontroversially, one important scientific task is identifying inhibitors of cancer. Assuming fundamental reality has abundant terminance relations like those found in models of paradigm theories of fundamental physics, there will virtually always be a determinate answer to the question, "What contextualized events fix a probability near x for the event E that is defined to represent coarsely a person's contracting cancer within the next t units of time?" There will also virtually always be a determinate answer to the question, "Which contrastive events inhibit someone's contracting cancer (coarse-grained as E)?" Unfortunately, the guarantee of a determinate answer provides little guidance to the scientist seeking a description of cancer inhibitors in practical terms, say by specifying a chemical formula or the design of a medical device. Quite to the contrary, even if we knew the correct fundamental laws and categories of fundamental material stuff, expressing the contrastive events in terms of these fundamental attributes would surely result in a fantastically complicated description. My metaphysics of causation is not intended to guide anyone in finding more practical descriptions. Given all the uncertainty that results from not knowing the correct fundamental laws, not knowing the precise arrangement of fundamental attributes, and not knowing how to derive the consequences of the dynamical laws for realistic arrangements of attributes, other techniques will be far more effective.

As we will see later in §8.2, we have intuitive heuristics that help us learn about promoters and inhibitors in practical terms. For example, our intuitions about preëmption help us to estimate which kinds of events can interrupt processes that are "going to" bring about *E* and thus identify event-kinds that are reasonable candidates for being inhibitors of *E*. But identifying useful techniques for characterizing inhibitors and promoters in a tractable language is an activity within the domain of the special sciences, not within the metaphysics of causation, because such heuristics can excel without a comprehensive set of rules that distinguish "genuine" cases of preëmption and satisfy STRICT standards of adequacy. My metaphysics of causation only supports the search for inhibitors in a more theoretical manner by ensuring that ultimately there is always a determinate fact of the matter as to which contrastive events inhibit which coarse-grained events. If you are seeking inhibitors or promoters and the fundamental laws are friendly enough, my account says that you can find them in principle if you are able to overcome your epistemological limitations by acquiring a sufficiently rich representation of fundamental reality. Thus, in this highly idealized (and surely impractical) sense, when fundamental reality is sufficiently rich in terminance relations, the ceteris paribus clauses and hedge clauses and non-interference clauses that routinely appear in the special sciences can be understood merely as a practical response to limitations in our epistemology and control. They need not signify any ontological fuzziness or that reality is dappled with laws that cannot be completely and relatively simply unified or that the actual world contains more than fundamental physics.

4.4 Temporally Extended Events

Returning our attention to prob-influence, there is a potential worry that needs to be addressed. So far, my examples of promotion have employed contextualized events that last only a single moment. Many causal claims, though, make reference to processes or events of non-zero duration. Contextualized events that are temporally extended do not in themselves pose any difficulty. One just assumes any pair of contextualized events one wishes to consider and then compares what probability each one fixes for the effect. A reasonable worry, though, is that some causal claims might not translate well into my formalism without begging questions about what follows nomically from what. Suppose someone is trying to evaluate the extent to which operating one hundred coal-fueled power plants promotes acid rain over a ten year span versus a combination of ninety coal plants and twenty wind farms. One could postulate a pair of contextualized events that span the ten years and instantiate the normal operation of the plants, but in order to do interesting work, the contextualized events must extend far out into space and be filled out in full microscopic detail for all ten years. The problem is that such events will trivially establish how much acid rain falls during their occurrence regardless of the dynamical laws. They will presuppose how much acid rain falls rather than deriving it from the laws.

The general solution I recommend is to start with contextualized events that occupy a single time slice, let the fundamental laws propagate a probability distribution over the future, and then conditionalize the resulting probability distribution on one's choice for the candidate cause and its contrast, both construed as plain coarse-grained events. To conditionalize a contextualized event $\overline{E}$ using an event E within E's region, remove any members of $\overline{E}$ that do not instantiate a member of E (in the appropriate location) and then renormalize the probability distribution. For example, one could consider some time at which the political decision to build the plants is made. Construct a contextualized event for everything happening at that time. Then, let it evolve according to the fundamental laws for at least ten years, and then conditionalize on one hundred coal plants operating regularly for the first ten years. Repeat for the contrast event, conditionalizing on ninety coal plants and twenty wind farms operating regularly for the first ten years. Assuming such contextualized events are nomologically possible, each will fix a non-trivial probability for acid rain over that period, thus establishing a determinate magnitude of prob-influence.

It is possible to impose by fiat a probability distribution on a temporally extended event that does not agree with the probability distribution that the dynamical laws would generate using its initial probability distribution. In such cases, the imposition can give rise to a nomological impossibility or it can place a very strong and frequently unnatural constraint on what happens temporally in between the initial and final parts of the event. For example, one could choose two entirely unrelated contextualized global states and stipulate that they are situated one second apart from one another. If there are no solutions of the fundamental laws consistent with this pair of global events, there is no contextualized event comprising them and any nomic conditionals requiring the existence of such a contextualized event are undefined. If there are possible solutions to the imposed constraints, it is still unlikely that the probability distribution one event extends over the other will match the distribution that was imposed by fiat. The best way to avoid such pitfalls is to impose a probability distribution only on an event at a single time—the earliest state being optimal—and then let the laws propagate that distribution to the rest of the event.

Conditionalizing a contextualized event using future coarse-grained events can raise worries about conflating evidential relations with causal relations as well as worries about Simpson's paradox. I will eventually address these issues in §6.4 and §5.7.1 respectively. The only observation I am making at this stage is that using promotion relations to represent causal generalities among temporally extended events does not result in trivial proclamations.

4.5 Idiomatic Differences between Promotion and Causation

The point of having promotion spelled out in terms of the formal machinery of the nomic conditional is for it to serve as a conceptual regimentation to take the place of claims of the form "*C*'s cause *E*'s." Although the important upshot of saying "*C*'s cause *E*'s" is supposed to be translatable into "$\tilde{C}$ promotes *E*," there are several discrepancies between 'promotes' and 'causes' in ordinary language. Many of these differences can be attributed to the fact that general causal claims are standardly construed as generalizations about claims of culpable causation. The sentence "*C*'s cause *E*'s" in ordinary language can often be interpreted as some sort of quantification over culpable causes, some statement roughly in the neighborhood of, "An event *c* of type *C* will (typically? generally? sometimes?) cause an event *e* of type *E*."

For further exegesis, I will now mention four ways our ordinary usage of 'cause' fails to match the technical usage of 'promote'. The important point to keep in mind is that these differences are artifacts of the way we talk and think about causation and do not count as counterexamples to the thesis that the important content of general causal claims can be entirely captured by relations of promotion. In an empirical analysis of the metaphysics of causation, the task is not to take for granted a collection of recognized truths of the form "*C*'s cause *E*'s" and to seek truth conditions for them. Rather, the goal is to seek a structure that serves as a universal scheme for precisifying general causal claims. Claims of the form "*C*'s cause *E*'s" can then be interpreted as an informal gloss on these more precisely characterized structures, these prob-influence relations.

First, promotion is not factive. Because promotion issues from a contrastive event (which serves as a pair of event types), $\tilde{C}$'s promotion of *E* can obtain regardless of what events are actually instantiated. By contrast, in ordinary speech, when we say that *C*'s cause *E*'s, that seems to imply (at least as a conversational implicature) that at least some instances of *C* and *E* exist.

Second, promotion exists in degrees ranging from just above zero to exactly one, but in ordinary language causal relationships are frequently simplified into all-or-nothing relations. The simplification is evident in cases where the probability of an effect is greatly raised from an extremely small number to just a very small number. We can all agree that buying a lottery ticket promotes one's winning the lottery, but ordinary folk do not claim that buying a ticket causes one to win the lottery, even for the eventual winner. The reason, I think, is that even though the purchase greatly increases one's chance of winning when measured as a multiplicative factor, it does not raise it greatly in absolute terms. We are often reluctant to claim "*C* causes *E*" when the absolute magnitude of promotion is not high and there are other candidate promoters that are stronger, in this case, the fact that the ping pong balls happened to bounce in the particular way they did.

Third, our intuitions about culpable causes are responsive to an instinctive distinction between foreground and background causes, and that infiltrates notions of general causation. We do not say oxygen *causes* people to live, even though oxygen certainly promotes life. Instead, we say oxygen allows or enables life to continue. Similarly, we often focus on a more narrow collection of salient partial causes when making general causal claims. For example, we routinely draw a distinction between proximate causes, which happen just before the effect, and distal causes, which are temporally remote, and in many cases we attend more to proximate causes than distal causes. These and other similar distinctions play a role in deciding whether it is appropriate to say that *C*'s cause *E*'s, but they do not need to be interpreted as metaphysical distinctions.

Fourth, sometimes we claim that *C*'s cause *E*'s even when they inhibit *E*. For example, it is fair to say that automobile airbags cause injuries even though they reduce the probability of injuries. Such a case might be viewed as a counterexample to the claim that causation can be analyzed in terms of probability-raising, but there is no deep conflict here. The counterexample can be explained away by recognizing that there are two different relations of prob-influence corresponding to two different effects, one of which is more finely grained than the other. Airbags inhibit injuries on the whole, gathered under the type *E*, but there is a more narrowly characterized type of blows to the head, E', that airbags promote. In general, we may be tempted to say that *C*'s cause *E*'s because *C*'s cause E''s and E' is a special case of *E*, but our convention of inferring from "*C*'s cause a special case of *E*'s" to "*C*'s cause *E*'s" does not reflect any metaphysical principle that we need to hold.

Another consideration bearing on the discrepancy here is that although in general, having an airbag device in your automobile lowers your chance of injury, there do exist circumstances where their presence raises your chance of injury. Sometimes we say "*C*'s cause *E*'s" to mean something roughly like, "There exist some background conditions such that in those conditions, *C* raises the probability of *E*." In the language of contrastive events, such claims would be vindicated by the existence of some actual state, vast in size and fuzzed a bit if you like, such that when some occurrence of *C* in that state is hypothetically replaced by a contextually relevant non-occurrence of *C* and the fundamental laws are allowed to operate on both the actual and altered states separately, the resulting probability of *E* fixed by the actual state is higher than the probability fixed by the contrast state.

The reason I have discussed the four discrepancies in this section is to signal that I am well aware that it is easy to find cases where a statement of the form, "*C* causes *E*" is not equivalent to a claim of promotion. However, I do not believe these differences are important to the metaphysics of causation. They simply result from our idioms and conventions regarding the appropriate application of the word 'causes'. In the next chapter, I will clarify the empirical content of general causal claims and how they are captured by promotion relations. That will provide the justification for my invocation of promotion (and more generally prob-influence) relations.

In the next six sections, I will discuss several features of promotion. The primary purpose of each section is to clarify the promotion relation further, but a secondary purpose is to illustrate how the promotion relation helps to make sense of our ordinary causal claims. According to empirical analysis, remember, it is of no importance whatsoever that common-sense causal claims turn out to be explicitly true, meaning true under the most straightforward literal interpretation. However, it is important that an empirical analysis of causation makes some sense of the utility of our ordinary way of thinking about and talking about causation. Although not all of our pre-theoretical commitments regarding causation correspond to metaphysical features of causation, the ones discussed in the rest of this chapter—aspect causation, causation by omission, contrastivity, transitivity, continuity, and shielding—do represent aspects of causation that are situated within the scope of metaphysics.

4.6 Aspect Promotion

Sometimes we attribute the label 'cause' not to the occurrence of some event but instead to an aspect of some event or object.[38] For example, the sourness of cranberries causes children to reject them. Aspects can accurately be said to cause because they sometimes promote. A contrastive event $(\overline{C}_1, \overline{C}_2)$ can be chosen so that the difference between its pair of contextualized events is merely that $\overline{C}_1$ instantiates some aspect that $\overline{C}_2$ lacks. Let $\overline{C}_1$ be the actual state of the world right now and let $\overline{C}_2$ be just like $\overline{C}_1$ except that cranberries are sweet rather than sour, implemented by altering their sugar content and the genetic code of all cranberry shrubs and vines to maintain their future sweetness. The fundamental laws will presumably ensure that the probability of later children rejecting cranberries is higher in the $\overline{C}_1$ worlds than in the $\overline{C}_2$ worlds. Thus the sourness promotes children rejecting cranberries. Describing causes as aspects is our way of communicating one's intended contrast. If one said, "Cranberries cause children to reject them," that would suggest an implicit use of a contrast where the relevant cranberries do not exist. Mentioning their sourness as the cause lets people know that you are comparing actual circumstances (contextualized to encompass prototypical encounters with cranberries) with a contrast where the cranberries exist but are not sour.

---

[38] See Paul (2000) for discussion and further references.

4.7 Promotion by Omission

Philosophers like to puzzle over how to make sense of the conjunction of (1) the seeming reasonableness of numerous causal claims that cite non-existing entities as causes with (2) the fact that if something does not exist, it cannot cause anything.[39] Causation by omission is when the culpable cause of some effect is properly describable as something that does not exist, or is an omission of some action, or is a failure of some process to take place, or something similar. I think it is easy to see how non-existents can cause because it is easy to see how non-existents can promote. Promotion by omission is just a special case of aspect promotion. When we correctly claim that a failure to feed one's dog at the appropriate time causes hunger in the dog, *E*, we superficially attribute causation to the non-occurrence of an event. But if one's dog is not fed and becomes hungry, there are in fundamental reality a multitude of fine-grained events that individually termine the dog's hunger, each one of which instantiates an absence of anyone feeding the dog. Thus, citing the omitted feeding is a correct way of describing the fundamental events that bring about the dog's hunger. What's more, the omitted feeding is a promoter of the dog's hunger. Take any state when the dog is supposed to be fed, coarse-grain it a bit if you like, and call it $\overline{C}_1$. Hypothesize a contextualized event $\overline{C}_2$ that is just like $\overline{C}_1$ except that someone is dispensing food into the dog's bowl in a prototypical way. Presumably, the fundamental laws are such that the value of $\overline{C}_1 \mathrel{\Box\!\!\Rightarrow} E$ is high and the value of $\overline{C}_2 \mathrel{\Box\!\!\Rightarrow} E$ is low, so the lack of feeding promotes the dog's hunger. So far as I can tell, there is nothing to promotion by omission that is not a special case of aspect promotion.

4.8 Contrastivity

Causal claims are often contrastive (Hitchcock 1993, 1996a,b; Maslen 2004; Schaffer 2005, Northcott 2008, Steglich-Petersen 2012). In particular, many attributions of culpable causation implicitly or explicitly relativize the connection between cause and effect to contrary-to-fact alternatives. What is controversial is how best to make sense of the presence of contrastivity in causal claims, especially whether it is best explained by contrastivity in the metaphysics of causation or instead in terms of pragmatics layered on top of a non-contrastive metaphysics. If contrastivity is best accommodated in the metaphysics, there is a further question as to whether it is fundamental or derivative. In my account, the contrastivity is built into the derivative metaphysics of causation.

There are two appearances of contrastivity: in the 'cause' slot and in the 'effect' slot. The first kind of contrastivity is illustrated by, "The presence of a cup of salt rather than sugar in the preparation of the dessert causes a culinary abomination."

---

[39] See Paul (2009), Beebee (2004), and Schaffer (2000b) for discussion and further references.

The second kind is illustrated by, "The presence of spice in the main dish causes the food to be tasty rather than bland."

It should already be clear how my account incorporates contrastivity in the cause slot. There is nothing contrastive in fundamental reality; contributors band together to bring about other events without regard to what non-actual contributors would do. But prob-influence is inherently contrastive because it is a form of difference-making that requires the contrast to be stipulated. Unlike other models of causation based on counterfactual dependence, prob-influence leaves the contrast as a fundamentally arbitrary parameter, so the contrastivity is irremovable and manifest. Models of counterfactuals based on world-similarity tend to hide the contrastive character of an individual counterfactual comparison from in causal judgments. For example, in Lewis's (1973b, 1986) account, when we evaluate whether *C* is a cause of *E*, we are supposed to consider what would have happened if *C* had not occurred, and if the theory of counterfactuals were to work correctly, the intuitively more distant possibilities would be discarded as irrelevant to whether *E* would have occurred. Thus, the full range of possible contrasts—ways *C* could have not occurred—are not encompassed by his semantics of the counterfactual conditional. His (2000) account remedies this problem not by changing the account of counterfactual dependence but by evaluating causation in terms of a multitude of counterfactuals. On my account, the contrastivity in the 'cause' slot is unfiltered by semantic rules; one can fill in the contextualized events any way one chooses, and the associated nomic conditional should be assessed only using that choice together with fundamental laws (and other related contingencies in fundamental reality such as the arena topology or the values of fundamental constants).

In practice, we humans cannot cognitively process beliefs about the entire infinite range of contrastive events. So, we tend to focus on contrasts that are most useful to us or are ingrained in our psychology. On some occasions, we do this by simply replacing the event of interest with a neutral background condition, and often this background is understood in terms of what is normal or what should happen, either in a normative sense or in the sense of what is to be expected. I do not have my own psychological theory to offer about how we adopt certain contrasts as appropriate, but because the contrasts in my theory are fundamentally arbitrary parameters, it is adaptable to a wide variety of psychological theories of contrast selection.

My account, as presented so far, includes no explicit contrastivity in the effect slot, but it can be easily accommodated. When someone makes a claim of the form "*C* causes *E* rather than *E′*," *E′* is meant to indicate what the intended contrast fixes. The contrastivity in the 'effect' slot plays a distinct role from that in the 'cause' slot. The contrast in the 'cause' slot is a parameter one chooses as part of one's stipulation of the intended promotion relation one wishes to consider. The contrast in the 'effect' slot is not stipulated but is an output of theory; it is whatever event the contrast in the 'cause' slot fixes. For example, I might say, "Adding a dash of salt to the dish rather than no salt causes the dish to be tasty." The presence of no salt in the dish (when situated within a suitably large background) fixes a later event that assigns a high probability to the dish being bland. I can make this result explicit by stating the contrast in the effect slot: "Adding a dash of salt to the dish rather than no salt causes the dish to be tasty *rather than bland*." If I had used a non-default contrast in the 'cause' slot, like "Adding a dash of salt to the dish *rather than a cup of salt* causes the dish to be tasty," we get a statement whose natural 'effect' slot contrast can be made explicit by appending the words, "rather than salty." That is because it is apparent to everyone that a cup of salt would fix a high probability for the dish being salty. This way of modeling the contrast suggests that when people are thinking correctly about causal claims with contrasts, the correct contrast in the 'effect' slot ought to correspond to the contrast in the 'cause' slot.

There is a convenient way to make the contrastivity of promotion more explicit by instead modeling effects as contrastive events. Recall from §3.7 that for any contrastive event, $\tilde{C}$, there exists a unique maximal contrastive event $\tilde{E}$ that it fixes. One can read off of $\tilde{E}$ any coarse-grained events that are promoted by $\tilde{C}$ by checking what probabilities are assigned to them by $\tilde{E}$. For example, if the first contextualized event in $\tilde{E}$ differs from the second by assigning a probability 0.9 to their being a stench in space-time region R rather than 0.7, then $\tilde{C}$ promotes a stench in R to degree 0.2.

4.9 Transitivity

Terminance was shown in §2.8 to be transitive in a suitably defined sense, and now we can consider how relations in the middle conceptual layer of causation inherit a form of transitivity from the bottom layer.

The weakest form of transitivity holds uncontroversially:

Weak Transitivity of Fixing: If $\overline{E}$ fixes a contextualized event throughout region R, any event fixing $\overline{E}$ also fixes a contextualized event throughout region R.

So long as shielding (from §2.7) holds in fundamental reality, a stronger form of transitivity holds for fixing relations that are chained together going in the same temporal direction:

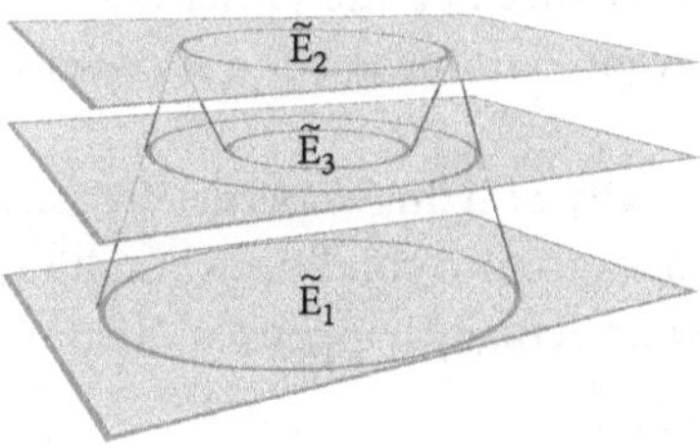

FIGURE 4.5 *$\tilde{E}_1$ fixes $\tilde{E}_2$ and $\tilde{E}_3$. $\tilde{E}_2$ fixes some $\tilde{E}'_3$ at $\tilde{E}_3$'s location that is not necessarily consistent with $\tilde{E}_3$.*

Unidirectional Transitivity of Fixing: For any $\overline{E_1}$, $\overline{E_2}$, and $\overline{E_3}$, if $\overline{E_1}$ fixes $\overline{E_2}$ and $\overline{E_2}$ fixes $\overline{E_3}$ and $\overline{E_2}$ is intermediate between $\overline{E_1}$ and $\overline{E_3}$, then $\overline{E_1}$ fixes $\overline{E_3}$.

The unidirectional transitivity of fixing helps to impart a transitive character to causal generalities.

We can drop the requirement that the fixing relations go in the same temporal direction to obtain the following:

Strong Transitivity of Fixing: For any $\overline{E_1}$, $\overline{E_2}$, and $\overline{E_3}$, if $\overline{E_1}$ fixes $\overline{E_2}$ and $\overline{E_2}$ fixes $\overline{E_3}$, then $\overline{E_1}$ fixes $\overline{E_3}$.

Strong transitivity of fixing does not hold in absolutely full generality because if we chain an instance of future-directed fixing together with an instance of past-directed fixing while ignoring intermediate events, we might not get the same contrastive event that $\tilde{E}_1$ fixes via a temporally direct route. However, when continuity and shielding both hold, $\tilde{E}_1$ will fix an event occupying the entire region from it to $\tilde{E}_2$ and $\tilde{E}_3$. This makes it reasonable to regiment our concepts so that $\tilde{E}_1$'s fixing of $\tilde{E}_3$ through $\tilde{E}_2$ must agree with what $\tilde{E}_1$ fixes for $\tilde{E}_3$ directly. I will defend this regimentation later in Ch. 6.

Although fixing relations obey unidirectional transitivity, nothing ensures that the contrastive event, $\tilde{E}_2$, that $\tilde{E}_1$ fixes will match the contrastive event we naturally tend to pick out when thinking about what actually happens at the intermediate time. So when translating from ordinary language, we should expect to find cases that superficially violate unidirectional transitivity because of how we select contrastive events.

In one example, unidirectional transitivity superficially fails because the default contrast picked out by an ordinary language characterization of the intermediate contrastive event does not match what is fixed by the initial contrastive event.

> Potassium salts are put into a fire at $t = 1$, which successfully promotes the existence of a purple fire at $t = 2$. The existence of the purple fire at $t = 2$ successfully promotes a marshmallow melting at $t = 3$. Yet, the potassium salts being put into the fire do not promote the marshmallow's melting.

This might appear to represent a counterexample to the unidirectional transitivity of fixing, but only because the reasoning conflates two distinct contrastive events at $t = 2$. Adding the salts fixes the existence of a purple fire rather than a yellow fire, which does not promote marshmallow melting. The contrastive event that promotes the marshmallow's melting is the existence of a fire rather than no fire at all, but adding the salts does not fix that event. So, this is not a successful counterexample to unidirectional transitivity.

In a second example, the unidirectional transitivity of fixing superficially fails because of a mismatch in the contrastive events' background conditions resulting from constructing the intermediate event using what actually happens at the time it occurs.

> Jane and Jill are camping where bears tend to eat unprotected food. In the morning, Jane removes her dinner from the bear box that protects the food, and she places the food out in the open. Jane's action inhibits E, the existence of Jane's dinner that evening. But Jane's action also promotes Jill's seeing the food at noon and storing it back in the bear box. Suppose Jill does in fact see the food at noon, and she stores it safely in the bear box. That action promotes E. So the event in the morning inhibits E but also promotes an event at noon that in turn promotes E, seemingly contrary to unidirectional transitivity.

Properly construed, this is not a counterexample to the unidirectional transitivity of fixing. Let $\tilde{M}$ be the morning contrastive event where Jane removes the food rather than leaving it in the bear box. Let $\tilde{E}$ be the contrastive event in the evening fixed by $\tilde{M}$. One can read off of $\tilde{E}$ that Jane's dinner is less likely to be available to her because the food was left out where a bear could get it. Let $\tilde{N} \equiv (\overline{N}, \overline{\neg N})$ be the contrastive event at noon fixed by $\tilde{M}$. One can read off of $\tilde{N}$ that Jill is more likely to see and then store the food, and $\tilde{N}$ does fix $\tilde{E}$. But there is another contrastive event $\tilde{N}'$ representing Jill's seeing and then storing the food. Let $\tilde{N}' \equiv (\overline{N}', \overline{\neg N}')$ represent what we naturally think of as Jill's storing the food rather than not storing the food in the actual environment occurring at noon. The contrastive event, $\tilde{N}'$, does promote the coarse-grained event E, but $\tilde{N}'$ is not the same as $\tilde{N}$. Even though $\tilde{M}$ makes Jill's storing the food more likely and $\overline{N}'$ makes E more likely than $\overline{\neg N}'$ makes it, $\tilde{M}$ makes E less likely because what $\tilde{M}$ fixes includes other possibilities not countenanced by $\tilde{N}'$, for example, that a bear eats the food before noon.

What makes this scenario seem like a counterexample to transitivity is that it is a counterexample to a crudely constructed alternative formulation of transitivity. Recall that promotion can be modeled in two ways. The first applies to effects that are plain coarse-grained events where we say a contrastive event $\tilde{C} \equiv (\overline{C_1}, \overline{C_2})$ promotes E iff $p_{\overline{C_1}}(E) > p_{\overline{C_2}}(E)$. If we try to chain together two instances of this relation, the formalism does not allow us because the relata that are to be linked together need to be the same kind of event. Both examples above illustrate how transitivity can fail when one uses contingencies from the actual state at the intermediate time to inform one's translation from the coarse-grained event to the contrastive event that it is supposed to be chained to. The second formulation upholds unidirectional transitivity by applying to effects that are construed as contrastive events. The fixing relations that bind contrastive events together can be linked in chains without falling prey to the above counterexamples, largely because fixing relations depend on no contingencies in the universe's material content.[40]

Another counterexample to transitivity will appear in §9.7.

---

[40] My discussion in this section parallels well-known examples in the philosophical literature such as McDermott (1995), Hall (2000), Ramachandran (2004), Maslen (2004), and Schaffer (2005).

Although it has been frequently argued[41] that causation is not transitive, the mere denial of transitivity leaves us with no explanation of why chaining together sequences of causal regularities is so frequently useful for bringing about effects. The resources of my account allow us to say more about why it is handy to think of causes in terms of chains. Fixing is a causal relation that is unidirectionally transitive. If $\tilde{E}_1$ fixes a later $\tilde{E}_2$ that fixes a still later $\tilde{E}_3$, then we know automatically that $\tilde{E}_1$ fixes $\tilde{E}_3$. That tells us that if we bring about protrast of $\tilde{E}_1$ in order to promote the protrast of $\tilde{E}_2$, we thereby promote the protrast of $\tilde{E}_3$.[42] Insofar as we are planning causal strategies, it is acceptable and useful to employ transitivity. One must just keep in mind that it is the fixing of contrastive events, not culpable causation among mundane events, that satisfies unidirectional transitivity. Causal culpability is not a transitive relation, and that explains why—when we do a retrospective examination of processes and coarse-grain events in ways that seem natural to us—we sometimes get events that do not stand in a transitive relation of culpable causation.

4.10 Continuity

Another significant feature of promotion is that it exhibits continuity whenever terminance is continuous. Assuming the fundamental laws are such that terminance is continuous, as discussed in §2.7, several corollaries hold regarding relations among derivative events.

These definitions make reference to a fixed intermediate, whose previous definition in §2.7 can be extended in the obvious way to apply to contextualized events:

> A contextualized event $\overline{I}$ is a *fixed intermediate* on the way from $\overline{C}$ to $\overline{E}$ (or E) iff the region R occupied by $\overline{I}$ is intermediate between $\overline{C}$ and $\overline{E}$ (or E) and $\overline{I}$ is the unique maximal contextualized event fixed by $\overline{C}$ for R and $\overline{I}$ fixes $\overline{E}$ (or a probability for E).

- *Continuity of Probability-fixing*: If a contextualized event $\overline{C}$ fixes a probability p for some E and there exists some region R intermediate between $\overline{C}$ and E, then there exists a unique maximal contextualized event $\overline{I}$ that occupies R, is fixed by $\overline{C}$, and fixes a probability p for E. (This $\overline{I}$ is a fixed intermediate on the way from $\overline{C}$ to E.)
- *Continuity of Fixing*: If a contextualized event $\overline{C}$ fixes a contextualized event $\overline{E}$ and there exists some region R intermediate between $\overline{C}$ and $\overline{E}$, then there exists a unique maximal intermediate contextualized event

[41] See (Hall 2000) for discussion and further references.

[42] Recall that the protrast is the first member of the ordered pair that constitutes the contrastive event.

$\overline{I}$ which occupies R and is fixed by $\overline{C}$ and fixes $\overline{E}$. (This $\overline{I}$ is a fixed intermediate on the way from $\overline{C}$ to $\overline{E}$.)

- *Continuity of Prob-influence*: If a contrastive event $\tilde{C}$ prob-influences some E to degree d and there exists some region R intermediate between $\tilde{C}$ and E, then there exists a unique maximal intermediate contrastive event $\tilde{I}$ which occupies R and is fixed by $\tilde{C}$ and prob-influences E to degree d.

The continuity of prob-influence has important consequences for learning about causal regularities and exploiting them in order to accomplish goals. If we come to believe that C promotes E, we can infer that the promotion is transmitted by way of intermediate promoters. If we are able to find other prior events that promote or inhibit aspects of the intermediate protrast, we can often use these events to promote or inhibit E.

4.11 Shielding

If the fundamental laws obey shielding, as discussed in §2.7, then there are shielding principles that apply to derivative events as well, three of which are listed below. The first applies to contextualized events fixing a probability for coarse-grained events, the second applies to contextualized events fixing contextualized events, and the third principle applies to contrastive events prob-influencing coarse-grained events.

None of these principles reveal anything surprising; they are just the natural extension of fundamental shielding to derivative events. These versions are based on the corollary to the Shielding of Terminance, and can be strengthened. Refer back to Fig. 2.6 for an illustration.

- *Shielding of Probability-Fixing*: For any contextualized event $\overline{C}$ that fixes a probability for a coarse-grained event E and for any contextualized event $\overline{I}$ occupying region Q that is a fixed intermediate on the way from $\overline{C}$ to E (so that it is fixed by $\overline{C}$ and fixes a probability for E), then for any region R that lies entirely within $\overline{C}$'s domain of terminance and contains no points on a c-path going from $\overline{I}$ to E, the contextualized event $\overline{J}$—defined as whatever $\overline{C}$ fixes for R ∪ Q—fixes the same probability for E that $\overline{I}$ fixes for E.
- *Shielding of Fixing*: For any contextualized event $\overline{C}$ that fixes a contextualized event $\overline{E}$ and for any contextualized event $\overline{I}$ occupying region Q that is a fixed intermediate on the way from $\overline{C}$ to $\overline{E}$ (so that it is fixed by $\overline{C}$ and fixes $\overline{E}$), then for any region R that lies entirely within $\overline{C}$'s domain of terminance and contains no points on a c-path going from $\overline{I}$ to $\overline{E}$, the contextualized event $\overline{J}$—defined as whatever $\overline{C}$ fixes for R ∪ Q—fixes $\overline{E}$ (just like $\overline{I}$ does).

- *Shielding of Prob-Influence*: For any contrastive event $\tilde{C}$ that exerts a well-defined degree of prob-influence over event E and for any contrastive event $\tilde{I}$ occupying region Q that is a fixed intermediate on the way from $\tilde{C}$ to E (so that it is fixed by $\tilde{C}$ and exerts a defined degree of prob-influence over E), then for any region R that lies entirely within $\tilde{C}$'s domain of terminance and contains no points on a c-path going from $\tilde{I}$ to E, the contrastive event $\tilde{J}$—defined as whatever $\tilde{C}$ fixes for $R \cup Q$—prob-influences E to the same degree that $\tilde{I}$ prob-influences E.

These principles are noteworthy because they play a role in the proof of causal directness in chapter 6.

4.12 Partial Influence

Prob-influence is the version of influence that plays the primary role in my metaphysics of causation, but there is a third, less discriminatory form of influence that is useful to have in our conceptual toolbox. In this section, I will construct the liberal notion of influence called 'partial influence'.

To set this up, let us first define a *contrastivization* of a plain coarse-grained event. To contrastivize C is to construct a certain sort of contrastive event using C as a scaffold.

> A *regular contrastivization* of a coarse-grained event C is a contrastive event $(\overline{C_1}, \overline{C_2})$ such that all three of the following conditions hold: (1) for every member of $\overline{C_1}$, there is a member of C agreeing with it throughout the region where they overlap; (2) none of $\overline{C_2}$'s members agrees with any of C's members where their regions overlap; and (3) $\overline{C_1}$ and $\overline{C_2}$ agree with each other everywhere outside C's region.

The purpose of contrastivizing an event C is to garner a representation of C that is big enough to fix a pair of probabilities for any plain coarse-grained events of interest but has its foreground localized to C's region.

We are now in a position to define partial influence.

> An event C *partially influences* an event E iff some contrastivization of C prob-influences E to a non-zero degree.

Partial influence gets its name because whenever C partially influences E, C's region is the foreground *part* of a contrastive event that prob-influences E (to a non-zero degree).

An examination of partial influence reveals that if we restrict contrastivizations of C to just those that occur "at the same time" as C by extending C in space-like directions, the resulting partial influence relations would be nearly equivalent to the notion of contribution. For the most part, C would make a difference to E just in case E's region overlaps with C's domain of influence.[43]

When $\tilde{C}$ can be construed as happening at a single time, it is a convenient cognitive simplification to think of $\tilde{C}$'s promotion of E as implemented solely by what happens in the foreground, C, within the environment formed by the background of $\tilde{C}$. At the end of §4.3, I cautioned that in some cases, it is potentially misleading to think of $\tilde{C}$'s promotion of E in this way, and now it is time to expose the source of the potential misinterpretation of promotion.

C's contrastivizations are not restricted to those that occur at the same time as C. One could define a contrastivization $\tilde{C}$ with a background consisting of a global state occupying a single time slice and whose foreground is a later event—disconnected in the arena from the background—whose protrast and contrast are both nomologically possible given the existence of the global state. The existing definition of 'regular contrastivization' has no problem making sense of the prob-influence exerted by such events.[44]

For more detail, consider the possibility that the fundamental laws are that of the toy theory of particle decay from §2.10.2. Recall that the arena is Minkowski space-time and that the laws incorporate fundamental chanciness directed toward the future. Imagine a contextualized event $\overline{S}$ that occupies a global time slice and is fully filled in. To the future of $\overline{S}$ there is a mundane event, C, and three other mundane events, E_1, E_2, E_3, that are located respectively to the past of C, space-like to C, and to the future of C. Let $\overline{S\&C}$ be the composition of $\overline{S}$ and C with $\overline{S}$'s probability distribution extended over C using the fundamental laws and conditionalization. Let $\overline{S\&\neg C}$ be just like $\overline{S\&C}$ except that the material contents of C are replaced with material contents that do not instantiate the original event C. Let $\tilde{C}$ be $(\overline{S\&C}, \overline{S\&\neg C})$.

There are many events E_1, E_2, and E_3 one can choose such that $\tilde{C}$ prob-influences all three of them. The difference between $\overline{S\&C}$ and $\overline{S\&\neg C}$ consists only

---

43 For the technically minded, a few discrepancies are worth noting for the record. First, it is conceivable for the fundamental laws to be such that (1) some full event c is a non-trivial minimal determinant of some later event e and (2) if any nomologically possible material contents were substituted for c's material contents, the altered event would still determine e. In other words, fundamental reality could be such that all possible full specifications of what happens in c's region determine the same event e. In such a case, c contributes to e but does not partially influence it. Such laws are probably too weird to take seriously in physics, but the logical possibility exists. Second, the fundamental laws might allow contribution relations that do not implicate probability. In §2.11.2, we noted that laws might have some fundamentally mental or theological mode of evolution that does not work in terms of determination or probability-fixing. Third, the laws might dictate that (1) the presence of an electron at p makes nomologically possible a maximally fine-grained electron decay process that has probability zero, and (2) the absence of an electron at p makes that decay process strictly impossible. If such a decay occurs, the electron's presence at p contributes to the decay but does not partially influence it because the probability was zero either way.

44 For further discussion of causation in the presence of background regions with unusual shapes, readers may be interested in Hitchcock (2001) and Yablo (2002).

in what happens at C. So, C counts as the foreground of $\tilde{C}$ and $\overline{S}$ counts as the background. Thus, C partially influences E_1, E_2, and E_3.

Note that C's partial influence on E_1 demonstrates that a form of past-directed influence can exist even when the fundamental laws permit no past-directed contribution.

C's partial influence on E_2 also demonstrates that partial influence can exist even though C does not contribute to E_2. This means partial influence is an extremely permissive notion of influence—in ways, more permissive than contribution. Alyssa Ney (2009, p. 743), when discussing accounts like mine that posit relations among physical causes to serve as a basis for difference-making relations, suggests that such accounts should maintain that "an event cannot count as making a difference to an effect unless it is at least a physical cause of that effect." What the example in this section demonstrates is that even Ney's rather weak (and prima facie plausible) constraint can be violated without undermining the overall thesis that difference-making is founded on physical relations. By counterfactually altering what happens at C's location, a probabilistic difference to E_2 is made without C's instance being a contributor to E_2.

The reason the above definition of a contrastivization was specified as a *regular* contrastivization is that it is unnecessarily restrictive by countenancing only contextualized events that are nomologically possible. A good illustration of this shortcoming occurs when the fundamental laws are deterministic and the background is a (fundamental) global state s at time t with the desired foreground occurring entirely after t. The intended protrast could be the state s with a certain light switch being on ten seconds after t and the intended contrast could be the state s with that same light switch being off ten seconds after t. Although such a comparison is a natural one to consider, the formalism developed so far would fail because one of these two contextualized events cannot exist. The deterministic laws only permit s to be compatible with one possible fundamental future. Thus, either the intended protrast or intended contrast would be a nomologically impossible event. Because contrastive events can only be constructed using possible events, the resulting degree of prob-influence would be undefined.

However, it is easy to extend the notion of a contrastivization to include a contrast or protrast that is nomologically impossible given the existence of the background. The remedy I will adopt is to be more permissive in our definition of a contrastivization by assigning zero probability to events that are disallowed by the fundamental laws. We can thereby have a well-defined measure of prob-influence that applies to a wider range of conditions without relying on the concept of an impossible event.

Consider the following definition.

An *irregular contrastivization* of some coarse-grained event C is an ordered triplet $(C_1, C_2, \overline{B})$ where $\overline{B}$'s region does not overlap with C_1 or C_2, and C_1

is identical to C (but with its location relative to $\overline{B}$ added) and where C_2 is a coarse-grained event located in the same region as C_1 (relative to $\overline{B}$) but without any members that are members of C.

Then, we define prob-influence to agree with the above definition in all cases where the referenced events are defined, but also to assign zero probability to the contrast (or protrast) when the background is well-defined but renders impossible the coarse-grained event used to construct the contrast (or protrast).

An *irregular contrastivization* $(C_1, C_2, \overline{B})$ prob-influences E to the degree $p_{\overline{B}}(E|C_1) - p_{\overline{B}}(E|C_2)$.

Assuming we are following the instructions for handling temporally extended events established in §4.4, this definition makes the degree of prob-influence exerted by an irregular contrastivization equal to that of the corresponding regular contrastivization when both of its contextualized events are well-defined. It is easy to verify that in the special case of a deterministic theory with a background $\overline{B}$ that is a (trivially contextualized) global state, if either of the later events C_1 and C_2 are incompatible with $\overline{B}$, its associated value (appearing as one term in the above formula for prob-influence) is zero rather than undefined.

One reason I have introduced the idea of partial influence is that it captures a sense in which we can remotely influence distant events. The partial influence that C exerts on E_2 is a form of space-like influence because while $\overline{B}$ is held steady, counterfactual twiddling what happens in C's region gives rise to a difference in the probability of E by virtue of the fundamental laws.

Recall that all the fundamental laws discussed in chapter 2 obey non-spatiality, the principle that space-like terminance is forbidden. Non-spatiality implies that a contrastive event $\tilde{C}$ can prob-influence a coarse-grained event E only if every subevent of E is c-connected to $\tilde{C}$. This raises the possible worry that space-like partial influence might conflict with non-spatiality, but the threat dissipates when the entire contrastivization is taken into account. For example, the prob-influence that $\tilde{C}$ exerts on E_2 in Fig. 4.6 is consistent with non-spatiality because all of E_2's subevents are c-connected to $\overline{S}$.

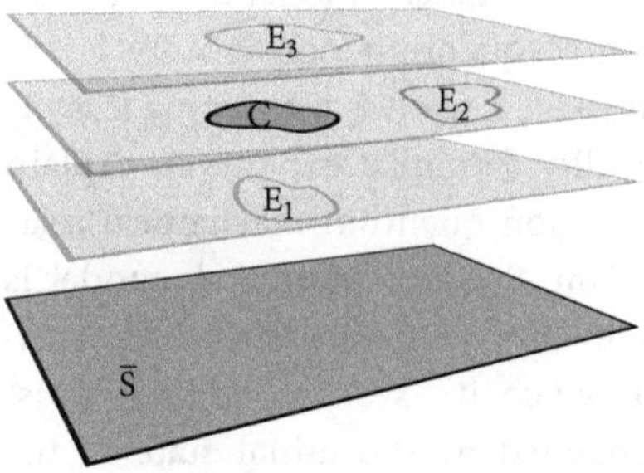

FIGURE 4.6 *C partially influences E_2 by virtue of its being a part of $\overline{S\&C}$ even though it does not contribute to e_2 or prob-influence E_2.*

This space-like partial influence also does not conflict with the kind of locality commonly imputed to relativistic theories because partial influence is not a fundamental relation. However, space-like partial influence might seem to conflict with a common sentiment that according to relativity, what I do makes no difference to the probability of anything that happens at space-like separated regions. I will challenge this sentiment later when discussing backtracking influence, but for now it suffices just to flag the existence of space-like partial influence and to note that my theory does not write off such influence as merely epistemic or evidential. It is correct to interpret the prob-dependence of E_2 on $\tilde{C}$ by saying that the state $\overline{S}$ prob-influences E_2 regardless of C and that all C does (in effect) is to conditionalize the probability distribution that $\overline{S}$ fixes for what happens at E_2, and that that conditionalization can be interpreted as being evidential in character. Specifically, one can say that the reason C's existence makes a difference to the probability of E_2 is that it provides evidence about the intervening physical development, which provides a constraint on how $\overline{S}$ eventually evolves toward E_2. All that is true, yet it is still true that C makes a difference to E_2 causally because $\tilde{C}$ prob-influences E_2 to a non-zero degree. I understand that a causal interpretation of this probabilistic relationship is not standard, but its existence follows from the choice to define prob-influence without imposing any restrictions on the scope of the events involved. Further justification of the legitimacy of that interpretation must be postponed until chapters 6 and 7.

Just to note one philosophical application, we can briefly examine Roderich Tumulka's (2006) relativistic model of the new-fangled flash-based spontaneous collapse interpretation of quantum mechanics from §2.13.3 that incorporates the signature quantum non-locality. His theory assumes the existence of an initial global state (expressed as a wave function) together with some initial flashes. The material content according to the model includes point-like flashes scattered throughout space-time and includes the wave function part of the ontology only at the initial boundary of the universe. Crucially, the theory would not work if it specified a determinate (collapsed) quantum state for later times. The model's fundamental laws define the probability of a flash existing in some chosen region R conditional on the initial quantum state together with a specification of flashes that lie to the past of R or are space-like from R. The only non-trivial terminance relations in Tumulka's model are the kind illustrated in Fig. 4.6, and its implementation of quantum non-locality is exactly what I have called 'space-like partial influence'. I see two interesting lessons to be gleaned here. First, the kind of non-locality Tumulka exploits to explain the peculiarly quantum effects *already exists* in non-quantum-mechanical relativistic theories like relativistic electromagnetism. Second, Tumulka's model is in fact relativistically *local* in the sense that it obeys non-spatiality in a Minkowski space-time. All the flashy events for which it fixes probabilities must have (and do have) their past light-cones spanned by the initial state of the universe.

4.13 Summary

In this chapter, I have distinguished three theoretically defined notions of influence: contribution, prob-influence, and partial influence. The purpose of the contribution concept is to capture the idea of a partial cause insofar as it exists fundamentally. The purpose of the prob-influence concept is to capture a notion of difference-making that quantifies how probable one event makes another; this is a derivative relation that will prove its value in explaining the empirical phenomena associated with effective strategies. Partial influence is another derivative relation, and it is a cruder and more indiscriminate conception of difference-making. Partial influence by itself has little value as a construction material for understanding effective strategies, but it provides a contrast with contribution that will be explored later in this volume. Specifically, in almost any reasonable fundamental theory, there will be instances where an event partially influences E even when it does not contribute to the instance of E; this raises important questions about how to understand causal locality and how to distinguish properly between causal relationships and evidential relationships, such as those exhibited in cases of a common cause of two effects.

The most important observation made in this chapter, however, is that any general causal claim can be made more precise as a claim about promotion, which is a claim about prob-influence. I surveyed some characteristics of the promotion relation to show how it helps to make sense of some of the features we associate with causation: aspect causation, causation by omission, and transitivity. In the next chapter, we will see how promotion relations are well suited for the explanation of empirical phenomena that we would ordinarily describe as evidence of causal generalities.

{5}

The Empirical Content of Promotion

One of the three main tasks I have set out for my metaphysics of causation is to help explain why a wide range of conditions exist where one kind of event is useful for bringing about another kind of event. We can now begin to uncover an account of this fact by specifying a type of experiment whose results constitute the empirical content of the claim that an event of type C is useful for bringing about an event of type E. The goal of this chapter is to demonstrate how the terminology developed in previous chapters facilitates the explanation of the results of such experiments. The other two tasks—explaining causal directness and causal asymmetry—will be addressed in the next two chapters. (In order to facilitate clarity, I will mostly restrict attention in this chapter to future-directed prob-influence. All the conclusions I draw, however, are fully compatible with the existence of past-directed prob-influence.)

Although the theoretical machinery developed in previous chapters was designed to help explain why C-events are good for bringing about E-events, the metaphysics of causation itself does not provide any explanations of particular causal regularities, nor does it provide a complete explanation of why there are lots of patterns that look causal, nor does it provide much help for special scientists when they attempt to discover causal regularities. Instead, it provides a general framework that helps to make sense of how causal generalities can obtain by virtue of an underlying fundamental structure that resembles paradigm fundamental theories of physics.[45] It does so by serving as an all-purpose explanatory backstop for the special sciences in the sense that any causal regularity can be characterized in terms of the probabilities that contrastive events fix. What justifies the value of my framework is that it is a *completely general* scheme for linking all causal regularities to (the presumably fully objective) fundamental reality through one's fundamentally arbitrary choices about how to characterize happenings in terms of derivative events.

---

[45] Recall that my account does not rule out the possibility of emergence or non-physical contributors. It does hypothesize that fundamental reality includes causation-like relations, e.g., terminance, among events at different times, but these events might instantiate more than just the attributes of fundamental physics and the fundamental laws might (for all that has been presumed so far) relate geological, psychological, or social events.

5.1 The Promotion Experiment

In conducting an empirical analysis of X, one should attempt to formulate experiments where the results constitute the empirical phenomena that motivate folk to have some concept of X. For my empirical analysis of the metaphysics of causation, the guiding idea is that we believe in causation because of the following kind of phenomena. We note that in situations when someone throws a rock at a calm unfrozen lake, a splash reliably occurs. We also note that in situations that are similar except that no one throws a rock at the lake, a splash reliably does not occur. This seems to indicate that the rock is making a difference as to whether a splash occurs. What's more, the basic structure of this example generalizes to all possible circumstances across all possible events.[46]

The task for this section is to generalize this example in terms of an experimental schema designed to indicate whether an event is a promoter of some effect E. The following experimental schema is suitable for any coarse-grained event E and any contrastive event $\tilde{C} \equiv (\overline{C}, \overline{\neg C})$ that precedes E and is spatially big enough and is filled in enough.

1. Identify or create a zillion separate instances of the initial conditions, each selected randomly from the members of $\overline{C}$ using its built-in probability distribution.
2. Identify or create another zillion separate instances of the initial conditions, each selected randomly from the members of $\overline{\neg C}$ using its built-in probability distribution.
3. Observe whether E happens in each separate run of the experiment.
4. Define $f_{\overline{C}}(E)$ as the fraction of $\overline{C}$-runs where E occurs and $f_{\overline{\neg C}}(E)$ as the fraction of $\overline{\neg C}$-runs where E occurs.
5. The observed value O is defined as $f_{\overline{C}}(E) - f_{\overline{\neg C}}(E)$.

The default prediction for any such experiment is that the observed value, O, will very nearly equal the predicted value, P, which is defined as the degree to which $\tilde{C}$ prob-influences E, namely $p_{\overline{C}}(E) - p_{\overline{\neg C}}(E)$.

For a simplistic illustration, consider the following. I fire a hundred cannonballs toward a distant pond by randomly selecting vertical angles between 20° to 30° with respect to the horizon. I calculate the likelihood of their landing in the pond by using my estimates of the cannonball velocity, the laws of physics, and the random initial conditions. I then observe how many actually land in the pond. I repeat the experiment using random angles from 30° to 40°. My

---

[46] In order to avoid minor complications, I will only consider nomological possibility in this volume, but it is a simple exercise for the reader to extend my account to more general possibilities by appending to each fundamental event a specification of whatever possible fundamental laws one wants to associate with that event.

calculations lead me to predict that the low angle makes it 40% likely for the cannonballs to land in the pond, but the high angle makes it only 15% likely. I thus predict that using the lower angle (rather than the higher angle) promotes a cannonball landing in the pond to degree 25%. My observations show 42 low-angled cannonballs landing in the pond and 16 high-angled ones. The observed value, $O = 42/100 - 16/100 = 26\%$, is reasonably close to the prediction, $P = 25\%$, so I conclude that my observations accord with my prediction.

There are several ways the magnitude of $\tilde{C}$'s prob-influence on E can fail to match the experimentally measured value, O. Any test of a probabilistic rule for nomic connections can fail just by virtue of accidental mismatches between actual frequencies and the probabilities set by the laws. Also, any finite number of tested initial conditions could constitute an unrepresentative sample of the members of a contextualized event. These kinds of failures for P to match O are standard in science, so I am assuming that the default prediction of $O = P$, if tested for a wide variety of promotion claims, will be falsified sometimes. The failure rate, though, should accord with what we know about standard sources of experimental error: inaccuracy of measurement, difficulty in creating or discovering the appropriate sample of initial conditions, brute chance, etc.

The $O = P$ prediction is not a trivial claim. It could turn out that there are fundamental laws dictating objective chances for how material stuff evolves but where the actual material layout does not routinely match what the chances indicate. Or it could turn out that there are no fundamental laws rich enough to account sufficiently for the correlations we observe in the material layout of the universe.

If experimental testing confirms that $O = P$ in some particular case, there is a simple explanation available. The contextualized event $\overline{C}$ adequately represents the weighted average of the initial conditions, and $\overline{C}$ fixes a probability $p_{\overline{C}}(E)$ for the occurrence of E among all the $\overline{C}$-runs. Similarly, $\overline{\neg C}$ fixes a probability $p_{\overline{\neg C}}(E)$ for the occurrence of E among all the $\overline{\neg C}$-runs. Because outcome frequencies tend to match objective probabilities in large test samples, O tends to be close to P.

Even though this explanation is framed in terms of events and does not say anything about strategies, it explains the existence of effective *strategies* because a strategy can be treated just like any other event. Every instance of a strategy is a fine-grained event, and any instance can be coarse-grained and contextualized in multiple ways to represent strategies in a more general way.

The success condition for a strategy is defined in terms of whatever event E counts as "what the strategy is trying to accomplish." To evaluate how effective a strategy $\overline{C}$ is for bringing about E, one should contrast it with alternative strategies. A strategy $\overline{C}$ is more effective than alternative strategies $\overline{C}_1, \overline{C}_2, \ldots, \overline{C}_n$ for bringing about a given effect E to the extent that $\overline{C}$ fixes a probability for E that is higher than the probabilities fixed by $\overline{C}_1, \overline{C}_2, \ldots, \overline{C}_n$. In practice, we ordinarily designate a strategy as effective only if the absolute probability it fixes is sufficiently high and if it raises the probability (relative to one's chosen contrast strategies) beyond a negligible amount. (When we attempt to measure the effectiveness of a strategy, the value is given in terms of the observed frequencies rather than the fixed probabilities, but such measurement raises the question of how to link our imprecise epistemic grip on the strategy being tested with the artificially precise contextualized events that represent strategies. This issue will be addressed in §5.2, §5.6, and §5.7.)

An important observation to make at this point is that although the exploration of causation in this book began with a preliminary focus on effective strategies, the empirical phenomena that bear on effective strategies and are captured by the promotion experiment have now been subsumed as part of a general pattern of promotion that draws no distinction between strategy-implementations and other kinds of events. This result is welcome because it ameliorates the worry that my metaphysics of causation would end up being too narrowly tied to agency and thus would not be suitable for understanding causation generally. Although our recognition that some strategies are more effective than others gives us a good reason to believe that reality includes causation, we can now conclude that there is nothing special about strategies or agency insofar as the promotion experiment is concerned. This counts as a notable advantage over manipulationist theories of causation,[47] which incorporate agency in their characterization of causation. The significance of effective strategies is not altogether expunged, however, because the promotion experiment does not capture all the empirical phenomena relevant to causation. In particular, the effectiveness of strategies and its implicit reliance on agency will end up playing a significant role in the explanation of causal asymmetry.

The rest of this chapter is intended to explore the relation between promotion and the promotion experiment in order to clarify the explanatory work done by an empirical analysis of the metaphysics of causation.

5.2 Insensitivity Considerations

Promotion is expressed in terms of contrastive events, which are much more determinate than what we intend when we make causal claims. When we think that smoking causes cancer, we do not have in mind a precisely delineated set of possible microphysical instances of some act of smoking embedded in its environment, nor do we precisely weight the possible instances with a probability distribution. If we choose a certain pair of contextualized events to represent the contrast between an instance of smoking and an instance of abstinence, there will be precise facts about how probable later lung cancer is. But what justifies our use of less precise expressions like "smoking promotes cancer" given that there are no identities

[47] See Woodward (2012) for a summary and further references.

between the contrastive events and what we mean by 'smoking'? The answer is that in many circumstances there are enough similarities among almost all the contextually relevant contrastive smoking events that there is only a negligible difference among their magnitudes of promotion, and what differences there are can be accounted for by understanding claims of promotion as holding only relative to a precisification of the promoter. To the extent that there are abundant similarities among almost all the contrastive events that instantiate smoking, it is not misleading to simplify the multitude of promotion relations by just saying, "Smoking causes cancer." That is, for practical purposes, it usually does not matter exactly how we contrastivize events. For convenience, I hereby introduce the umbrella term '*insensitivity considerations*' to refer to whatever set of principles vindicate this practice of abstracting away from fundamental reality without needing to do so in too precise a manner.

The fact that many contextualized events (that represent event-kinds people typically concern themselves with) are insensitive to slight modifications of their probability distributions (or the precise choice of their members) is intuitively plausible when dealing with some simple forms of behavior exhibited by macroscopic objects. It is a demonstrable regularity that any rock being held by an Earthling who releases it (with nothing else around that could support the rock) will lead to the rock's falling. We can explain this by noting that virtually any microscopic configuration that counts as a released rock will fix a high probability of the rock's falling. In theories with deterministic laws, there is virtually always some small set of possible events that develop in bizarre ways where the rock does not fall, but with any uncontrived probability distribution over the microstates, the troublesome microstates will have a negligibly small probability. (If the fundamental laws were to supply enough chanciness, we would not have to worry about the troublesome microstates, but it is better to remain agnostic about how much fundamental chanciness exists.)

A contextualized release-of-rock event, $\overline{C}$, fixes (for many mundane events) probabilities that are insensitive to the microscopic details of the rock. That is because the macroscopic motion of a released rock is largely insensitive to its microconditions. But there are also cases where the macroscopic facts are extremely sensitive to the microconditions, and yet we are still safe in being imprecise about how we contextualize events. A helpful example was discussed by Adam Elga (2007), drawing on work by Diaconis and Engel (1986). We can imagine someone throwing a sequence of darts at a dartboard covered with black and white patches. Suppose the person who painted the dartboard wanted to ensure a reliable regularity where if someone strikes the dartboard with a dart, there is a one-third chance of the dart landing on a white spot. A very good plan would be to make an extremely fine grid of alternating black and white spots, with the black spots taking up exactly twice as much area as the white ones. Although the probability distributions that describe a person's ability to place darts varies quite a bit (on the scale of inches) depending on the person's skill, people are never so good at darts that they can reliably strike a spot that is significantly smaller than a square millimeter. So, if the grid is very finely grained relative to the lumpiness of probability distributions that are reasonably attributed to a person's accuracy, any thrown darts will strike the white spots nearly one-third of the time, regardless of the player's skill. So, even though the location the dart lands on is highly sensitive to the throw, virtually any contextualized event $\overline{C}$ representing the person's ability in a reasonably smooth way will fix a probability of the dart's landing on white at nearly one-third and on black as nearly two-thirds. Thus, throwing a dart at the board (conditional on its landing on the board) promotes the dart landing on black to degree two-thirds, and such promotion is largely insensitive to how we represent the promoter.

Details relevant to insensitivity considerations have been discussed recently by Keller (1986), Strevens (1998, 2003, 2011), Goldstein (2001), Frigg (2009, 2011), Maudlin (2007b, 2011), Volchan (2007) among others, and there is a sizable technical literature dedicated to the design of concepts for dealing with such issues, mostly as part of the literature on the foundations of statistical mechanics. Owing to the many subtleties that come into play when attempting to make the underlying principles precise and justifying them, even a cursory review of the technicalities involved would be beyond the scope of this book. I will restrict my discussion to two brief points at this stage and a few more comments in the next section. Beyond that, I will mostly resort to bracketing the insensitivity considerations as one component of my overall account of causation and deferring to experts the task of spelling out the resources needed to make adequate sense of why some probability distributions are better than others for doing science and formulating adequate rules of thumb for when the precise probability distribution one uses is not critical.

First, Michael Strevens has provided the most extensive philosophical discussion of how probabilities in the special sciences can hold largely by virtue of deterministic laws. His (2003) attempts to show how probabilistic causal relations among macroscopic entities arise almost entirely from the character of the dynamical laws, with a supplementation discussed in (2011) using facts about actual frequencies. Much of the hard work required in his account comes from having to show how certain non-fundamental probabilities—in particular, what he calls 'enion probabilities'—can fix non-trivial probabilities that remain non-trivial even when one conditionalizes on (fundamental) states. My framework makes Strevens' task easier because the probabilities fixed by contextualized events are automatically (by stipulation) insulated from the fundamental events that instantiate them. Even better, because the legitimacy of enion probabilities, according to my framework, would be established by demonstrating their *utility* as abstractions from the fundamental details, one only needs to show enough of their connection to fundamental reality to justify using them as a practical conceptual tool. It is not clear to me what standard of rigor Strevens thinks an adequate account of non-fundamental probabilities needs to meet, but the framework of abstreduction is flexible enough to accommodate multiple depths of explanations for the utility of enion probabilities. Furthermore, the explanation Strevens provides unavoidably appeals to some imprecise hedging of the assumptions and conclusions. The probability of the dart landing on white will be *almost* one-third. The size of the black and white regions needs to be *small enough* for their ratio to be *nearly* constant over *suitably sized* regions, what Strevens calls 'microconstant'. The density of the darts' initial conditions needs to be *smooth enough*, what Strevens calls 'macroperiodic'. My account helps Strevens by making room for his sought-after enion probabilities in the top conceptual layer of causation, so that they are suitable for the RELAXED standards of theoretical adequacy that allow us to get away with using hedge clauses that are not explicitly spelled out. Although the kinds of considerations Strevens points to are surely a large part of an adequate justification of the utility of non-fundamental probabilities in the special sciences, nothing in my metaphysics or its application to the special sciences requires Strevens' specific notion of an enion probability or his specific argumentative strategy for their defense.

Second, my out-sourcing of the provision of a detailed account of insensitivity considerations does not render my theory faulty because an adequate treatment of these issues is needed in science regardless of what one wants to say about the metaphysics of causation. The problem of how best to make sense of probabilities in physics and in science generally is not at all specific to my own theory.

Third, I cannot see how out-sourcing this task could generate any plausible threat of circularity because the kinds of causal notions required for explaining insensitivity and the seeming objectivity of certain kinds of probability distributions, etc., should all be available from my account of fundamental causation from chapter 2. There is no plausible need for a notion of culpable causation, for example, in order to spell out how chaotic the behavior of matter is.

With the existence of acceptable insensitivity considerations taken for granted, we can now return attention to my contention that the important content of claims of the form, "*C*'s cause *E*'s," can be cashed out in terms of what contrastive events promote. If one fills in all the fundamentally arbitrary parameters needed to characterize $\tilde{C}$, there will be a determinate degree to which the effect is prob-influenced. But what makes our talk of causal regularities useful is not merely that there is *some* way of setting the parameters that implies a definite degree of prob-influence, but also that the degree is insensitive enough to the parameter settings one could have reasonably chosen, so that it depends much more on the laws of nature and the character of the events we are considering than on our choices about how to contrastivize. It is not enough to have a few contrastive smoking events that promote cancer. What vindicates the claim that smoking causes cancer is that an overwhelming majority of reasonable contrastivizations of smoking promote cancer.

5.3 Thermodynamics and Statistical Mechanics

In order to connect my invocation of insensitivity considerations to some formal physics, it is helpful at this stage to examine some thermodynamic behavior in terms of statistical mechanics. Statistical mechanics is the branch of physics that attempts to explain the macroscopic behavior of large quantities of particles in terms of their aggregate microscopic behavior. I will now recite in this section the standard statistical-mechanical explanation for why gasses tend to expand to fill the volumes of their containers, under the simplifying fiction that the simple theory of classical mechanics is the correct and complete theory of fundamental reality. Everything in this section is intended to be uncontroversial textbook material except that I will illustrate some concepts using contextualized events. By discussing this example, I will be able to introduce the concept of entropy, make a few more comments on insensitivity considerations, and prepare the stage for a discussion of what I call 'the asymmetry of bizarre coincidences'.

A useful mathematical device for understanding thermodynamic quantities and how they relate to fundamental quantities is the so-called phase space. A *phase space* is a mathematical space with enough dimensions, x_i, so that each possible state of the universe can be represented as a single point in the phase space. In order to facilitate discussion, I will conceive of phase space as including one more dimension, t, representing time, so that each possible state of the universe at time t can be represented as a single point in the time slice of phase space at t. That allows us to represent the full history of the universe as a time-like path in phase space. As we saw in §2.4.2, the relevant minimal determinants from the simple theory of classical mechanics are global states that specify all relative particle positions and velocities. Thus, a complete specification of the possible arrangements of N particles (as normally understood in terms of real numbers) requires $6N$ parameters, three for each corpuscle's position and three for each corpuscle's velocity. (We do not represent the intrinsic properties of the corpuscles like mass and charge using phase space but instead incorporate those properties into the laws that govern the evolution of the system.) Thus, the phase space for a universe with N particles has $6N + 1$ dimensions. Fig. 5.1 illustrates a pair of corpuscle world lines. At time t_0, corpuscle 1 has a position with coordinates (x_1, y_1, z_1) and velocity with coefficients $(\dot{x}_1, \dot{y}_1, \dot{z}_1)$ along unit coordinate tangent vectors. Similarly, the coordinate-based description for corpuscle 2 assigns it position (x_2, y_2, z_2) and velocity $(\dot{x}_2, \dot{y}_2, \dot{z}_2)$. Thus, the state of the two-corpuscle system is represented by a single point in the phase space, $(t_0, x_1, y_1, z_1, \dot{x}_1, \dot{y}_1, \dot{z}_1, x_2, y_2, z_2, \dot{x}_2, \dot{y}_2, \dot{z}_2)$.

Note that each possible history[48] is represented in phase space as a single path that intersects every hyperplane of constant t exactly once. Because virtually

[48] For simplicity, we ignore the possibility of corpuscles springing into and out of existence, and the possibility that a corpuscle could change its mass or charge, and the possibility that space-time could change its structure as time goes by.

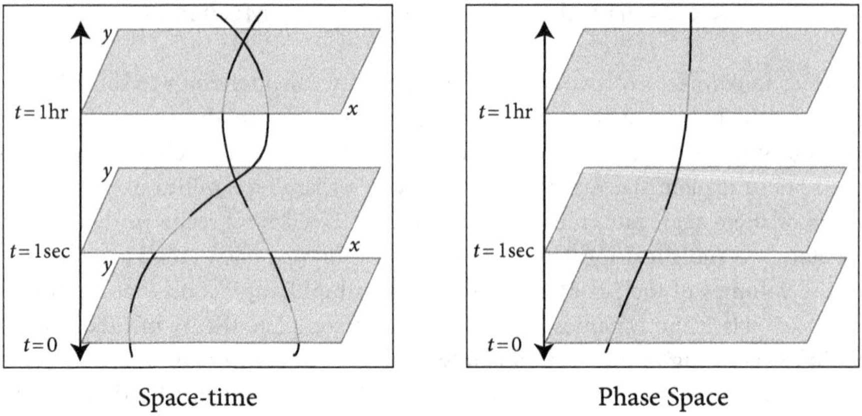

FIGURE 5.1 *Space-time, depicted on the left, contains two corpuscle world lines. This corresponds to a single state-path in a thirteen-dimensional phase space, depicted on the right.*

every realistic model of the simple theory of classical mechanics exhibits ubiquitous determination, we can ignore the worries about indeterminism discussed in §2.11.1 and assume that each point of the phase space determines a unique path of temporal development so that the paths never intersect.

Phase spaces facilitate visualizing a general feature of contextualized events. Because each contextualized event consists of a probability distribution over a set of fine-grained events, we can depict an instantaneous contextualized event, $\overline{C}$, as some patch of a plane with its probability density represented by shading in the patch more heavily where more probability is assigned. Then, because the fundamental laws tell us how each point of phase space is carried forward in time to a new point of phase space, the laws will dictate how the patch will evolve in phase space, and it will do so while preserving a well-defined probability distribution over the patch. So, the value of $\overline{C} \,\square\!\!\Rightarrow E$ can be reckoned by picturing $\overline{C}$ as a shaded patch at one time, letting the laws develop it forward in time as much as one likes and then identifying which subset of phase space counts as instantiating *E*. Because *E* is a plain coarse-grained event, it can be pictured as just a region in the phase space. If *E* is an instantaneous event at time t_2, the value of $\overline{C} \,\square\!\!\Rightarrow E$ is just the fraction of the shaded patch at t_2 (weighted by the amount of shading) that exists inside *E*'s region, as depicted in Fig. 5.2. In general, *E* can be temporally extended. In that case, the value of $\overline{C} \,\square\!\!\Rightarrow E$ is the proportion of $\overline{C}$'s points that evolve in a path through *E*'s volume.

Thinking about the nomic conditional in terms of phase space is helpful because there are some general features of paradigm fundamental theories that bear on how patches of phase space evolve. First, virtually any fundamental theory that adequately represents the motions of colliding subatomic particles, whether they are fundamentally corpuscular or field-like, will need to accommodate the experimentally recognized fact that particles (nearly) collide a lot and when they do, their

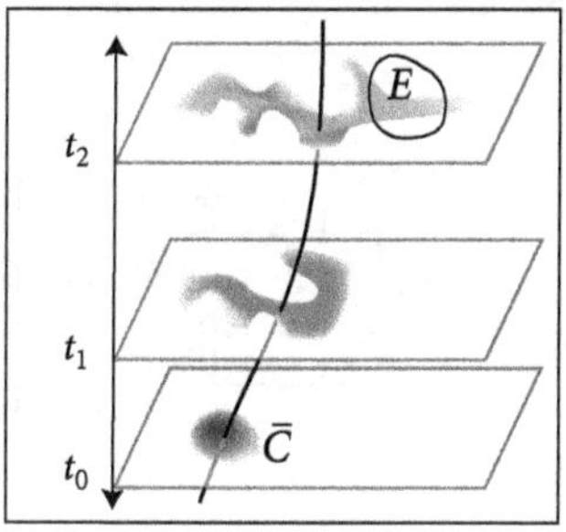

Phase Space

FIGURE 5.2 *The value of* $\overline{C} \;\square\!\!\Rightarrow E$*, namely* $p_{\overline{C}}(E)$*, is the proportion of* $\overline{C}$*'s points that develop in a path through E's region.*

post-collision trajectories often depend very sensitively on their pre-collision trajectories. This means that two points of phase space near each other at one time will typically evolve away from one another. (Talk of 'near' is a simplification. For most fundamental theories, there is always some neighborhood, usually fantastically small, where two points will not differ much in their future evolutions, but points in phase space that are roughly near each other generally tend to spread quickly apart as they evolve.)

For illustration, we can consider a simplistic model of an ideal gas. Standard practice in physics dictates that we assign a particular kind of probability distribution, called the microcanonical distribution, over all the possible configurations of gas particles. The microcanonical distribution has several formal properties that encourage us to think of it as a natural distribution. The one that is most important for the current discussion is that for a gas uniformly spread out in a sealed cubic box, the chance of any chosen particle being located in any chosen region V is equal to the proportion of the total volume of the box that V takes up. Any gas particle in the box has a one-half chance of being located in the left half of the box and a one-third chance of being located in the top third of the box. A similar feature holds for particle velocities, though there are some complications imposed by the conservation of energy. Also, the directions of particle velocities are randomly distributed. The microcanonical distribution counts as objective in the sense that the statistical predictions it gives rise to are confirmed by experimental data regardless of anyone's subjective degrees of belief about the gas.

Suppose there is a ten-liter vacuum tank with 10^{25} molecules uniformly spread inside a box as shown in Fig. 5.3 with a temperature of 300 K $\pm$ 1 K. The lid of the box has been opened quickly at time t_0. With that information, we are able to specify a phase space for the system of molecules inside the tank. Let T be the region of phase space corresponding to all possible configurations of the 10^{25} particles within the confines of the tank.

The *macrostate* of the gas is the condition of the gas as characterized by parameters like temperature, volume, and pressure. We can depict the gas's macrostate in phase space by just considering all the points of phase space at t_0 such that the corresponding particle configuration (in physical space) lies entirely inside the box

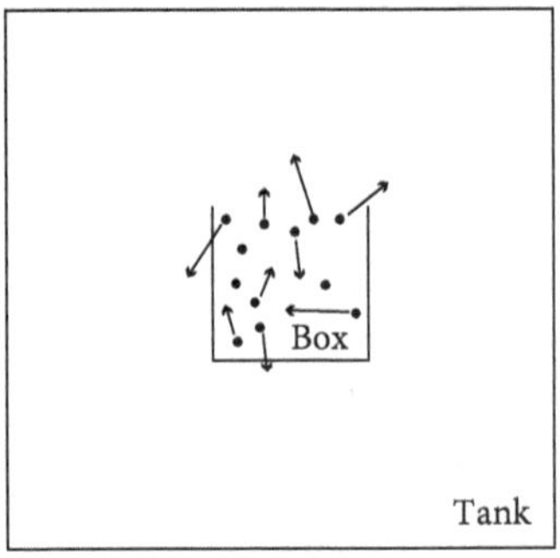

FIGURE 5.3 *The ideal gas initially occupies the box inside the tank.*

and has a temperature of 300 K $\pm$ 1 K. We can think of this initial state of the gas as a coarse-grained event $C(t_0)$. Then, we can apply the microcanonical probability distribution to $C(t_0)$ to form the contextualized event $\overline{C}(t_0)$. The depiction of $\overline{C}(t_0)$ in phase space is a uniformly shaded region. (See the left side of Fig. 5.4.)

It follows from the fundamental laws of classical mechanics that $\overline{C}(t_0)$'s region in phase space will evolve toward the future by maintaining its precise volume but stretching into a thin fiber that squiggles around to fill a vast hypersurface within T. (It would squiggle throughout all of T except that conservation laws restrict it, for example to a hypersurface of constant energy.) By saying that the later event $\overline{C}(t_2)$ squiggles throughout the hypersurface, I mean that it will wind around within the hypersurface so much that any given point of the hypersurface will be very near some part of the fibrillated $\overline{C}(t_2)$. As a result of the fibrillation, the vast majority of $\overline{C}(t_2)$'s members will correspond to the gas being spread throughout the tank, and only a fantastically small fraction of them are microstates that will re-contract inside the box. Thus, $\overline{C}(t_0)$ fixes a high probability for the gas expanding uniformly throughout the tank and staying that way. And that is more or less the standard statistical-mechanical explanation for why gasses expand to fill their containers, restated using the language of contextualized events.

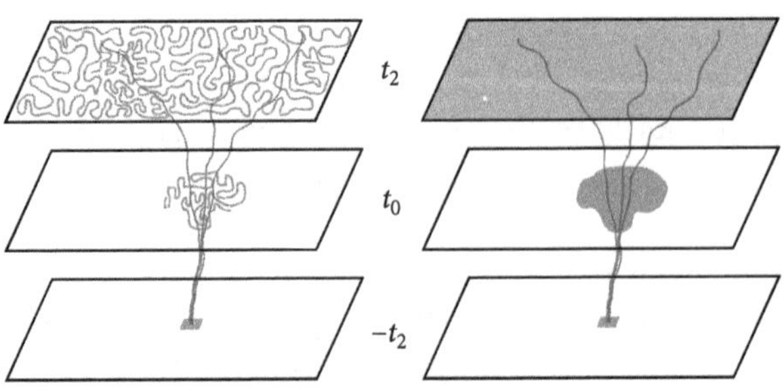

FIGURE 5.4 *$\overline{C}(t)$ in phase space is depicted on the left. The corresponding $P(\overline{C}(t))$ is on the right.*

We can now relate this explanation to the famous concept of entropy. To do so, it is convenient to partition the entire phase space (at one time) into sets of fundamental events that count as having the same macroscopic condition. Talk of "the same macroscopic condition" is to be cashed out in terms of (1) a set of fundamentally arbitrary parameters that specify macroscopic quantities like pressure and temperature in terms of the fundamental attributes of any fundamental event, and (2) a set of fundamentally arbitrary parameters that specify a precise range of these macroscopic quantities to be used for grouping fundamental events. For example, one might choose to group together all fundamental events whose temperature is within between 6.7C and 6.8C and whose pressure is between 103kPa and 104kPa. Assuming there exists an adequate set of fundamentally arbitrary parameters for exhaustively categorizing every relevant fundamental event into groups, we can construct several new theoretical structures to represent events in terms of their macroscopic groupings.

Let $M(e)$ be a function that outputs the *macrostate* of a fundamental event e, which is the set of possible fundamental events that are grouped together using one's chosen parameters. We can assume that we are working in a context where the fundamental events in question occupy a well-defined region so that $M(e)$ can be regarded as a coarse-grained event occupying a single region.

We can extend M's range of applicability by constructing a new function, $S(E)$ that takes a plain coarse-grained event as input and produces as output a set containing all and only the macrostates of E's members.

We can then extend S's range of applicability by constructing a new function, $P(\overline{E})$, that takes a contextualized event as input and produces as output a probability distribution over $S(\overline{E})$. $P(\overline{E})$ groups together all of the fundamental events in $\overline{E}$ that have the same macrostate, and it assigns them the very same probability that $\overline{E}$ assigns to them (as a group). $P(\overline{E})$ can be said to constitute a *pixelation* of $\overline{E}$. The redistribution of probability generated by $P(\overline{E})$ is somewhat analogous to how an digital image is pixelated by averaging the color-values for a block of neighboring pixels and assigning all those pixels this one average value. Any set of fundamental events in $\overline{E}$ that share the same macrostate will be assigned the same value by the pixelation, $P(\overline{E})$. A slight disanalogy is that the probability assigned to each macrostate is the *total* probability of its members in $\overline{E}$, not their *average* probability.

The *entropy* of an actual fundamental event e instantiating the ideal gas in the tank is defined as the volume of $M(e)$ measured on a logarithmic scale. Thus, as $\overline{C}(t)$ develops through time by fibrillating, $S(\overline{C}(t))$ will grow in size by filling up almost all the space around $\overline{C}(t)$, and the members of $P(\overline{C}(t))$ carrying the most probabilistic weight will expand to fill virtually all of the phase-space volume. That means the entropy of the gas will very likely increase until the gas fills the tank, at which time the entropy will remain constant. The increase of entropy is depicted on the right side of Fig. 5.4.

The statistical-mechanical explanation of gas expansion extends to handle more general phenomena, such as how temperature differences tend to even out over time, how liquids evaporate, etc. How it does so involves an abundance of technicalities. For the purpose of understanding promotion, we only need to note a few issues. One of the major issues concerns how we understand the probability distributions used in statistical mechanics. Some probability distributions are demonstrably better than others when it comes to quantitative predictions, and the appropriate distribution can depend on the materials involved. Bosonic material requires a different treatment from fermionic material. Furthermore, the structure of the probability distribution often depends on the underlying structure of the arena. In classical mechanics, one standardly uses probability distributions that respect symmetries of the space-time. In general relativity, there are problems even making sense of the structure of phase space because the space-time geometry keeps changing.

Insofar as promotion and contextualized events are concerned, it is acceptable for there to be no unique correct objective probability distribution that vindicates the practice of abstracting away from fine-grained events with contextualized events. There does need to be some sort of fundamental structure to make sense of why some ways of assigning probabilities are better than others, and this structure will also make some contextualized events count as more appropriate or more useful than others. Exactly how that is to be worked out in detail is a major technical problem that is the province of statistical mechanics and is intended to be addressed by the insensitivity considerations I mentioned earlier.

For current purposes, a proper account of causation only needs the insensitivity considerations to justify our treating bizarre evolutions as very rare. Bizarre evolutions were characterized in §3.4 as possible situations where things behave radically different from the way they normally do, like scenarios where objects spontaneously leap into the air, or where food ingredients spontaneously assemble themselves into an elaborate dessert. We are now in a position to redefine bizarre evolutions a bit more formally.

We can start with a fundamental event that is temporally extended and consider the special case where all its subevents are connected to each other and where it has an initial state c to which some appropriate statistical-mechanical probability distribution applies meaningfully. We then activate the apparatus of statistical mechanics to attribute a suitable probability distribution to a slight coarse-graining of c, giving us a contextualized event $\overline{C}$ that now represents the initial condition c with a bit of fuzziness. We then let e be everything else that happens in the chosen fundamental event, and we slightly coarse-grain it as E. If $p_{\overline{C}}(E)$ is fantastically close to zero (according to the scales of magnitude commonly appearing in statistical mechanics), then the evolution of c into e is bizarre. In providing this semi-formalization of bizarreness, I have deliberately employed imprecise terms like 'slight coarse-graining' and 'fantastically close' so that applications of 'bizarre' in particular cases are meant to be justified only to the extent that nothing of importance hinges on the precise boundary of 'slight' and 'fantastically close'. The proper usage of these terms, though, is intended to have the result that violations of thermodynamic regularities will count as bizarre, but that events like winning a lottery three times in a row will be merely very improbable, not fantastically improbable.

In order for me to convey the place of bizarreness in the conceptual landscape, it is valuable for us to consider its relation to another term that designates the overwhelmingly probable: 'typical'. The concept of typicality tries to capture the idea of "true in the vast majority of cases" without positing a specific probability or probability distribution. As Goldstein (2010, p. 53) puts it, typicality plays "the role of informing us when a set of exceptions is sufficiently small that we may in effect ignore it." In the simple example of the ideal gas being released from the box, the microcanonical distribution tells us that microstates that expand to fill the tank are typical and microstates that spontaneously contract from the full tank into the box are atypical. But in dealing with more general phenomena, it helps to be able to identify an event *E* as either fantastically improbable, fantastically probable, or of middling probability, even if we lack a specific probability distribution that would make sense of precisely how probable *E* is. That is the purpose of typicality. Refining and clarifying the concept of typicality is a research program in the foundations of physics extending beyond what I can reasonably engage in this examination of causation, and so typicality is meant to be included in the set of insensitivity considerations, which I have outsourced. Here, I will just illustrate the role played by typicality in enough detail to connect it to bizarreness.

As discussed by Tim Maudlin (2011), certain statistical frequencies can be designated as typical and thereby defended as legitimate objective probabilities without recourse to an objective probability distribution over the precise initial conditions. For example, the symmetry of coins together with some innocuous assumptions about the fundamental laws and what it means to toss a coin fairly will suffice for the typicality of a Bernoulli distribution of outcomes—basically that half will land heads, half tails, and the outcomes will be statistically independent. To say that the Bernoulli distribution is typical for fair coin flips is to say that virtually all possible states that determine a very large collection of fair coin flips will eventuate in nearly this distribution of outcomes, almost totally without regard to which particular probability measure is invoked to quantify the size of the "bad" set that leads to a pattern of outcomes not resembling a Bernoulli distribution. The key feature of a typicality-based defense of the objectivity of certain patterns of outcomes is that it does not require attribution of an objective probability to any particular coin outcome or small set of coin outcomes. Typicality, by construction, only applies legitimately in cases where the set of outcomes is sufficiently large.

Talk of typicality can be connected with talk of bizarreness as follows. Let us first assume that a typicality-based justification exists for the attribution of derivative chanciness to a reasonably wide range of happenings like coin flips, dice rolls, lotteries, organism survival, sexual reproduction, appearance of four-leaf clovers, lightning strikes, temperatures evening out, stars burning, and gasses diffusing. Otherwise, we would not need to concern ourselves with its connection to bizarreness.

We can think of each of these kinds of happenings in terms of an ordered pair $(\overline{C}_i, \overline{E}_i)$ where $\overline{C}_i$ is a contextualized event representing the initial condition (or what is often known as the chance setup), and $\overline{E}_i$ is a probability distribution over some chosen set of coarse-grained events that represent the *typical* pattern of outcomes for occurrences of $\overline{C}_i$. For example, if $\overline{C}_i$ represents the rolling of a die, $\overline{E}_i$ would consist of a set of six coarse-grained events corresponding to the six faces that can appear on top and a distribution that assigns each of these events a probability of nearly one-sixth.

We can take all of the situations where typicality can be invoked and bundle all of the ordered pairs together in a single set S and thereby gain a formal representation of what is typical for a wide variety of different circumstances. We can say of any sufficiently large and filled in (possible) fundamental event e (of roughly the size of a galactic cluster and lasting for many millions of years) that e is *future-bizarre* if it instantiates a very large number of instances of the initial conditions $\overline{C}_i$ that are present in S and the overall pattern of corresponding outcomes significantly deviate from what is typical for $\overline{C}_i$. Similarly, a possible fundamental event e is *future-typical* if it instantiates a very large number of the initial conditions in S but does not have significant deviations from what is typical. If an event does not instantiate enough instances of the initial conditions in S or if the degree of match is neither determinately close nor determinately not close, then it is inappropriate to attribute either label.

The purpose of introducing 'future-bizarre' is to designate events that include a vast pattern of coincidences of the kind that would strike us as overwhelming evidence for the violation of well-established rules of chance. We can also apply 'future-bizarre' to states that determine a future-bizarre evolution. For an illustration, suppose the fundamental laws are deterministic and resemble the laws of the paradigm fundamental theories. Then, imagine a device that has been constructed to select a random number every second in the range from one to a million. It is thoroughly tested, and to all appearances it works as designed. Then, the device is hooked up to a million bird perches so that if a bird lands on a perch, a signal representing the perch's number is sent to the random number generator, which then checks to see if the perch number matches the random number. If the numbers match, a signal is sent back to a bright light that reliably scares the bird away. Suppose that we operate this device and allow one bird at a time to fly to the perches, and that almost every time the bird lands on a perch the flash occurs. This would be a remarkable discovery, the kind of outcome that would immediately arouse suspicion that it is a hoax or a malfunction. But suppose we repeatedly test different setups using a wide variety of birds and a wide variety of random number selectors based on wholly different physical mechanisms none of which apparently has anything to do with birds, and we almost always get the same coincidence that the light flashes far more often than we would expect given that the device appears to select numbers randomly. If the world were like that, we might suspect that there is a special law of birds that overrides the behavior of random number devices so that the numbers they select are attuned to the future arrival of the bird on the perch. Because the example by stipulation has no such laws, the initial conditions of the universe must have just been in some freak initial condition leading to a series of coincidences involving birds and random number generators that continues to occur no matter how thoroughly we test it. If the world were like that, the pattern of events around Earth would count as future-bizarre, and the initial conditions of the universe would count as future-bizarre.

Defining bizarreness is a bit tricky because we do not want to entirely rule out the existence of seemingly chancy processes exhibiting long periods of deviation from whatever patterns of outcomes count as typical. Such deviations, after all, are implied by the rules of chance themselves. Suppose, for example, we consider a global state that determines the eternal existence throughout many locations in the universe of the kind of material interactions we are familiar with, so that the universe has an infinite number of chance outcomes (without everything running down and becoming an undifferentiated diffuse gas or getting sucked into black holes). Because such states imply an infinite number of coin flips or dice rolls or some other chancy phenomenon, the law of chance will imply the existence of many localized patterns of coincidences. My use of 'significantly deviate' in the definition of future-bizarre is meant to designate patterns of outcomes where the degree of mismatch between the actual frequencies and typical frequencies does not occur much more often than should be expected given the rules of chance and the overall number and distribution of outcomes. But remarkable patterns restricted to even a single kind of scenario, like the birds landing on the randomly selected perches, can be sufficient to designate an event as future-bizarre (or any initial conditions that determine such an event).

In summary, in order to vindicate the claim that smoking causes (promotes) cancer, we need the insensitivity considerations and the related defense of the explanatory value of typicality to substantiate a distinction between (1) a suitably large set of acceptable contextualizations of smoking that fix a fantastically low probability for bizarre evolutions toward the future and (2) a suitably small set of unacceptable contextualizations of smoking that fix a non-negligible probability for bizarre evolutions toward the future. Those resources will in turn justify the acceptability of certain kinds of probability distributions for use in contextualizing events with the desired result that the magnitude of an event's prob-influence is not terribly sensitive to minor jiggling of the "good" probability distributions and not terribly sensitive to the choices we make about which particular fine-grained events count as smoking and not terribly sensitive to the choices we make about how to precisify the background environment. It is an important feature of our causal claims that they not be too sensitive to our fundamentally arbitrary choices about how to represent them.

5.4 The Asymmetry of Bizarre Coincidences

I believe we have substantial evidence for the proposition that the actual world does not behave bizarrely as it develops toward the future. In other words, actual states of the world are future-typical. Across a wide range of circumstances, we have evidence that matter behaves in a chancy way without there being an outrageous streak of freak outcomes to suggest that some conglomerations of matter are misbehaving by violating the normal rules of chance. Of course, no amount of experimentation can conclusively confirm this proposition, but it is uncontroversial enough that I feel safe in assuming it holds.

However, everything is different with regard to how things look going back in time. If one visualizes the universe's states in a sequence or continuum going toward the past, a vast multitude of coincidences continually occur all across the universe where disparate microscopic configurations spontaneously spring into coordinated macroscopic action, depicted vividly by film projectors running in reverse. Because the world behaves bizarrely as it evolves toward the past, we can say that actual states of the world are *past-bizarre*.

Actual states of the world are future-typical and past-bizarre. This fact is what I call '*the asymmetry of bizarre coincidences*'.

In order to connect the asymmetry of bizarre coincidences with the previous discussion of statistical mechanics and typicality, it helps to revisit the example where our box of gas inside the vacuum tank has an open lid at t_0 under the assumption that the fundamental laws are deterministic toward the past as well as the future. If we just take the state at t_0 as a given and ignore everything that happened prior to t_0, we can contextualize the presence of the gas exactly as we did above and then use that contextualized event, $\overline{C}(t_0)$, to retrodict what likely happened in the past. When we do so, it turns out that $\overline{C}(t)$ develops toward the past in the same general manner that was discussed with regard to its future-directed evolution. The tidy rectangular shape of $\overline{C}(t_0)$ is spread out and fibrillated by the fundamental laws in a highly winding pattern. When we pixelate the contextualized event as $P(\overline{C}(t))$, the vast majority of its probability distribution comes from events that have much higher entropy than they had at time t_0. This represents that entropy very likely increases as we hypothetically develop $\overline{C}(t_0)$ back in time, which means that $\overline{C}(t)$ decreases in entropy from t_{-2} toward t_0. It is thus overwhelmingly likely, given the occurrence of $\overline{C}(t_0)$ at time t_0, that the gas spontaneously collapsed from the tank into the box. And that is a paradigmatic bizarre evolution.

When we ramp up the example by replacing $\overline{C}(t_0)$ with some reasonable contextualization, $\overline{A}$, of the current global state of the actual world, the same reasoning holds. It is overwhelmingly likely, according to the procedure just outlined, that the universe would have developed from a very high entropy state through some extremely bizarre evolution to arrive at the present condition $\overline{A}$. According to this inferred history, virtually everything we think is true of the past is incorrect. It

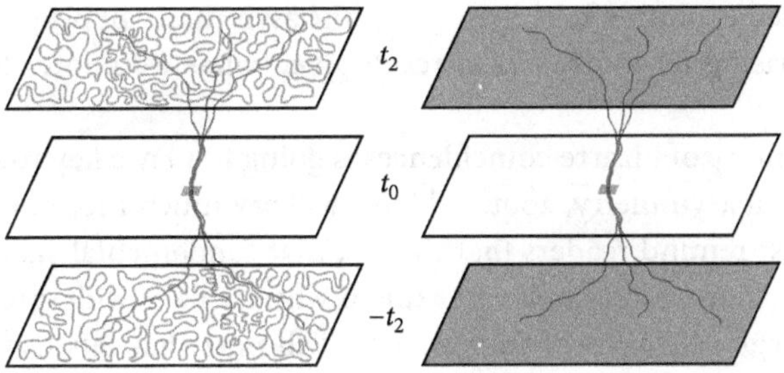

FIGURE 5.5 *$\overline{C}(t)$ in phase space is depicted on the left. The corresponding $P(\overline{C}(t))$ is on the right.*

would be overwhelmingly unlikely (according to what is fixed by $\overline{A}$) that dinosaurs existed, or that Rembrandt was around to paint, or that any plants were alive last week.

Because it is demonstrably true in the actual world that entropy has been increasing toward the future for billions of years, or at least we take it to be true on pain of skepticism, there must be something wrong with the procedure we just used to retrodict what likely preceded $\overline{C}(t_0)$. Indeed, the argument leading to this retrodiction is self-undermining in the sense that it implies that the evidence confirming statistical mechanics is likely spurious. To make a long story short, the standard diagnosis of what is faulty with this inferential procedure is that it is unable to generate the fact that the distant past was in some sort of state such that entropy increased toward the future of that state. Exactly how to incorporate this fact into our overall conception of reality is controversial. Perhaps the specialness of the early condition of the universe is just a brute fact, or perhaps there is something about the fundamental laws that opens up new possibilities toward the future so that global states early in the history of the universe that are unremarkable become seen as overwhelmingly improbable from a later standpoint. Perhaps some combination of the initial conditions and fundamental laws provides the best account of the steady increase of entropy, or maybe there exists some altogether different account of why the universe long ago was in the kind of condition that would naturally generate an expansive entropic asymmetry.

Whatever ultimately explains the asymmetry of entropy will almost surely explain the asymmetry of bizarre coincidences as well. In any case, the evolution of the world toward the past—that is, how the actual global state changes as one "goes back in time," not how the fundamental laws propagate it toward the past—is definitely a bizarre evolution in both the informal and semi-technical senses. As the previous example of the ideal gas demonstrates, the changes the actual world undergoes toward the past certainly violate the kinds of patterns that would be expected from what is fixed by any slight coarse-graining of the actual world with any reasonable probability distribution. The very existence of a non-equilibrium present condition that evolves non-bizarrely toward the future virtually ensures that its past-directed evolution is bizarre. Thus, the asymmetry of bizarre coincidences is not a surprising or controversial result given what we know about statistical mechanics.

The asymmetry of bizarre coincidences is going to play a key role in the explanation of causal asymmetry, about which I will say much more in chapter 7. For now, I will just remind readers that the fact that fundamental states evolve non-bizarrely toward the future helps to explain why we have widespread patterns of effective strategies toward the future, and I will note that this fact will also help to explain why we are unable to exploit our influence over the past to further our goals.

5.5 The Analogy to Thermal and Mechanical Energy

The rest of this chapter is dedicated to providing evidence that 'prob-influence' has been optimized for understanding the empirical phenomena related to causation, namely the results of promotion experiments. I will do so primarily by showing that prob-influence is an appropriately flexible concept that does not conflict with any well-founded principles from the special sciences. In particular, my treatment permits one to understand causal claims in the special sciences as claims of promotion that ultimately hold true by virtue of laws of fundamental physics. This picture of how causation works in the special sciences is not broadly accepted, but my task is not to demonstrate that this somewhat reductive approach is superior to alternatives that are less reductive. The goal is merely to clarify what this option consists of.

In particular, I want to remind readers that in §1.7, I emphasized that my account of the concepts in the middle conceptual layer of causation is modeled on how mechanical (or thermal) energy abstreduces[49] to the motions of corpuscles in classical mechanics. The point was that the fundamental attributes—the masses, relative positions, and relative speeds of the corpuscles—together with some fundamentally arbitrary parameters—a choice of rest and a choice of which corpuscles would be grouped together—suffice for how much thermal energy and mechanical energy each object possesses. That relationship constitutes an abstreduction of thermal and mechanical energy to the fundamental attributes (under the fiction that the simple theory of classical mechanics is the correct and complete theory of fundamental reality).

A key point I made in that section is that the mere existence of some quantity that can be derived from the fundamental attributes and fundamentally arbitrary parameters is unremarkable. One can cook up any function of the fundamental attributes one wants, but that doesn't make the resulting quantity scientifically interesting. What makes the derivation of thermal and mechanical energy worthy of

[49] See §1.7 for the definition of 'abstreduction' as a form of reduction.

interest is that (assuming the fundamentally arbitrary parameters have been chosen wisely) those quantities are useful for explaining and predicting macroscopic behavior in the special sciences and in everyday life.

I mentioned that in order to achieve a reductive explanation of thermal and mechanical energy to the fundamental attributes of classical mechanics, one can chain together two links: a precise link provided by the abstreduction of thermal and mechanical energy to the fundamental attributes and a fuzzy link provided by some sort of account of why thermal and mechanical energy are handy quantities to consider. All I want to add here is that the very same explanatory pattern holds for the concept of prob-influence and thus promotion. There is a precise link from the fundamental laws to the derivative relation of prob-influence once we select some (fundamentally arbitrary) contrastive event and some (fundamentally arbitrary) coarse-graining of the effect. There is also a fuzzy link to be provided in great part by the insensitivity considerations that I outsourced. I mention this analogy in order to emphasize that although I have provided a STRICT account of prob-influence, there is a further explanation of why prob-influence is a useful relation, and that explanation might involve a bit of hand-waving and imprecise terminology and tasteful judgment. It is not clear how one could ever specify principles in advance that would adjudicate every possible conflict that may arise in the fuzzy link of the explanation. Perhaps the possibility of conflicts will never threaten one's favored account of why there are recognizable ways of abstracting away from fundamental reality that are consistent with statistical mechanics. But in case there are any conflicts, my account of prob-influence (and thus promotion) is still just as STRICT as the relation from thermal and mechanical energy to the fundamental attributes of the simple theory of classical mechanics. And so, my reduction of promotion is just as reductive as the account of thermal and mechanical energy.

5.6 Broad and Narrow Promotion

Any successful empirical analysis needs to have its concepts be suitably flexible for their intended task so that they can maintain proper relationships with each other in ways that do not introduce artificial mystery in deviant or borderline cases. This was the principle of graceful degradation, introduced in §1.1.

One example of this principle is that it is better to avoid construing 'effective strategies' so literally that we require an adequate theory of causation to provide a rich, detailed account of what constitutes a genuine strategy. In §5.1, I argued that my account of causation exhibits the proper behavior by accommodating strategies as just a special case of events. Then, because we have an account of why *events* in general promote, we automatically have an account of why *strategies* promote with no need to decide precisely what counts as a strategy.

In order to demonstrate further how the concept of promotion exhibits graceful degradation, I will raise a potential problem for my account and then resolve the problem by showing how the connections among our concepts of promotion, ability, and control can be understood flexibly enough.

Consider a prototypical case of an agent having deliberate control over whether a certain future event E occurs, say whether his mouth is closed a second later. We can represent this control with a contrastive event. Let $\tilde{C} \equiv (\overline{C}, \overline{\neg C})$ be some reasonable contextualization of a deliberate decision to have one's mouth closed rather than a deliberate decision to have one's mouth open. It is easy enough to fill in the details so that $\tilde{C}$ greatly promotes E and that its reversal, $(\overline{\neg C}, \overline{C})$, greatly promotes having one's mouth open and thus $\neg E$. In many such cases, it is fair to say that the agent controls whether E occurs. For some philosophical purposes, construing control in this counterfactual way would be controversial, but it will do no harm here to focus on a conception of control based on the nomic conditional and prob-dependence.

In circumstances where an agent has deliberate control, we should normally expect the agent to be able to demonstrate the control by performing $\overline{C}$ and $\overline{\neg C}$ repeatedly on command and having the corresponding effect occur. If the desired effect is too narrowly circumscribed, say, by being a singular event like destroying the *Mona Lisa*, it will not be possible to bring about that particular effect repeatedly, so we need to focus on cases where the desired effect is general, like closing one's mouth. We often take an agent's inability to bring about or forestall E repeatedly as evidence that he does not have deliberate control over whether E occurs. Any particular agent may be trying to conceal his ability or might be temporarily disabled or might have an extremely bad run of luck, but in ordinary circumstances where we can be confident that the agent is making a good-faith effort to make E come about or prevent E and does not succeed often enough, we have reason to believe he cannot control whether E occurs.

However, I contend that these heuristics for inferring the presence or absence of deliberate control should be understood as defeasible. It is coherent and helpful to conceive of control so liberally that an agent can count as having control over E despite the guarantee of a widespread pattern of failures to bring about E. In more generality, it is consistent that $\tilde{C}$ promotes E significantly even though every actual instance of $\overline{C}$ is guaranteed to be unaccompanied by E. Because promotion abstracts away from fundamental reality, the possibility always exists that what happens in a widespread pattern of attempts to bring about E mismatches what is expected given one's action (or strategy implementation) $\overline{C}$. I will now illustrate with three examples.

For the first example, imagine a universe much like ours except that particles only come about through supernova-like explosions. Each current galaxy is composed of matter from a single explosion, with a bit of intergalactic mixing from stray particles or galactic collisions. Imagine the laws of nature are such that when any of these explosions occur, the masses of the created up-quarks all match a stochastically determined quantity. (For readers who are less familiar with physics, all you need to know is that an up-quark is a fundamental particle that is a common component of ordinary matter.) Because the galaxies and the star systems within them are formed from the detritus of these explosions, virtually all the up-quarks have the exact same mass as their neighbors. Yet, throughout the universe as a whole, the masses vary randomly. Suppose that for most macroscopic processes, the variation in quark mass makes little difference in causal regularities, yet there are some rare chemical reactions that sensitively depend on it. For concreteness, let the reaction of white roses to the presence of silver in the soil be one such reaction. Given the statistical distribution of quark masses fixed by the laws, planting a silver coin under the roots of a white rosebush makes highly probable that its blooms turn blue. However, with the very specific improbable value of the quark mass in our galaxy, the silver has no effect on the color.

There exist both a narrow and a broad way of thinking about the prob-influence of the silver on the blooms turning blue. If we coarse-grain narrowly, only using states that have the precise value of the quark masses in our local environment, then blue blooms will be extremely improbable regardless of the presence of the coin, leading to no significant promotion. If we coarse-grain broadly across instances with different quark masses, then the probability of blue blooms will be high with the coin and low without the coin, leading to promotion over the color of the roses. Both versions of promotion are technically legitimate, but the narrow version has a superior connection to actual statistical regularities in the local environment. The broad version provides a better measure of how likely the roses will turn blue in general circumstances across the universe. It is only because we are stuck in our own galaxy that the probability fixed by the broad coarse-graining is a poor guide to the outcomes we observe when we run the promotion experiment in our galaxy. Because up-quarks on Earth always have the same mass, any testing one does on the effects of silver on roses will reveal statistics matching the probability fixed by the narrow coarse-graining.

Because testing the promotion in our local environment reveals no effect, it might seem like an objective fact about our local environment that the presence of silver in the soil does not affect the color of the rose blooms. However, even if we assent to the "objectivity" of this fact, we do not need to hold that there is a complete lack of promotion. It is better to be more flexible and maintain that in general, there are multiple legitimate precisifications of prob-influence. After all, any way of coarse-graining fundamental reality is going to allow possible mismatches between what happens fundamentally and what the coarse-graining indicates, and allowing multiple ways of coarse-graining is going to allow possible mismatches among what the various coarse-grainings indicate. In the particular example chosen to illustrate this possibility, the silver narrowly prob-influences the blue color to degree zero but broadly prob-influences the blue color positively. The broad version is legitimate even though only the narrow version matches the correlations in our galaxy between silver and the color of the roses' blooms. We can say the broad promotion exists but is unexploitable by Earthlings for achieving the goal of turning the blooms blue because of the uniform exemplification of the special value for the up-quark mass throughout our galaxy.

Previously, I stated that my default prediction for any promotion experiment is that the observed value O will nearly equal the predicted value P, the degree to which $\tilde{C}$ prob-influences E. What I am now pointing out is that the default prediction can be overridden if there is reason to believe the environment in which the promotion is being tested fits into an atypical special case. If there is some contingency that affects the effectiveness of some strategy $\overline{C}$ throughout the environment of its testing, then that contingency can be taken into account without falsifying the hypothesis that $\tilde{C}$ promotes E to degree P. In the example, what makes our galaxy atypical with respect to silver's promotion of blue roses is the unusual up-quark mass that was randomly selected by nature at the infancy of our galaxy.

In order to avoid trivializing the $O = P$ prediction, one must be selective about the kinds of facts that are permitted to rescue a seemingly falsified claim of promotion. I suspect that adjudicating which facts can be appealed to legitimately is not a task that can be specified with precise and explicit rules. Although one could point to examples like the one above for guidance, making distinctions between broad and narrow prob-influence relies on the same good taste that is required generally in science to weed out ad hoc hypotheses.

A second example is the phenomenon of nuclear-spin echo, as presented by Brewer and Hahn (1984). A sample of glycerin is placed in a constant magnetic field and a machine successively sends two kinds of electromagnetic signals to it, which we can call 'ALIGN' and 'FLIP', and then the machine waits to receive an electromagnetic wave back from the sample, called 'ECHO'. When the machine broadcasts ALIGN at time $t = 0$, the spin states of the atoms in the sample become aligned in parallel. After the alignment occurs, the spins begin to precess around the magnetic field lines, and because the particles precess at different rates, after a few thousandths of a second the pattern of spin directions appears to be as randomized as it was before the ALIGN signal. After a span of time Δt, a few hundredths of a second, the machine sends the FLIP signal, and that flips every particle's spin state. After the flip, because all the particles are still in the same magnetic field, they each continue to rotate at the same rate but in the opposite direction. When another Δt of time has passed, the particle spins in the sample naturally become aligned, which generates an electromagnetic ECHO signal going from the glycerin to the machine.

What is interesting about nuclear-spin echo for understanding the promotion experiment is that it provides an example where even slightly coarse-graining an event is enough to make it misleading with regard to tests of its promotion. At time $t = \Delta t$ just before the FLIP signal is sent, if one contrastivizes the sending of FLIP as $\tilde{C}$ by coarse-graining its background so that the relative spin orientation of the glycerin molecules are not held fixed, then $\tilde{C}$ will not promote E, the ECHO signal. In all normal circumstances, where the glycerin molecules' spins have not been pre-configured in the special arrangement by ALIGN, sending the FLIP signal does not promote E. However, when the FLIP signal occurs Δt after an ALIGN signal, there is a pattern of E's produced. Tests of whether $\tilde{C}$ promotes E will show that it does: FLIPs are regularly followed by ECHOs even though the predicted degree of promotion is zero, given that we have coarse-grained away the fragile spin configuration.

Fortunately, no additional theoretical machinery is needed to handle this case. One can just leave $\tilde{C}$ as the FLIP signal with the coarse-grained glycerin sample and define $\tilde{C}'$ to be a temporally extended event defined by the following procedure. Start with a reasonable contextualization of the ALIGN signal, propagate it toward the future using the fundamental laws, conditionalize on the existence of a FLIP signal, and then discard everything before FLIP. Although $\tilde{C}$ does not promote E, the described $\tilde{C}'$ does. In brief, the FLIP signal does not broadly promote E, but it does narrowly promote E.

The point of the nuclear-spin echo example is to highlight a qualification to the default rule from §5.2 that a promotion claim is only useful if it "is insensitive enough to the parameter settings one could have reasonably chosen." The motivation for the default rule is that in normal circumstances we do not want to allow people to select their coarse-graining and contrasts in gerrymandered ways that allow them to vindicate ostensibly false causal claims. Yet in the case of the nuclear-spin echo, the correct claim that $\tilde{C}'$ promotes E is vulnerable to the same criticism that it encodes an atypical condition in its background. After all, the part of $\tilde{C}'$ that encodes the condition of the glycerin sample has to be extremely detailed regarding the spin orientations of its molecules.

The resolution to this problem is to draw a distinction between a gerrymandered coarse-graining and a special coarse-graining. Even though $\tilde{C}'$ encodes a special background condition, it is not chosen ad hoc to defend some dubious claim. It is a condition that can be regularly brought about by sending an ALIGN signal to the sample. So we should not discount the ability of $\tilde{C}'$ to promote the ECHO signal just because it requires a very specific background condition to work.

The upshot is that what makes a contextualization representative can be subtle, allowing ample room to distinguish between claims of promotion that are robustly testable and claims that abstract away from the experimental data in an ad hoc manner.

For the third example, consider our seeming inability to thwart ubiquitous determination. Under ubiquitous determination, any actual global state of the world before the year 3000 determines f, the full, maximally determinate, historical evolution of everything after the year 4000. Suppose I take an action a in the year 2000 with the goal of making f not happen. Let a be contextualized as $\overline{A}$ using a (continuous) probability density function. All the members of $\overline{A}$, we may assume, differ with respect to what occurs in the region actually occupied by a, and $\overline{A}$ is large enough to fix an event occupying the whole of the arena after the year 4000.

Owing to the supposed deterministic laws, all of the members of $\overline{A}$ (except the one actual state that instantiates a) determine that f does not occur. Thus, there is at most one member of $\overline{A}$ that determines f. Because $\overline{A}$ includes a continuous probability distribution, the set of members that determine f has measure zero. Thus, $\overline{A}$ fixes a probability of zero for f's occurrence. Thus, it is overwhelmingly likely that if I were to take action $\overline{A}$, f would not occur. Furthermore, if I attempt to prevent f by instantiating $\overline{A}$ multiple times, each one will "make" the occurrence of f fantastically improbable, yet f will still occur. There is a seeming conflict between the claim that I have the ability to make f not happen and the claim that no matter how many times I attempt to make f not happen, f is guaranteed to happen.

The reason f occurs despite my numerous attempts at prevention is just definitional; 'f' was stipulated to designate the actual future. What this shows is that we should not adopt a conception of promotion such that the guaranteed repeated failure to make f not happen counts as an inability to inhibit f. We can have a looser conceptual link whereby when we promote some kinds of future events, that only justifies a defeasible inference to a pattern of future outcomes. Normally, if the observed frequencies do not match what we think the fixed probabilities are, we have reason to revise our judgments about the fixed probabilities. But not so if the effect has been identified using a description that ensures it will happen regardless of the particular character of the laws or circumstances. We can instead declare that the inhibition of such effects exists but is *unexploitable*.

These three examples are intended to demonstrate that we need to be nuanced in how we treat the relation between promotion and the empirical data that promotion is intended to help explain. Promotion implies defeasibly that there will be a corresponding pattern observed when running the promotion experiment. But there can be reasons, when testing $\tilde{C}$'s promotion of E, for the experimental results to mismatch the degree to which $\tilde{C}$ prob-influences E. But when there is a mismatch, there is often another promoter that does match the empirical data and is not contrived in an ad hoc manner. In cases where we have a good understanding of what is responsible for the mismatch, we can continue to maintain our belief in the fundamental laws and that $\tilde{C}$ promotes E.

An important application of this lesson appears when we consider causal asymmetry in chapter 7. Briefly stated, the argument will go as follows. By virtue of the pervasive asymmetry of bizarre coincidences in our universe, the evolution of any actual global state toward the past will be bizarre (in the sense introduced in §3.4 and made semi-formal in §5.3). Thus, it will not obey the familiar patterns exhibited by future-directed causal interaction among macroscopic objects. This plays a key role in explaining why past-directed promotion can exist (in a broad sense) but never reveal itself when we attempt to test for its existence. The past-bizarreness of the actual world counts as a non-ad-hoc fact that defeats attempts to promote effects that would be expected if the actual world did not routinely evolve in bizarre ways toward the past.

5.7 Inferences from Empirical Data to Promotion

The promotion experiment was described using the language of random initial conditions selected from either $\overline{C}$ or $\overline{\neg C}$. However, in real tests of promotion, one does not have the ability to select a random fine-grained event from some prescribed set and instantiate that one fundamental event. For one thing, we never have perfect control over all the microscopic details. At best, we can attempt to instantiate something close to the desired initial condition and check afterward to gather information about which initial condition we created. Then, as we get enough experimental runs, we can adjust the relative weights of the instances to better match the probability distributions encoded in our chosen contextualized events.

In practice, our epistemic access to promotion is even more limited, for we often have no way to prepare the initial conditions; we just sift through what we know about the past in order to find patterns among various events. Think here of astronomy or paleontology. This raises an important issue. The promotion experiment is formulated in a way that presumes we start with some $\tilde{C}$ of interest and then test it with particular instances, whereas often we collect data and then seek an interpretation of them in terms of promotion. This inference from a collection of initial conditions to the contrastive event that best describes them is common in scientific practice, perhaps more so than the reverse inference where we choose a contrastive event and set out to verify whether it promotes a chosen effect.

Inferring an appropriate contrastive event from sets of initial conditions is a task fraught with difficulties, but such problems occur generally in science and do not pose any special problem for understanding promotion or an empirical analysis of the metaphysics of causation. The inference from sets of initial data to contrastive events is just like the problem of how best to fit a curve through a collection of data points on a graph. No matter how large the number of data points one graphs, there are always an infinite number of curves that will fit the data. However, it is common scientific practice—vindicated in experience—to think of some curves as more useful representations of the data than others vis-à-vis prediction and explanation. In cases where abundant testing continually adds more data points that all very nearly fit a simple figure such as a line or bell curve, it is typically reasonable, though defeasibly so, to infer that the data are best represented by that simple figure. The point of having the promotion experiment framed using a zillion instances of $\overline{C}$ and another zillion instances of $\overline{\neg C}$ is just to ensure that there are enough data to provide a basis for a reasonable inference when possible. Although gathering a trillion data points does not in principle get one any closer to a unique contrastive event than gathering a thousand, in many circumstances, as the data points grow in number, an ever more limited set of contrastive events remain reasonable representations of the data.

The following point deserves emphasis. The empirical analysis of the metaphysics of causation in general places no constraints on scientific practices concerning how to abstract away from collected data to a pattern or formula representing the data. My account in particular provides little or no help to the practicing scientist about how to draw appropriate inferences from empirical data to claims about promotion. One practical implication of my account, perhaps, is that it is conducive to a pluralistic approach toward promotion. There can often be different kinds of promotion and inhibition mixed up in the same phenomena without there being a unique correct way to interpret the data in terms of prob-influence. For further clarification of how this could be, it is helpful to investigate a well-known special case: Simpson's paradox.

5.7.1 SIMPSON'S PARADOX

One particular form of the problematic inference from empirical data to claims of promotion is the example known as Simpson's paradox, which refers to a certain statistical relationship that sounds counterintuitive. For a stock illustration, consider prospective students applying for admission to the philosophy and English departments at some college. The philosophy department happens to be small and more popular among male applicants. The English department happens to be large and more popular among female applicants. Suppose that one year, the data concerning admission is as listed in Table 5.1. One can easily inspect that the frequency of males who are admitted overall is 14% and the frequency of females who are admitted overall is 9.6%. But the frequency of males who are admitted to philosophy is 20% compared to 25% for females, and the frequency of males who are admitted to English is 8% compared to 9% for females. The paradox is that when the data are viewed at the department level, it seems males have a harder time being admitted than females, but when viewed at the college level, it seems females have a harder time being admitted than males.

Sometimes this example is accompanied by the following story. Initially, when outsiders look at the college-level statistics alone, they begin to suspect discrimination against females. Then, further investigation shows that admission decisions are made by the individual departments, so that it is the department-level data

TABLE 5.1 Frequencies Illustrating Simpson's Paradox

	Accepted		Rejected	
	M	F	M	F
Philosophy	5	1	20	3
English	2	9	23	91
Total	7	10	43	94

alone that is causally relevant to the assessment of bias. The seeming anti-female bias in the college-level statistics is then interpreted as non-causal in character, merely the result of the (perhaps) pro-female departmental level causation combined with the tendency of women to apply more often to the English department.

Someone might think that Simpson's paradox poses a problem for my account because my account treats all promotion relations as causal regardless of whether they characterize events at a department level or college level. So, one might worry whether my account has the resources to privilege the department-level statistics in such cases as indicative of genuine causation.

My response to this kind of worry is to deny that the college-level statistics are any less representative of real causal relations. They just provide evidence of different prob-influence relations than the ones based solely on the department-level statistics. To see this, let us contemplate three contextualized events that we can use to build the two contrastive events corresponding to two different ways in which being female rather than male can affect one's chance of admission.

- $\overline{C_f}$ is some reasonable contextualization of the actual state of the world at a time just before applications are submitted. It incorporates the institutional structures that govern admissions such as the policy of evaluating the applications at the department level and limits on how many students will be admitted. It also instantiates the 154 applicants such that their sexes and interests match the data in Table 5.1. For example, 104 out of the 154 applicants instantiated by $\overline{C_f}$ are female, and 100 out of those 104 are interested in applying to the English department.
- $\overline{C_m}$ is just like $\overline{C_f}$ except that one of the women applying to the English department, Laura, is hypothetically replaced by a man who has the very same interests and abilities and submits an application that is identical to Laura's except that any information that would identify her as female is altered to identify the applicant as male.
- $\overline{C_{m'}}$ is just like $\overline{C_f}$ except that we hypothetically replace a randomly selected female with a new male whose interests are representative of the general pattern among the 50 males in the population. For example, $\overline{C_{m'}}$ assigns an equal probability to his being interested in applying to philosophy and being interested in applying to English. Furthermore, the quality and character of the application he submits is assigned a probability distribution weighted to match the actual distribution of application quality among the male applicants.

There are two contrastive events we can consider. Let $\tilde{C}_d$ be defined as $(\overline{C_f}, \overline{C_m})$ in order to represent Laura's being female rather than male, while holding her abilities and interests fixed. Let $\tilde{C}_c$ be defined as $(\overline{C_f}, \overline{C_{m'}})$ in order to represent the application by a female with interests and abilities representative of female applicants to the college as a whole rather than a male with interests and abilities representative of male applicants.

Suppose for the sake of argument that there is a pro-female bias in the department-level decisions that matches the data so that the discrepancy between the number of females admitted and males admitted is purely a result of the application reviewer seeing the applicant's sex and then slightly favoring women over men. Given that this inclination of the reviewers is built into all the contextualized events and that there are no other factors, $\tilde{C}_d$ promotes Laura's acceptance into the English department. This promotion represents how merely being female improves one's chance of being accepted. One can form a similar contrastive event to show that a philosophy applicant's being female improves her chance of being accepted relative to having the same abilities and interests but being male.

Under the very same assumptions about the institutional structure, we can evaluate how much $\tilde{C}_c$ prob-influences a woman's chance of acceptance. To do that, we imagine how events would evolve if a randomly selected female were hypothetically replaced by an average male, where this average is weighted to match the frequencies in Table 5.1. The hypothetical man would be much more interested in applying to philosophy than the average among women, and because it is easier to get into philosophy, he would be more likely to gain acceptance. In fact, we can just read from the data that his overall chance of being accepted to the college would be 7/50, which is larger than the 10/104 chance a representative woman has. Being female and having interests representative of women thus inhibits a woman's chance of acceptance into the college relative to a male with interests and abilities representative of men. But that is no surprise because it incorporates the fact that women's preference for English over philosophy makes them less likely to gain admittance. A similar argument can demonstrate that for Laura in particular, being female and applying to the English department rather than a representative male inhibits her being admitted to the college.

So, there are two different promotion relations. $\tilde{C}_d$ represents a condition that promotes Laura's acceptance and is reflected in the department-level data. $\tilde{C}_c$ represents a condition that inhibits female's acceptance and is reflected in the college-level data. Both of them are easily understood in terms of the hypothesized pro-female bias in the departmental admissions and the general inclination of women to apply to the department where acceptance is more difficult to achieve. Both of them represent real causal tendencies, but not in any way that raises a worry about consistency. So, I conclude that Simpson's paradox does not pose any problem for construing causal claims in terms of prob-influence.

In the end, my metaphysics of causation provides a lot of flexibility for inferring prob-influence relations from statistical data. As far as I can tell, the metaphysics itself does not provide much help to the working scientist concerning which causal inferences to draw from experimental data. At best, it can work in tandem with auxiliary hypotheses about insensitivity considerations to suggest cases where a promotion (inhibition) relation is unlikely to be helpfully reinterpreted in a Simpson's reversal as the result of some more finely grained inhibition (promotion) relation. We don't expect Simpson's reversals for thermodynamic regularities, for example, because the kind of contextualized events required would likely be extremely contrived and fantastically difficult to manufacture. The lesson here is that even though positive statistical correlations can be the result of a combination of negative statistical correlations, we often have auxiliary reasons in particular cases to suspect that interpreting positive correlations in terms of inhibitors is not going to lead to tractable relations of probabilistic influence.

The original worry that Simpson's paradox posed was motivated by the idea that there are some probabilistic relations that are representative of causation and others that are a mixture of causal and evidential relations. Although I demonstrated that Simpson's paradox can be resolved even if the two different levels of statistical data are both representative of prob-influence and are thus genuinely causal, my account is committed to the thesis that all causation is ultimately the result of fundamental causation-like relations. A pluralistic attitude toward prob-influence—allowing the justified inference of different (and perhaps even conflicting) prob-influence relations from the same data—is compatible with (and can be justified in terms of) a non-pluralistic attitude toward the fundamental relations like terminance that account for the way nature develops.

5.8 Why There Are Effective Strategies

The concept of promotion is intended to provide a conceptual structure optimized for explaining the existence of effective strategies, and I have already provided a schematic explanation for *particular* causal regularities in §5.1, but there is a further question that an empirical analysis of causation should help to answer. Why is our universe *generally* hospitable to so many different kinds of causal patterns (including effective strategies) across a wide range of circumstances? Any fully adequate answer to such a question involves several components, many of which are easy to cite in general terms but extremely difficult to clarify in any significant detail. Because the task of actually providing the full explanation is far beyond my ken, I will restrict attention to the more limited task of sketching several components of the explanation and mentioning how the theoretical structures provided by my account relate to them.

First, in order for there to be instances of effective strategies as we know them, we need a space-time with macroscopic objects. One could imagine possible worlds where there are effective strategies without any material objects or without space-time, but such possibilities are so remote from the explanation of the kind of causation present in the actual world that they are hardly relevant. Much of my discussion of causation has focused on fundamental theories where space-time is part of fundamental reality, but nothing in my metaphysics requires space-time to be fundamental. Furthermore, all the contents of space-time can be metaphysically derivative insofar as my account of causation is concerned.

Second, there are many quantities like the electron mass and the fine structure constant, α, that currently have no explanation. Perhaps they are fundamental constants, and their magnitudes have no explanation. Perhaps they are related to each other in a system that makes sense of their absolute or relative magnitudes. If the existence of atoms and molecules depends on the contingent constant α having a value within some circumscribed range, then part of the explanation for the prevalence of the kinds of effective strategies we are familiar with involves showing how the promotion relations would be different or would fail to exist if α were outside this range. For example, if it could be shown that water can exist only if the value of α lies between 1/138 and 1/136, then part of the complete explanatory story of why swimming is an effective method of travel for fish includes citing the existence of water which in turn is partially explained by the value of α. Providing explanations of how effective strategies depend on fundamental constants is far beyond anyone's capacity at present.

Third, another part of the explanation involves some account of why there exist macroscopic objects, matter that is clumpy at roughly human scale with at least some stability. The existence of effective strategies certainly requires that the universe not be just a soup of undifferentiated gas at thermal equilibrium, and an explanation of this fact presumably requires accounting for the entropy asymmetry in terms of statistical mechanics. But another part of the explanation of macroscopic stability involves a lot of chemistry and solid-state physics to fill in the details of why some matter sticks together and other kinds of matter do not. Also, a world where macroscopic objects change too haphazardly too often will not have effective strategies as we know them. The explanation of this component needs to posit enough structure in the laws or initial conditions to explain why (toward the future) we do not have widespread occurrences of bizarre processes like rocks and trees changing into clowns and waterfalls at the slightest disturbance and thereafter quickly morphing into a glob of peanut butter. This explanation, I believe, can be given partially in terms of the insensitivity considerations I mentioned in §5.2, though I have deferred to current and future experts on this issue instead of providing my own account.

Fourth, effective strategies require the existence of agents with everything agency requires of its environment. The explanation for this component in part calls on previous components, but it presumably can also be given an evolutionary explanation. Given the kinds of conditions that have existed on Earth, and given what we now know about evolution, it is perhaps not too surprising that organisms would develop sensory organs, motor skills, and cognition permitting rational goal-oriented behavior. The kind of agency that is needed to make sense of effective strategies is fortunately a very weak conception. I think it counts as a significant virtue of the conceptual design that agency can be treated as a special case of event promotion without requiring any special metaphysics of agency. So, if agency is constituted merely by appropriately functioning matter, we can make sense of why some strategies and actions are more effective than others. Because the explanation of effective strategies gracefully degrades as one considers agents that are ever less agent-like, there is no need to characterize agency formally, especially no need to engage in the kinds of debates over the character of agency prevalent in action theory.

This list is certainly not intended to be exhaustive. The point is merely that a full explanation of why there is a wide spectrum of seeming causal relations presumably requires both an account of esoteric fundamental physical facts as well as an account of more mundane phenomena. The concept of promotion is useful for incorporating fundamental and derivative explanatory components because promotion from any $\check{C}$ to any E exists partially by virtue of fundamental laws and partially by virtue of abstracting away from the details. That should make it easy to see how the existence of effective strategies can depend on basic physical parameters and yet be insensitive enough to the fundamental details to represent general conditions that are likely to repeat themselves even when fundamental arrangements of particles do not repeatedly occur.

5.9 Mechanistic Theories of Causation

One subset of the philosophical literature on causation emphasizes the importance of mechanisms in a scientific understanding of causation. Stuart Glennan (2009) provides a helpful survey of the relevant literature. The account of causation I have presented does not by itself say anything about mechanisms, but in §2.7 I flagged the auxiliary hypothesis that fundamental laws govern the evolution of the present state of the universe continuously into the future. If that assumption is correct, there is in effect a universal mechanism and process underlying all instances of causation, namely the fundamental laws that propagate any given state to other regions in the arena. Philosophers often like to distinguish a productive notion of causation that contrasts with a difference-making notion. The development of nature under the action of the fundamental laws is a version of this production notion, and any complete temporal sequence of terminants counts as a causal process. Thus, my account accommodates notions of process and production as part of the fundamental causation-like relations.

Many notions of mechanism appearing in the recent philosophical literature appear to be quite a bit more restrictive than the universal productive mechanism posited by paradigm theories of fundamental physics. Salmon (1984), for example, focuses on processes capable of bearing marks, Fair (1979) and Kistler (1999) on transference of energy, Dowe (2000) on world-lines of conserved quantities, Bechtel and Richardson (1993), Glennan (1996, 2002), and Machamer, Darden, and Craver (2000) on "systems of interacting parts," so these authors appear to be appealing to a different conception of mechanism.

Mechanism-based theories of causation are not competitors to my own theory because they are not empirical analyses in the sense I identified in §1.1, and they do not relate mechanisms to fundamental reality in the sense I identified in §1.6. Although advocates of mechanistic theories of causation intend their work to make sense of appeals to causal explanation in the special sciences, this literature does not at present include any specification of the empirical phenomena to be explained by a mechanistic theory of causation. There is no attempt to characterize an experiment like the promotion experiment, nor do any of the advocates of mechanistic approaches commit to the principle that explaining such experiments is the proper aim for a theory of causation (of the sort they are providing). Because I have insufficient understanding of what these authors' methodological commitments are and because I am unable to ascertain what criteria they are presuming one should use to assess theories of causation, I am not in a position to say much about their theories. I hope to be able to contrast their work with mine in the future once it becomes clearer to me what metaphysical task their projects can accomplish.

In the meantime, I can make two brief comments. First, if someone were to agree with me that the results of promotion experiments constitute the primary empirical content of causation, then we could go on to debate whether the kind of mechanisms mentioned in this literature could play a useful role in explaining the results of promotion experiments. I suspect that there is no prominent role for the kind of processes invoked by Salmon (1984), Fair (1979), Kistler (1999), and Dowe (2000) because their conception of a singular cause is far too liberal to be of much help in explaining the results of the promotion experiment. For example, if you want to know the causes of cancer, these accounts will tell you to screen the candidate singular causes by starting at every instance of cancer, E_i, and tracing back in time along paths where energy is conserved or transferred (or where some other quantity like charge is conserved or transferred or where there is some mark-bearing process). Suppose for the sake of argument that the relevant physical quantities or processes exist in paradigmatic cases of causation. Because every instance of cancer is instantiated by molecules that have been banging into many other molecules in the environment and those have been banging into many others, we quickly find that E_i is causally linked to just about everything on Earth that happened more than a few seconds previously. In order to find out how to cure cancer, we need to have a much more restrictive conception of the *relevant* causes, so that we do not waste time trying to prevent cancer with magma flows or anteaters. Like Woodward (2003), I have not yet seen a compelling account in this tradition of how to incorporate a more restrictive set of causes that could help to explain the existence of effective strategies and macroscopic causal regularities. In my own theory, the concept of a promoter is designed to make precise this narrower conception of cause, but I have so far been unable to figure out how one is supposed to explain, using the transfer of conserved quantities and the like, that radiation and viruses are among the kinds of partial causes that should be targets of cancer mitigation efforts, as opposed to the vast number of other potential partial causes. There are many approaches one could take to address this issue, such as using counterfactuals or laws governing the relevant physical processes. Because I cannot survey the entire range of conceivable theories, I will just state my suspicion that such strategies will be effective only if the added machinery is doing most of the heavy lifting.

Second, there are pervasive empirical phenomena of the following sort, for which talk of mechanisms is especially appropriate. When scientists begin to examine some empirical phenomenon, such as biological reproduction or photosynthesis or snowfall, they can start with something close to a "black box" functional description of the phenomenon. In the case of reproduction, for example, we may come to know on the basis of observation that having few apple trees in a field promotes the existence of more apple trees nearby a decade later. One of the things scientists can learn about this promotion is that the decade-long process always instantiates the continuous path of a seed from one tree's fruit into the ground, whence a new tree grows. After further investigation, we can observe a lack of promotion when we destroy all the seeds, and increased promotion when we protect the seeds from harm. Eventually, we interpret these data by concluding that the seed is part of a mechanism for reproduction. These kinds of discoveries are commonplace in science, and we can be fairly confident in the utility of the following rule of thumb for the special sciences. Wherever there is a case of promotion, there is likely an underlying mechanism.

In what way is my account of causation tailored to explain this? A big part of the answer is that the fundamental laws themselves provide the ultimate mechanism for all causation. So my theory, by isolating the fundamental causation-like aspects of reality, has stripped away many of the irrelevant features of macroscopic causation like its asymmetry in order to isolate a purified conception of mechanism and production that operates universally. Recognizing the existence of this universal mechanism, though, does little to explain why there are so many discoverable and manipulable mechanisms like seed dispersal that are presumably not fundamental. My answer to this question is largely the same as the answer provided in the previous section. There are a multitude of explanatory components, most of which are currently poorly understood, including the values of fundamental constants, the prevalence of materials that clump together at the right distance scales, etc. Although my metaphysics of causation cannot answer these complex scientific questions, it is able to provide additional resources for understanding the more restrictive conception of mechanisms employed in the special sciences. Although I have too little space to spell out the details here, a medium-length explanation of my views on this topic can be found in (Kutach 2011b).

There are two observations from (Kutach 2011b) that bear on our practice of seeking causal mechanisms. The first is merely that some ways of coarse-graining fundamental reality are better than others for the purposes of the special sciences. Some contextualized events are more general than others and less contrived and better suited to modeling the kinds of interactions we typically find in our local environment. There is no scheme built into my metaphysics of causation for identifying which coarse-grainings are best; this is a task for the special sciences themselves. My account merely provides a language for representing these non-fundamental mechanisms in terms of fundamental events. This allows us to treat all causation as operating through a single productive mechanism, the fundamental dynamical laws, while also allowing for multiple ways to abstract away from this universal mechanism to better capture phenomena of interest to more limited domains like botany. This result supports the scientific acceptability of multiple mechanistic explanations of the same phenomena, even when the cited mechanisms are incommensurable or inconsistent with one another in the languages of the special sciences.

The second observation is that when we more finely grain some promoter that we do not yet fully understand by discarding any of its members that do not instantiate the kinds of matter we find in our local environment, we will often find that the more narrowly characterized event promotes other effects that we are not initially aware of. For example, we humans quickly learned that the liquid that flows off glaciers satisfies thirst and is transparent, but it took a lot of work to discover that when you pass electricity through it, two gasses are generated that explode with one another when hot. The mystery is why the same stuff that promotes the transmission of light and the quenching of thirst also promotes the formation of a pair of explosive gasses. The answer is that when you more finely grain that which fixes a high probability for the transmission of light and quenching of thirst in our local environment, you end up with contextualized events that (when altered to incorporate electricity) fix a high probability of generating a pair of gasses that can react with each other. Consult my (2011b) for more details.

{6}

Backtracking Influence

The fundamental causation-like relations and derivative causation-like relations based on them provide an excellent framework for interpreting general causal claims, but most of my advocacy for the framework has pressed the claim that it is flexible enough to represent any causal generality without unreasonably constraining the investigation of causation in the special sciences. For two examples, my metaphysics of causation allows one to evaluate probability-raising relative to any counterfactual contrast one chooses to consider, and it does not impose any significant restrictions on how scientists should infer causal relations from statistical correlations.

A more impressive advantage of the framework, however, is that it can also enforce some highly non-trivial constraints on causation. If one takes for granted certain relatively uncontroversial principles governing the fundamental causation-like relations, it is possible to derive remarkable principles governing the derivative causation-like relations. In this chapter, I will finally draw upon the power of the principles governing terminance that were introduced in chapter 2 to demonstrate an important principle governing causation in our world: causal directness. Causal directness, you may remember from way back in §1.2.1, is the claim that backtracking nomic connections between two events never do anything beyond what they do by virtue of the temporally direct nomic connection between them. Phrased simplistically, influence does not zigzag through time.

To make the principle of causal directness more precise, let us define a *backtracking fixing relation* to consist of an event $\overline{E_1}$ that fixes an event $\overline{E_2}$ that counts as "happening at a different time" than $\overline{E_1}$, which in turn fixes an event $\overline{E_3}$ that counts as "happening at a different time than $\overline{E_2}$ in the opposite temporal direction." A backtracking fixing relation can go from $\tilde{E}_1$ toward the future to $\tilde{E}_2$ and then backtrack toward the past to $\overline{E_3}$, or it can go from $\tilde{E}_1$ toward the past to $\tilde{E}_2$ and then backtrack toward the future to $\overline{E_3}$. Note that 'backtracking' does not mean the same as 'past-directed'. (Philosophers often use 'backtracking' to designate past-directed difference-making.)

Backtracking fixing relations can also be chained together so that they backtrack multiple times, and we can call any of the events in the interior of a chain of backtracking fixing relations a *turnaround event*. Furthermore, we can let the final event in the chain be a plain coarse-grained event E, and the accompanying final relation be a relation that fixes a probability for E.

Backtracking prob-influence is defined in the obvious way as existing whenever a contrastive event $\tilde{E}_1$ fixes an event $\tilde{E}_2$ that counts as 'happening at a different time' than $\tilde{E}_1$, which in turn fixes a pair of probabilities for an event E_3 that counts as 'happening at a different time than $\tilde{E}_2$ in the opposite temporal direction'. The degree of backtracking prob-influence in such a case is equal to the probability difference that $\tilde{E}_1$ fixes (through the mediation of $\tilde{E}_2$) for E_3.

We can now formulate causal directness in terms of backtracking prob-influence:

> *Causal Directness*: Any backtracking prob-influence that $\tilde{E}_1$ exerts on E_3 (by fixing some turnaround event $\tilde{E}_2$ that fixes a pair of probabilities for E_3) is equal in value to the prob-influence that $\tilde{E}_1$ exerts directly on E_3 (ignoring $\tilde{E}_2$).

I have heard of people who believe that all influence is future-directed. If it were true that fixing relations and thus prob-influence were only future-directed, backtracking prob-influence would not exist and causal directness would trivially follow. So, before discussing backtracking prob-influence any further, I will first defend the idea that it is reasonable to keep our minds open to the possibility of past-directed influence in all its forms, including past-directed (pure) contribution, past-directed prob-influence, and past-directed partial influence. I will not try to argue in this chapter that there is anything wrong with the hypothesis that all influence is future-directed. I will just suggest that it is unwise when exploring the nature of causation to exclude the possibility of past-directed influence out of hand. If past-directed influence is to be rejected, it should be done with a full appreciation of its consequences. Afterward, I will demonstrate why—even if past-directed terminance exists—backtracking prob-influence is always superfluous. Thus, regardless of whether we reject or accept past-directed terminance, all backtracking prob-influence will be superfluous.

Near the end of this chapter, I will attempt to isolate the empirical phenomena that motivate us to disbelieve in influence that is routed through a backtracking nomic connection. Interestingly, it will turn out there is no single experimental schema corresponding precisely to this platitude. Instead, there turn out to be three different schemas, two of which demonstrate a lack of non-redundant backtracking influence and another that demonstrates the existence of a genuine non-redundant form of influence that superficially looks like it involves backtracking. This form of influence turns out not to be genuinely backtracking, but more important, it turns out to be unexploitable for furthering one's goals. A proper analysis of this situation, however, is only possible after a thorough examination of causal asymmetry, which is the subject of the next chapter.

6.1 The Direction of Influence

Whether there is a component of fundamental reality corresponding to a distinction between the two directions of time is controversial. In order to avoid begging any questions about whether time has a fundamental direction, I will implement a neutral convention for 'future' and 'past'. There are many empirically accessible temporal asymmetries in the material layout of the universe that are widely distributed and causally integrated. For example, seeds grow into plants in the same temporal direction as burning wood transforms into ashes and in the same temporal direction in which perfumes diffuse and in the same temporal direction in which objects slow down due to friction. Let us say that 'future' refers (de re) to the direction of time in which seeds grow into plants, wood transforms into ashes, etc., and that 'past' refers to the other direction of time. The direction of the future and the past can thus be located in terms of the local distribution of matter and is extended throughout the arena by parallel transporting this locally defined temporal orientation.[50] I presume that paradigmatic influence is future-directed, where the future direction can be identified using some subset of the numerous de facto material asymmetries.

The hypothesis that there is a fundamental direction of time can take many forms. Some models incorporate a direction in which time flows or passes. Others identify it with the direction in which the block of reality grows. It is also possible to equate the fundamental future as the only temporal direction in which influence is exerted. One could even formulate metaphysical theories where different kinds of fundamental temporal directions exist but do not align with each other in the way common sense takes them to, but in order to keep the discussion manageable, I will just assume that if there is a fundamental future, it is unique and that it at least plays the role of ensuring that there is no past-directed contribution. Let us say that if contribution exists only toward the future, the future counts as the *fundamental future*. If there is no fundamental future, then the future counts as the *derivative future*. The *fundamental past* and *derivative past* are defined similarly.

In §4.12, I discussed how our ordinary conception of influence can be thought of as an extension of our notions of control and manipulation. Here is a recap of the lesson. We initially take for granted that we are able to control whether certain future events occur, and all of these instances of control also count as instances of influence. However, what we influence is not restricted to what we control but also includes events where our degree of control is too weak or chaotic to be of any practical use. How far does the range of our influence extend? There is seemingly

[50] If there is no globally well-defined temporally orientation, talk of future and past will need to be restricted to more localized regions.

an excellent rule. The fundamental laws dictate the precise range of our influence. If there is a fundamental interaction that works like Newtonian gravitation, transmitting forces across infinite spatial distances with unbounded speed, then our influence is arbitrarily fast. If the fundamental dynamical laws are like Maxwell's laws for the electromagnetic field, then one could argue that influence extends throughout the future light cone but does not extend any faster than the speed of light.

This way of thinking about the relation between influence and the fundamental laws makes it natural to infer that if the fundamental laws are deterministic in both temporal directions, then we influence the past as well. The underlying reasoning would be that we try to determine the fundamental laws as best we can, and then—without tacking on a fundamental direction of influence to vindicate our prejudice against past-directed causation—we check the laws to see whether the fundamental laws can propagate states toward the past and whether counterfactual alterations to the present thereby imply counterfactual differences in the past. To spell out this procedure a bit more formally, just consider any state s that determines its future and its past. Then, consider a counterfactual alteration to s by letting s' be just like s except that it has some localized region R where the positions of some corpuscles are shifted or the strength of some fundamental fields are altered. Just about any s' that is so defined will determine a future and past different from what s determines. It follows that the counterfactual alteration to s that makes it into s' implies counterfactual differences in the past. The temporally unbiased way to interpret this counterfactual dependence is just to accept it at face value as a form of past-directed difference-making influence.

Some people find past-directed influence objectionable, but there is no evidence anyone has ever presented for its non-existence. As I will attempt to make clear in the next chapter, I believe there is substantial evidence for the lack of *exploitable* past-directed influence, but influence extends over a much greater range than exploitable influence. Given the central role our conception of time plays in practical reasoning and the apparent lack of utility in trying to affect the past, it is easy to understand how we might mistake the uselessness of past-directed influence for its non-existence. Before Newton suggested that the motions of celestial bodies and the trajectories of earthly objects are governed by the same laws, it was easy to mistake our lack of any exploitable influence over the moon for a total lack of influence. Perhaps, in the end, it will turn out that the best way to think about influence is to posit a fundamental future and fundamental past, but it is worth explicitly weighing the relative merits of explaining the temporal asymmetry associated with influence in terms of a fundamental asymmetry versus a derivative asymmetry.

I will continue to forgo any stipulations that c cannot termine a previous event e or that $\tilde{C}$ cannot promote a previous event E, and for terminological convenience, I will continue to maintain the equivalence between prob-dependence and prob-influence regardless of temporal direction.[51] Thus, in full generality, $\tilde{C}$ prob-influences E iff E prob-depends on $\tilde{C}$.

Once the possibility of past-directed prob-influence is perceived as a cogent hypothesis, it is not difficult to infer that future-directed and past-directed prob-influence might be linked together in a chain. In §4.9, I argued that it is plausible that fixing relations obey unidirectional transitivity, and in §2.8, I noted that terminance itself is transitive regardless of temporal direction. So, it might be possible under some circumstances to have non-trivial backtracking prob-influence.

Although ordinary counterfactual reasoning might motivate someone to believe in the truth of some backtracking counterfactuals, people are usually aware that the causal regularities that seem to underlie the reasonability of backtracking counterfactual inferences are not normally exploitable for causal purposes. This is illustrated in mundane cases of two distinct effects arising from a common cause. When I acquire an infection that makes me feverish and itchy, I might think that if I were not to have the itch, I would not have contracted the infection, and thus I would not have the fever. But I would be demonstrably wrong to think I can prevent or eliminate the fever by spreading anti-itch cream on my skin (except to the extent the cream has a future-directed curative effect on the fever, say by chemistry or placebo effect). Our inability to exploit such *common-cause patterns* is seemingly general. It is apparently very difficult if not impossible to manipulate an event in the present and thereby influence an event in the past and thereby influence something in the future. The task of this chapter is to unpack this apparently plausible principle using the methodology of empirical analysis. It will turn out that so long as a few general principles governing terminance hold true, then genuine backtracking prob-influence is guaranteed to be worthless.

After discussing the general disutility of backtracking prob-influence, I will call attention to a legitimate kind of prob-influence that is easily confused with backtracking prob-influence and permits a very limited channel through which common-cause patterns can make a difference. One effect of the common cause can non-trivially affect the other effect by virtue of the existence of the common cause (rather than some direct nomic connection between them). Even though this genuine form of influence defies common interpretations of causality, I will eventually argue that it can never be exploited.

6.2 Proof of Causal Directness

The conclusion to be established in this section is that every relation of backtracking prob-influence is made redundant by a temporally direct relation of prob-influence. Thus, any backtracking prob-influence a contrastive event exerts

[51] Remember that prob-dependence is counterfactual dependence as understood using the nomic conditional from §3.4. See also §3.1.

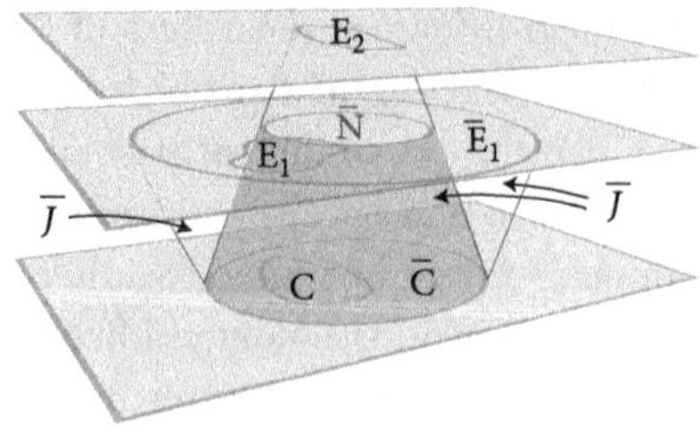

FIGURE 6.1 *$\overline{E_1}$ shields the fixing it exerts back to $\overline{C}$ and then forward to E_2.*

does not differ in magnitude from the non-backtracking prob-influence it exerts. This renders backtracking prob-influence superfluous.

Readers who find the details in this section challenging are invited to examine a simpler version, (Kutach 2011a), which includes some additional commentary.

Consider the situation depicted in Fig. 6.1. The starting point is an event e_1, coarse-grained as E_1, contextualized as $\overline{E_1}$, and contrastivized as $\tilde{E}_1 \equiv (\overline{E_1}, \overline{\neg E_1})$ so that $\tilde{E}_1$'s foreground is e_1's region and its background extends outward in space-like directions. The argument in this section only concerns whether the contrastive event, $\tilde{E}_1$, can prob-influence events in a non-redundant backtracking fashion and does not directly concern what the plain coarse-grained event, E_1, partially influences,[52] but I will return later to discuss this issue in §6.4.

I will initially assume $\tilde{E}_1$ wholly temporally precedes E_2 and consider whether $\tilde{E}_1$ can prob-influence E_2 by way of some event to the past of $\tilde{E}_1$. After presenting the proof, I will return to discuss how the argument applies to cases where E_2 comes before $\tilde{E}_1$ or at the same time. Because no temporal asymmetries are introduced in the proof, it automatically applies to the case where the turnaround event is to the future of $\tilde{E}_1$ and E_2, as occurs in the temporal reverse of the situation depicted in Fig. 6.1.

The proof does not require that the events are instantaneous, but it does require that $\overline{E_1}$'s region include a connected space-like subregion Q that spans $\overline{E_1}$ in the sense that every point in $\overline{E_1}$'s region is c-connected to Q. This requirement is imposed merely to ensure that even if e_1 occupies a temporally disconnected or convoluted region, then the chosen contextualization, $\overline{E_1}$, will envelop e_1 in a more manageable region. This maneuver does not constitute any loss of generality because in any situation where e_1's region is too complicated for the proof to apply directly, the proof will still apply to its subevents, and that will suffice for the general conclusion.

The turnaround event in the backtracking is instantiated by c. (If there were no event c to serve as a turnaround event, then there would be no backtracking prob-influence at all.) In order for $\tilde{E}_1$ to prob-influence E_2 by way of what happens in c's region, there needs to be some contrastivization of c lying entirely to the past of $\tilde{E}_1$ that prob-influences E_2 and is fixed by $\tilde{E}_1$. Without loss of generality, we

[52] See §4.12 for the definition of partial influence.

can assume for the sake of simplicity that this event is a regular contrastivization, $\tilde{C} \equiv (\overline{C}, \overline{\neg C})$. The goal of the proof is to show that the prob-influence $\tilde{E}_1$ exerts on E_2 by virtue of what $\tilde{E}_1$ fixes for *c*'s region is no different from $\tilde{E}_1$'s prob-influence on E_2 existing by virtue of how $\tilde{E}_1$ prob-influences E_2 directly (by virtue of what the fundamental laws dictate about events to the future of $\tilde{E}_1$).

Before giving the formal proof, let me sketch the intuitive idea behind it. In order for $\tilde{C}$ to prob-influence E_2, it needs to be large enough for E_2 to fit within its domain of terminance. Thus $\tilde{C}$ needs to span at least the past light-cone of E_2. If the fundamental laws are non-relativistic, as in the case of classical gravitation or non-relativistic quantum mechanics, then $\tilde{C}$ needs to span an entire time slice.[53] Similarly, in order for $\tilde{E}_1$ to fix $\tilde{C}$ it needs to be large enough so that $\tilde{C}$ fits within its domain of terminance. Thus $\tilde{E}_1$ needs to at least span the future light-cone of $\tilde{C}$. If the fundamental laws are non-relativistic, then $\tilde{E}_1$ needs to span an entire time slice. The intuitive idea behind the proof is that in order to get the backtracking prob-influence going, $\tilde{E}_1$ has to be so big and have the specification of its material contents so filled in that any influence it has on *c* is not able (as it goes forward in time) to zigzag around $\tilde{E}_1$ or go through $\tilde{E}_1$ or skip over $\tilde{E}_1$ to have some bearing on the probability of E_2.

The ingredients we need for the proof were identified in §2.14, §4.10, and §4.11. They are reproduced here for easier reference.

- Definition of a c-connection: A c-path is an everywhere differentiable path whose tangents are nowhere space-like and are well-defined and non-space-like in any mathematical limits along the path. Two points *p* and *q* are c-connected iff $p = q$ or a c-path exists between *p* and *q*. Two regions are c-connected iff some point in one region is c-connected to some point in the other. Two events are c-connected iff their regions are c-connected.
- Definition of an intermediate region: A region of the arena R is intermediate between *c* and *e* iff (1) every point of R is c-connected between some point of *c*'s region and some point of *e*'s region, and (2) there is a connected space-like subregion Q of R such that every c-path from a point in *c*'s region to a point in *e*'s region intersects Q.
- Weak Transitivity of Fixing: If $\overline{E}$ fixes a contextualized event throughout region R, any event fixing $\overline{E}$ also fixes a contextualized event throughout region R.
- Non-spatiality: The fundamental laws disallow space-like terminance (and thus space-like fixing).
- Continuity of Probability-fixing: If a contextualized event $\overline{C}$ fixes a probability *p* for some *E* and there exists some region R intermediate

[53] In the quantum-mechanical case, one might need to include a complete specification of the corresponding state in configuration space.

between $\overline{C}$ and E, then there exists a unique maximal contextualized event $\overline{I}$ that occupies R, is fixed by $\overline{C}$, and fixes a probability p for E. (This $\overline{I}$ is a fixed intermediate on the way from $\overline{C}$ to E.)

- Shielding of Fixing: For any contextualized event $\overline{C}$ that fixes a contextualized event $\overline{E}$ and for any contextualized event $\overline{I}$ occupying region Q that is a fixed intermediate on the way from $\overline{C}$ to $\overline{E}$ (so that it is fixed by $\overline{C}$ and fixes $\overline{E}$), then for any region R that lies entirely within $\overline{C}$'s domain of terminance and contains no points on a c-path going from $\overline{I}$ to $\overline{E}$, the contextualized event $\overline{J}$—defined as whatever $\overline{C}$ fixes for R ∪ Q—fixes $\overline{E}$ (just like $\overline{I}$ does).

Here is the proof. Consider a region, R′, consisting of $\overline{E}_1$'s region unioned with an inextendible space-like extension of E_1's required space-like subregion so that $\overline{C}$ and E_2 lie entirely on opposite sides of R′. Let R be the intersection of R′ with all the points that are c-connected between $\overline{C}$ and E_2. By construction, R is intermediate between $\overline{C}$ and E_2.

By continuity of probability-fixing, there exists some contextualized event, $\overline{N}$, occupying R, that fixes a probability for E_2 and is fixed by $\overline{C}$. $\overline{E_1}$ must contain this region R because by weak transitivity of fixing, $\overline{E_1}$ must fix some event occupying R, and if R were to occupy any part of the arena outside of $\overline{E_1}$, that would result in a violation of non-spatiality. So we can conclude that there is a fixed intermediate, $\overline{N}$, located in $\overline{E_1}$'s region on the way from $\overline{C}$ to E_2.

Remember that we are not concerned with what $\overline{C}$ by itself fixes for $\overline{N}$ or what probability it fixes for E_2. The fixing under consideration is what $\overline{E_1}$ fixes by going back to $\overline{C}$ and then backtracking through $\overline{N}$ to E_2.

Let $\overline{J}$ be whatever event $\overline{E_1}$ fixes for the region consisting of all the points c-connected between $\overline{E_1}$ and $\overline{C}$, excluding $\overline{E_1}$'s region. Shielding implies that $\overline{J}$'s fixing of $\overline{N}$ (and thus its probability-fixing of E_2) is shielded by $\overline{E_1}$, which implies that the probability fixed by backtracking from $\overline{E_1}$ through $\overline{C}$ to E_2 is $p_{\overline{E_1}}(E_2)$, which is the same as the probability we get from the nomic connection going straight from $\overline{E_1}$ to E_2.

The argument can be repeated to show that the probability fixed by the nomic connection backtracking from $\overline{\neg E_1}$ through $\overline{\neg C}$ to E_2 is the same as the probability fixed by the direct connection from $\overline{\neg E_1}$ to E_2. Thus the degree of prob-influence we get from backtracking through c's region is $p_{\overline{E_1}}(E_2) - p_{\overline{\neg E_1}}(E_2)$. Thus, $\tilde{E}_1$'s backtracking prob-influence on E_2 through the region occupied by c amounts to nothing more than its temporally direct prob-influence on E_2. Q.E.D.

I will now consider how the proof applies if $\tilde{E}_1$ does not precede E_2, but c is still situated to the past of both. If E_2 is entirely to the past of $\tilde{E}_1$, then whatever backtracking prob-influence eventually reaches E_2 will have to pass through E_2 on the initial stage of the backtracking. Because there will be a fixed intermediate just prior to E_2, the backtracking prob-influence going through c's region will be rendered superfluous. If E_2 happens at exactly the same time as $\tilde{E}_1$, then E_2's region will be a subregion of $\tilde{E}_1$'s and thus the pair of probabilities fixed for E_2 will be trivially fixed by $\tilde{E}_1$ without regard to any laws. If E_2 is temporally extended and is partly after $\tilde{E}_1$ and is partly at the same time and is partly before, then the proof above and the two previous arguments in this paragraph address each of these three subevents of E_2 separately.

Even though the assumptions of the proof are meager, one might wonder how solid our evidence is that the actual laws obey the principles assumed in the proof. Although one can never be sure about what principles the actual fundamental laws obey or even that there are fundamental laws, the four principles I have employed can be defended as plausible even though we know that none of the paradigm fundamental theories is a correct and complete theory of fundamental reality.

The transitivity of terminance is the least controversial of all the principles. The transitivity of determination is rather trivial and uncontroversial. Furthermore, in §4.9, I defined weak transitivity of fixing to ensure it would not be controversial when applied to contextualized events.

Continuity and shielding work together to rule out the existence of influences that hop across time without leaving an impact on intermediate states. Continuity is a plausible working hypothesis given how it holds for a wide variety of realistic theories beyond the paradigm fundamental theories including arenas that have a discrete temporal structure. Shielding is perhaps less secure. Because general relativity (GR) permits solutions with closed time-like curves, CTCs, as discussed in §2.12.2, there can be local circumstances that look like Fig. 6.1, but where a c-connection exists between E_2 and $\overline{C}$ that bypasses $\overline{E_1}$ through the CTC. Imagine a wormhole in space-time, for example, that tunnels around $\overline{E_1}$. This highlights the possibility that if there are CTCs in the actual world, for all GR tells us, they might allow past-directed prob-influence over the future and future-directed prob-influence over the past. Although there are numerous technicalities involved in sorting out what prob-influence would be like in such situations, it is uncontroversial that the reality of CTCs might call into question the applicability of our ordinary conception of how causation works, and the same goes for fundamental laws that build in explicit rules where terminance skips over intermediate states, for example, laws where the activation of a magic wand determines that a rabbit springs into existence a year ago and then vanishes one day after that. Even though CTCs are possible according to GR, we do not have any evidence that they exist. And even if they exist *somewhere* in the universe, that alone does little to help someone who is trying to exploit backtracking nomic connections in our locale (where there are presumably no CTCs of the right size to reveal their presence). I am not claiming that an interesting form of backtracking influence would exist if there were CTCs, only that they at least crack the door open for that possibility. Arguing against its plausibility would require too much of a digression.

Another way to evade the shielding assumption is for the fundamental laws to allow different kinds of terminance that run in opposite temporal directions. A theory resembling the classical unified field theory of §2.5.3 could exist where

$\tilde{E}_1$ is specified in terms of only electromagnetic properties and fixes $\tilde{C}$ solely through the electromagnetic interaction. If $\tilde{C}$ is composed of stuff that has both electromagnetic and weak properties, it could then use the weak interaction to prob-influence E_2 by going straight through $\tilde{E}_1$, which does not respond to the weak interaction. Even though such possibilities are worthy of exploration, I doubt that the kinds of matter we are familiar with can circumvent the principle of shielding for the two reasons mentioned in §2.5.3. Namely, we have evidence that all the known interactions are linked together because (1) in quantum field theory, the probabilities for particle interactions include subtle adjustments that take into account all possible interactions, and (2) gravitation is a universal interaction. According to what we know from general relativity, any kind of matter that can possess energy, even massless corpuscles, will serve as a gravitational source and every sort of matter is affected by gravity. Unless there is some surprising reason to think gravity is not comprehensive, there will be no terminance of merely an electromagnetic sort or merely a weak sort. A third reason one could add is that each of the fundamental interactions seems to work in the same way as the others toward the future and toward the past. There is no evidence that the purely electromagnetic $\tilde{E}_1$ fixes only toward the past while the weak subevent of $\tilde{C}$ fixes only toward the future. It is plausible that if one kind of fundamental attribute takes part in fixing toward the future, then the other does as well, and the same holds for any past-directed fixing. If so, there would be no distinct routes of influence that could give rise to non-redundant backtracking prob-influence.

Non-spatiality is obeyed by all the paradigm fundamental theories, but it may not have overwhelming intuitive appeal. Given that there exists arbitrarily fast contribution in the theory of classical gravitation, §2.4.4, and that quantum mechanics reliably exhibits space-like correlations that are arguably causal in some sense, it is not too much of a stretch to posit fundamental theories that violate non-spatiality. However, if one considers the role of non-spatiality in the proof, it becomes apparent that the job it performs is just to block the possibility of having a nomic connection issuing from $\tilde{C}$ that zigzags around the edge of $\tilde{E}_1$ to get to E_2. Perhaps such a nomic connection could exist if there were a temporal asymmetry in the speed of influence. One could imagine, for example, that space-time is structured so that at each point there is a double light cone structure: the normal light cone plus a super-light cone that represents the possible paths of super-photons traveling at twice the speed of light. Then, the fundamental laws might say that past-directed contribution relations stay within the normal light cone but that future-directed contribution can exploit the super-light cone to allow super-photons to zigzag around $\tilde{E}_1$ so that the prob-influence of $\tilde{C}$ on E_2 is not shielded by $\tilde{E}_1$. However, it is hard to see a plausible model where that could work. One implausibility is that it appears to require two different kinds of interactions: one for the past-directed terminance restricted to the normal light cone and a second for the future-directed terminance that exploits the super-light cone. The interaction linkage discussed in the previous paragraph suggests that the laws forbid the existence of such unlinked forms of terminance. If $\tilde{C}$ prob-influences events outside the regular light cone toward the future, then events outside the regular light cone ought to be needed as part of the background for $\tilde{E}_1$'s past-directed fixing of $\tilde{C}$.

One way to motivate non-spatiality is just to recognize that influence (in the sense of 'pure contribution') spreads out as time goes by (whether toward the future or past), as illustrated by the domain of influence in Fig. 2.5. The spreading of such influence corresponds with the shrinking of terminance. In relativistic electromagnetism, for example, influence spreads at the speed of light, and as a result, any finite-sized event can only termine events at other times that are smaller than itself. The size of the termined events decreases at the speed of light. (The shrinking of terminance is hidden in models using a classical space-time because the relevant terminants are superevents of global states, which makes the terminance shrink down to every event at every other time.) One could certainly imagine laws that violate non-spatiality by having terminants expand as time goes by, so that $\tilde{E}_1$ fixes a larger event, $\tilde{C}$, going toward the past, which then fixes an even larger event going back toward the future to reach points that are space-like related to $\tilde{E}_1$. But these laws would have an exceedingly strange character because the probabilities $\tilde{C}$ would fix for later events would not depend on stuff located barely outside of $\tilde{C}$. The stuff just beyond the edge of $\tilde{C}$ would either fail to have spreading influence or there would have to be a conspiratorial redundancy in the probabilities that such events fix.

Another way to motivate non-spatiality is to consider that if non-spatiality were generally violated, then a deliberate act like placing a couch against a wall might fix a probability for what objects exist at the same time on the other side of the room. But if we could learn about such laws, what would prevent us from repeatedly placing objects on the other side of the room in contravention of the lawful probabilities? This is not intended to be anywhere near a conclusive argument for non-spatiality but is meant to suggest that laws that violate non-spatiality would need to be made consistent with our apparent ability to arrange our furniture pretty much how we like on a macroscopic scale, and it is unclear how that consideration would best be accommodated.

Perhaps there is some other way of exploiting violations of non-spatiality, but I know of no viable options.

I suspect the best reason to question the proof's assumptions is that space-time may not be the arena of fundamental reality.[54] The assumptions of the proof included reference to c-connections, which are defined using terms like 'space-like', which applies to space-time arenas as well as quantum configuration space. If it turns out that fundamental reality has some arena whose structure does not support the kind of distinctions that Galilean and Minkowski space-times make when

---

[54] For reasons I cannot disclose here, I believe that fundamental reality has an arena but that it is not space-time.

they distinguish space-like relations, then the principles used in the proof would need to be reformulated, and it is uncertain whether a modified proof could be found. Also, I believe it is plausible that the actual arena could turn out to be some higher-dimensional space where terminance relations do not correspond to (or even imply) relations among (metaphysically derivative) events in space-time. It is hard to know what to say about this possibility from the perspective of current physics because it is speculative enough to warrant skepticism about how much we can reasonably infer about the nature of time in such a model, or probabilities, or even matter. However, given that our current theories obey transitivity, non-spatiality, continuity, and shielding, it could turn out that fundamental physics in some higher-dimensional arena justifies the application of these principles to events located in space-time, thus changing the metaphysical status of the principles to derivative. Whether that would suffice to rule out non-redundant backtracking prob-influence entirely, I cannot say.

Still, in the end, I think the superfluousness of backtracking prob-influence is a conclusion that is on fairly safe footing despite our ignorance of the actual fundamental laws (or lack of fundamental laws).

6.3 A Search for Empirical Phenomena

Backtracking influence (or causation) is any sort of influence (or causation) that holds by virtue of an initial event influencing (or causing) in one temporal direction an event that in turn influences (or causes) a third event in the opposite temporal direction. In the previous section, I provided a disproof of non-redundant backtracking prob-influence. In this section, I am bracketing this conclusion and investigating whether *some* form of backtracking influence could exist. The backtracking connections that I will be identifying as 'past-directed then future-directed' and 'future-directed then past-directed' represent hypothetical routes by which some form of influence might be conveyed. They do not stand for prob-influence relations but are unarticulated place-holders for some conjectured sort of influence that behaves in a temporally directed and localized enough manner in order to vindicate talk of its existing along routes. My aim in this section is to explore whether any experiment reveals evidence for such influence.

Across a wide variety of circumstances we find event-kinds that satisfy the three following conditions: (1) Each instance of event-kind C is regularly followed by an instance of event-kind E_1 and an instance of event-kind E_2. (2) Each instance of event-kind E_1 is regularly preceded by an instance of event-kind C. (3) Instances of event-kinds E_1 do not appear to cause instances of E_2 directly in any recognizable way. When a triplet of event-kinds $\{C, E_1, E_2\}$ satisfies these conditions, let us say that the event-kinds exemplify a *common-cause pattern*. Instances of such common-cause patterns include an infection of the body followed by a distinctive itch and a fever, and an electrical disturbance among clouds being followed by a bright flash of light and then a loud rumbling noise. I am deliberately defining 'common-cause pattern' imprecisely to avoid presupposing the conceptual apparatus of probabilistic relations among coarse-grained events usually employed in formal definitions of a common cause. It will not matter that the definition of 'common-cause pattern' does not match the standard definition of a common cause because its only purpose is to focus the discussion on certain paradigmatic instances of what we understandably interpret as causation involving a common cause. In particular, we can understand the word 'regularly' used in conditions (1) and (2) above liberally to include robust positive statistical correlations, not merely exceptionless co-instantiation. And we can interpret 'cause' and 'directly' in condition (3) as referring to some (presumably localized) process extending from E_1 to E_2 that people would interpret as causal.

Our experience might seem to indicate that for any common-cause pattern, E_1 does not influence E_2 through the mediation of C. The task of this section is to explore what evidence we have for thinking that such influence is impossible.

Remember that in conducting an empirical analysis of the metaphysics of causation, one is supposed to isolate the empirical phenomena that motivate our possession of (something in the neighborhood of) our actual concept of causation. The lack of influence via common cause is historically central to the distinction between genuinely causal probabilistic relationships and non-causal statistical correlations. Such cases are among those that motivated Nancy Cartwright's (1979) original distinction between effective strategies and ineffective strategies, and I agree with her that one good reason to draw some sort of distinction between causal and non-causal connections is because of how they relate to effective strategies.

If we are following the method of empirical analysis, however, we should not presume our pre-theoretical grasp of this distinction is adequate. Instead, we should attempt to formulate an experiment whose results give us a good reason to disbelieve in cases of influence via common cause. One might suspect that I would then try to explain these results using the causal directness principle I just defended. However, as it turns out, several problems arise in formulating an adequate experimental schema, and I suspect that the deficiencies cannot be remedied as a single schema. According to my investigation, there turn out to be no empirical phenomena corresponding precisely to the seemingly plausible conjecture that common-cause patterns are not routes for influence. Instead, the empirical phenomena are better organized in terms of three distinct experimental schemas.

This motivates a reappraisal of our conception of what kinds of causal relations exist. It will turn out, on my account, that common-cause patterns can be the source of bona fide difference-making influence between the events we ordinarily think of as effects of the common cause. However, I will also argue that this influence cannot be *exploited* in the sense relevant to explaining the empirical phenomena that give us a good reason to distinguish between effective and ineffective strategies. The upshot is that the distinction between effective and ineffective strategies should not be cashed out in terms of causal versus non-causal relations or influence versus lack of influence, but instead in terms of exploitable versus non-exploitable influence. Exactly what exploitation ultimately amounts to is going to be expressed in terms of an experimental schema.

At the end of this chapter, I will begin the task of identifying the empirical phenomena that will serve as the basis for suitable principles to substitute for the initial platitude that influence via common cause is impossible, and I will complete the discussion in the next chapter. The three substitutes will vindicate the utility of our simplistic belief that common causes cannot serve as routes for influence but will correspond more closely with what is empirically testable. Furthermore, it is amenable to scientific explanation, which I will also provide in the next chapter.

Revising the target of an empirical analysis in light of discoveries we make when trying to formulate experiments is an important feature of the methodology behind empirical analysis that I emphasized in §1.1. This section thus provides a valuable lesson about empirical analysis. The difficulty of defining an experiment to capture the intuitive content of an initially plausible principle guides an empirical analysis toward novel principles that improve the conceptual architecture.

The initial goal in this section is to consider a special case of the promotion experiment from §5.1 where we test whether a common-cause pattern allows one effect to influence the other probabilistically through the common cause. To set this up, we need to run some preliminary checks to ensure we have correctly identified a common-cause pattern. To do this, we identify some C, contrastivized as $\tilde{C}$, that we think promotes two distinct events E_1 and E_2. For the sake of clarity, we can focus discussion on cases where the locations of E_1 and E_2 relative to $\tilde{C}$ are such that none of these three events overlap each other spatio-temporally. By running promotion experiments, we can empirically confirm whether $\tilde{C}$ promotes E_1 and E_2 and also confirm the lack of any plausibly manipulable form of influence going directly from E_1 to E_2. We do not, strictly speaking, need to verify whether instances of E_1 are reliably preceded by C because to the extent that a nomic connection going from E_1 back in time toward C is weak or not robust under alterations of the background conditions, it will be harder to detect. But because we are presumably more likely to detect backtracking influence if E_1's are reasonably often preceded by C, we should initially try to investigate such event-kinds. For discussion's sake, let us assume we have found suitable event-kinds that meet these criteria so that we are reasonably confident that we have a common-cause pattern.

For the main experiment, we want to evaluate whether someone can verify the existence of backtracking routes of influence by making a difference to the probability of E_2 through the use of E_1's nomic connection to their common cause, C. We can now specify the *backtracking experiment*, whose purpose is to detect backtracking routes of influence.

Consider initial conditions instantiating an agent who is able to fiddle directly with E_1 and is motivated to affect the probability of E_2, perhaps by knowing there is a cash reward that depends on whether E_2 occurs. Let the event A be an agent wanting E_2 to occur, and let $\tilde{A} \equiv (\overline{A}, \overline{\neg A})$ be a contrastivization of A representing the agent's wanting E_2 to occur rather than wanting E_2 not to occur. It helps to focus on a case where $\tilde{A}$ includes enough structure in its background for C to have a reasonable possibility of occurring and a reasonable possibility of not occurring, and also that the agent knows that C is often a common cause of E_1 and E_2, and the agent is capable of affecting what happens around where E_1 might occur.

Then, instantiate a zillion instances of $\overline{A}$ selected randomly using the probability distribution built into $\overline{A}$ and observe the fraction of runs in which all three events, E_1, C, and E_2, occur. Call that value $f_{\overline{A}}$. Then, in a second set of runs, instantiate a zillion instances of $\overline{\neg A}$ and observe the fraction in which all three events occur, $f_{\overline{\neg A}}$. The observed degree of prob-influence is $f_{\overline{A}} - f_{\overline{\neg A}}$, which in theory should match the theoretically predicted value, $p_{\overline{A}}(E_1\&C\&E_2) - p_{\overline{\neg A}}(E_1\&C\&E_2)$.

The backtracking experiment, as formulated so far, has several grave problems providing a decent measure of an agent's ability to affect the probability of E_2 by virtue of a lawful connection going from E_1 to C to E_2. I will just note four deficiencies, numbered in Fig. 6.2. First, it does not account for the possibility that the agent can directly affect E_2 without going through E_1. Second, it does not distinguish between ways of instantiating E_1 that have a direct influence on E_2 (but were previously overlooked) and those influences that go through C. Third, the experiment does not adequately distinguish between attempts to instantiate E_1 directly from attempts to instantiate E_1 by instantiating C. These three problems are depicted in Fig. 6.2.

Fourth, the experimental design incorporates two concatenated instances of backtracking. The agent presumably is acting to create E_1 in an ordinary future-directed process, which then (we are supposing) might go back in time to C, and then forward in time to E_2. This prevents the experiment from distinguishing 'past-directed then future-directed' influence independently from 'future-directed

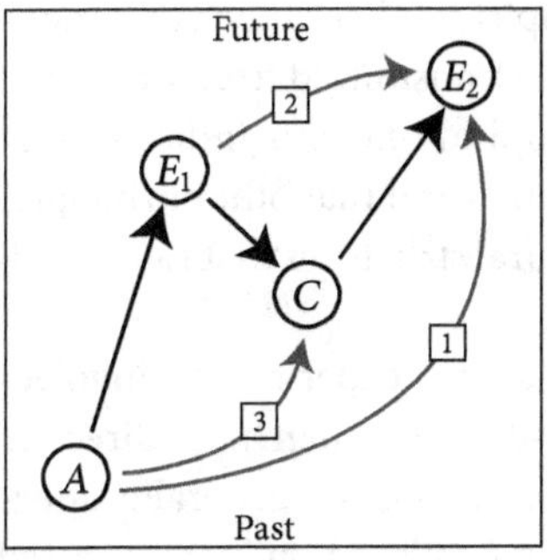

FIGURE 6.2 *The backtracking experiment fails to rule out the three numbered routes of influence.*

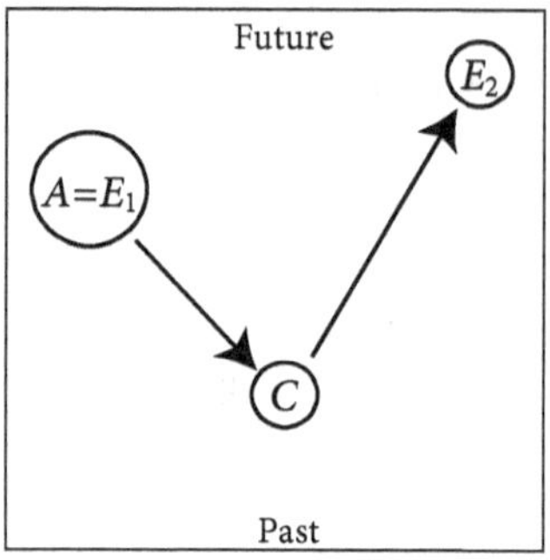

FIGURE 6.3 *A conceivable case of 'past-directed then future-directed' influence.*

then past-directed' influence. Perhaps only 'future-directed then past-directed' influence is unexploitable, and if so, this way of characterizing the empirical phenomena will lead us to mistakenly overlook evidence for 'past-directed then future-directed' influence.

In exploring how to respond to deficiencies in the design of the backtracking experiment, I think it is best to focus on the fourth problem by distinguishing between two cases. In the first case, depicted in Fig. 6.2, the agent's action A is external to the common-cause pattern the agent is trying to exploit. In the second case, depicted in Fig. 6.3, the agent's action A *is* the effect E_1, so the action itself is part of the common-cause pattern the agent is trying to exploit.

My aim in this chapter and the next is to demonstrate that agents in the first case will not affect the probability of E_2, but that for agents in the second case, the situation is more nuanced. There is one sense in which agents can affect the probability of E_2 and another sense in which they cannot. This result will suggest that the backtracking experiment should be replaced with three different experiments, two of which demonstrate the inability of agents to affect the probability of E_2 and one of which demonstrates an ability of agents to affect the probability of E_2 by virtue of the common cause. In the end, I do not believe there is a unique backtracking experiment that can be adequately formulated.

The replacement of one experimental design with three more precise designs illustrates an important feature of empirical analysis. When we require ourselves to formulate an experiment corresponding to the empirical phenomena that motivate us to believe a platitude like "influence cannot be transmitted from one event to another through a common cause" and we fail to find one, this indicates that the principle is probably defective or at least suboptimal and that other principles should be sought that identify more precisely the natural structures that make such a platitude useful as folk wisdom.

My plan for distinguishing these three backtracking experiments involves deferring exploration of the possibility of 'future-directed then past-directed' influence to the next chapter, where I will show that it is quite generally unexploitable. My attention in the rest of this chapter will be placed on 'past-directed then future-directed' influence, which reveals the ambiguity that arises when one of the common cause's effects is the agent's decision.

6.4 'Past-directed then Future-directed' Influence

There are many simple devices that reliably exhibit a common-cause pattern. Think, for example, of a brake pedal that slows down four separate wheels or a switch that turns on both a light and a fan. Although the activity of our brain is more complex, there is no apparent barrier in principle to its being a source of common-cause patterns. In particular, if we adopt the view that our brains obey the same laws of physics as everything else, it might be possible for the molecules constituting our brains to be correlated with a mechanical device such that some external effect, E_2, is reliably correlated with our future action, E_1. Call such a device a 'brain correlator'. A brain correlator is a limited mind-reading device that is capable of detecting enough about our brain structure to establish a reliable correlation between one kind of future behavior—an agent's taking action E_1 rather than action $\neg E_1$—and the output of the device, E_2 rather than $\neg E_2$. It is a decision predictor of the kind discussed in Robert Nozick's famous (1969) discussion of Newcomb's problem. If a brain correlator were successfully implemented, it might warrant ordinary language claims like, "If the agent were to take action E_1, then E_2 would probably occur, and if the agent were to take some alternative action, then E_2 would probably not occur." Given the close connection in my account between counterfactual dependence and prob-influence, it is worth exploring whether such a counterfactual can be cashed out in terms of prob-dependence and thus indicate the existence of prob-influence. That is, it is worth exploring whether an agent's action E_1 can influence E_2 by way of a lawful connection that "goes through" or at least exists by virtue of a brain correlator.

One can easily formulate principles governing fundamental reality that make the existence of a brain correlator unlikely or even impossible. One hypothesis is that there is so much fundamental chanciness in the brain's behavior that it is virtually impossible for the brain correlator to correlate with what future decisions an agent will make. Another hypothesis is that there exist fundamental volitions that serve as causal contributors to the physical development of the world and cannot be detected beforehand by any physical device. Even though one could reject 'past-directed then future-directed' influence on the grounds that a brain correlator cannot exist, I will continue to explore the consequences of brain correlators for two reasons. First, it is plausible that the fundamental laws permit brain correlators, and we need to know what to say about the possibility of their instantiating backtracking influence. Their existence is not an outrageous hypothesis because unless the sources of decisions takes a special metaphysical form, like some variants of Cartesian dualism, the correlator is not relevantly different from ordinary correlation-establishing devices like barometers and thermometers.

Second, everything I will claim about brain correlators can be said of more mundane common-cause patterns involving an agent's action such as the existence of an alcohol addiction promoting an agent's accepting free booze and exhibiting tremors. The focus on the brain correlator merely helps to keep the relevant events spatio-temporally localized for the sake of clarity.

To add some detail, let us say that the brain correlator only correlates with a single action the agent can take: a deliberate choice between yellow and green. The brain correlator is set up so that it scans the agent's brain and relays an electronic signal to a separate room where a flag is raised so that the color of the flag corresponds to the color that the agent will choose. If the agent does not choose a color, no flag is raised. It is convenient to imagine the flag room being epistemically inaccessible to the agent because that helps to prevent interaction effects that might undermine the reliability of the brain correlator. If the flag were to rise before the agent's choice so that the agent could see the flag's color and afterward deliberately select the opposite color, that would tend to result in the brain correlator not working as advertised. That possibility is a legitimate way to defeat the successful operation of the brain correlator, but it would demonstrate nothing about whether the brain correlator can be reliable in cases where the agent is unable or unwilling to countermand the reading of the brain correlator. We can focus on the favorable cases because the question we are trying to address is whether it is possible to influence along a 'past-directed then future-directed' route, not whether it is possible to ensure that the brain correlator works in all circumstances.

Suppose we run a promotion experiment with one zillion randomly selected instances of the initial condition, $\overline{S}$, which instantiates an agent and a properly operating brain correlator before they interact with each other and establish a correlation. We can check the experimental runs to verify that the brain correlator did not break down for mundane reasons and to verify that the agent did not avoid interacting with the brain correlator. We simply ignore data from any such faulty runs. Then we observe the remaining runs to see how many times the agent chose yellow, how many times green, and how many times the color of the raised flag matched what the agent chose. If the brain correlator operates as advertised, we will observe a strong correlation between the agent's choice of color and the actual color of the raised flag.

Such correlations are perfectly understandable without positing backtracking prob-influence or past-directed promotion. The agent's action is, after all, a consequence of prior conditions that the brain correlator was able to detect. One could think of the agent's choice in such a case as an indication of some prior physical condition. This hypothesis might grate against some people's intuition that a proper agent is able to choose independently of the prior conditions, but we are considering realistic agents under the assumption that their atomic constituents are obeying the same laws as other atoms.

What I will consider now are two possible ways of modeling the relationship between E_1 and E_2 in terms of prob-influence, the first of which results in E_1 not being able to affect the probability of E_2, and the second of which allows E_1 to affect the probability of E_2.

First, we can construct a regular constrastivization of E_1 that wholly occurs "at the same time" as E_1. Let us construct $\overline{E_1}$ by first extending E_1 in space-like directions and selecting a localized region for E_2 so that E_2 lies entirely to the future of $\overline{E_1}$. For contrast, $\overline{\neg E_1}$ differs from $\overline{E_1}$ by replacing every instance of E_1 with an instance of $\neg E_1$. The resulting contrastive event $\vec{E_1} \equiv (\overline{E_1}, \overline{\neg E_1})$ will have a foreground whose region is the same as E_1's and a background whose region occurs everywhere else nearby "at the same time."

The experimental design is meant to rule out the agent somehow changing the color of the flag after it has been set by the brain correlator. And my proof of the redundancy of backtracking prob-influence from §6.2 ensures that $\vec{E_1}$ can only prob-influence E_2 to the same degree it does in a purely future-directed manner. Thus, $\vec{E_1}$ cannot promote E_2.

Second, we can construct an alternative form of prob-influence. Recall from §4.12 that events have no restrictions on their size and shape and are thus allowed to be composed of two disconnected subevents that occur at different times, as shown in Fig. 6.4. In particular, an irregular contrastivization of E_1 can be formed by adopting $\overline{S}$ as the background and $\neg E_1$ as the contrast event whose region is identical to E_1's. It follows from the way the initial conditions were set up that the resulting irregular contrastivization $(E_1, \neg E_1, \overline{S})$ promotes E_2. In this way, we can understand the relation between the two effects of a common cause as a genuinely causal relation. E_1 affects the probability of E_2 in the sense that it partially influences E_2. E_1 occupies the foreground of a contrastivization, $(E_1, \neg E_1, \overline{S})$, that prob-influences E_2 to a non-zero degree.

Even though $(E_1, \neg E_1, \overline{S})$ exerts a non-zero degree of prob-influence on E_2, it does not count as *backtracking* prob-influence because neither E_1 nor $\neg E_1$ termine any events toward the past that can in turn termine events toward the future non-redundantly. One should instead understand this example as a case of future-directed prob-influence that exists by virtue of the future-directed fixing that issues from $\overline{S}$ together with the conditionalization on E_1 (contrasted with the conditionalization on $\neg E_1$). It requires no past-directed terminance relations. Because this kind of prob-influence nevertheless corresponds to some people's ordinary

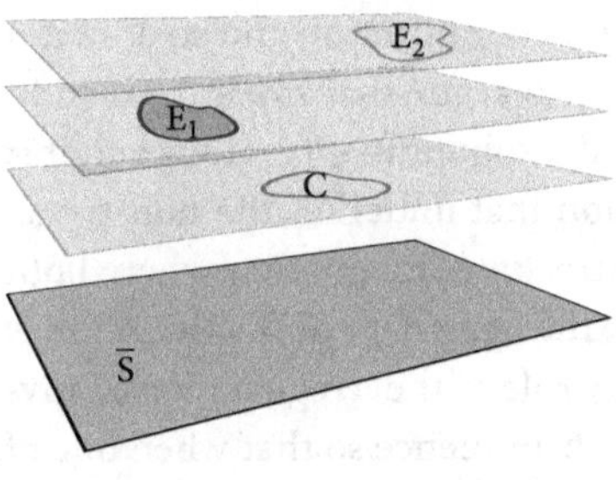

FIGURE 6.4 *Prob-influence exerted by a spatio-temporally disconnected event.*

use of backtracking counterfactuals, I will hereby dub it '*pseudo-backtracking prob-influence*'. This pseudo-backtracking prob-influence constitutes a legitimate form of influence that holds between two effects of a common cause by virtue of their common cause, although it would be misleading to call it bona fide *backtracking* influence.

In the end, whether E_1 affects the probability of E_2 depends on whether you render it (1) as a contrastive event at a single time or (2) as part of a temporally disconnected contrastive event or irregular contrastivization. There is no unique answer to the question of whether E_1 affects the probability of E_2.

My main reason for discussing this example is that it uncovers a hidden ambiguity in the idea of an effective strategy. When the agent chooses yellow, is she effective at bringing about the raising of the yellow flag? According to my theory there is no univocal answer. There is one sense in which the agent's choice of color is effective at raising the flag of that color. Whenever events start off like a typical instance of $\overline{S}$ and the agent chooses yellow, the flag that is raised is yellow more often than green. There is another sense in which the agent's choice of color is ineffective. If you hold everything fixed at the time when the agent chooses except what is happening in her brain, then her choice will make no difference as to which flag is raised. Furthermore, if you randomly choose a color (after the brain correlator has acted) which the agent as a rule recites unthinkingly, then her choice will again make no difference as to which flag is raised. These three possibilities will be converted into experimental schemas in §7.4.

That there is no ultimate fact of the matter about whether the agent's choice affects which flag is raised is not a problem for my account of causation because the distinction is unneeded for explaining the difference between effective and ineffective strategies, and the method of empirical analysis advises us not to worry about having a unique correct answer in cases where the empirical data does not demand it.

A good reason for treating the probabilistic relationship between E_1 and E_2 as causal is simply that it follows straightforwardly from the definition of partial influence. Even though the irregular contrastivization of E_1 that is responsible for the partial influence occupies a disconnected region, it implies probabilities for other events by virtue of the fundamental laws and so should count as genuinely prob-influencing. Furthermore, keeping it on equal footing with contrastive events that occur at a single time generates no significant deficiency in the account of causation nor any suboptimality in the conceptual organization and so does not warrant contrivances to rule it out. One might try to argue that my introduction of the idea of an irregular contrastivization is itself a suboptimality. However, it is not the invocation of an irregular contrastivization that underlies the non-trivial 'past-directed then future-directed' pseudo-backtracking. In any case where both E_1 and $\neg E_1$ are nomologically compatible with $\overline{S}$, the irregular contrastivization is equivalent to a regular contrastivization. The only role of the irregular contrastivization construction is to extend the notion of prob-influence so that when one of the coarse-grained events is nomologically incompatible with $\overline{S}$, it will have zero probability rather than an undefined probability.

I will now address two potential objections to my assertion that whether the probabilistic relation from E_1 to E_2 is causal depends on how you contrastivize E_1. The first objection claims that there needs to be a fact of the matter as to whether E_1 affects the probability of E_2 because of some issue beyond explaining empirical phenomena. A good example is the debate between causal and evidential decision theory. Decision theory is in the business of evaluating and modeling what is rational. It is not in the business of making empirical predictions about causal behavior. In order for the debate between causal and evidential decision theory to make sense, though, there needs to be some distinction between which relations are causal and which are not. Although my account does not draw the line between causal and non-causal in the way it is traditionally conceived in discussions of causal decision theory, nothing in my account absolutely demands abandonment of the traditional distinction. One can, if one likes, just define a decision-theoretic notion of causation in terms of what events prob-influence when they are contrastivized in ways that do not incorporate events at other times. Then, the relation between E_1 and E_2 will count as non-causal. Personally, I see no reason to maintain the traditional distinction, and I think the natural approach toward decision theory from the standpoint of my metaphysics of causation involves accepting that at least some correlations that have traditionally been construed as merely evidential are secretly causal after all. Consequently, the distinction between causal and evidential decision theory needs to be revised in light of the more flexible conception of causation provided by my account. Because I do not have the space here to initiate an empirical analysis of decisions, I will just remind readers that strictly speaking, my account of causation is neutral on issues of causal decision theory.

The second objection claims that there needs to be a fact of the matter as to whether E_1 affects the probability of E_2 in order to explain the empirical phenomena associated with effective strategies. The worry is that if my account is unable to identify the relation between E_1 and E_2 as non-causal, then it cannot adequately substantiate the distinction between effective and ineffective strategies. After all, a common-cause correlation was the example originally proposed in Cartwright's (1979) essay to illustrate the difference between effective and ineffective strategies. Fortunately, I have a way of deflecting this criticism. There is an alternative way to draw the distinction between effective and ineffective: a certain distinction between exploitable and unexploitable influence. My strategy in the next chapter, in §7.4, is to demonstrate that 'future-directed then past-directed' influence is generally useless for advancing goals and thus is unexploitable. That constitutes an explanation for why an agent cannot *intervene* in nature in order to exploit a common-cause pattern that is external to that agent. Furthermore, it will turn out that in the same sense of 'exploitation', a common-cause pattern that incorporates the agent's own action as one of the effects is also unexploitable. So, in general, common-cause patterns are not exploitable routes of influence (from one effect to another). However, the explanation I provide is compatible with my conclusion that when an action is itself one of the effects of a common cause, the agent does probabilistically affect other effects of the common cause by virtue of the agent's fundamental nomic connection to the common cause. In order to address this topic adequately, though, we must first tackle the problem of causal asymmetry.

{7}

Causal Asymmetry

Causation is associated with a multitude of asymmetries,[55] one of which is the idea that causal relations are never directed toward the past. In this chapter, I will attempt to analyze this conjecture about causation using the method of empirical analysis. Remember that the ultimate goal of an empirical analysis of *X* is to assist in the scientific explanation of whatever empirical phenomena motivate our having a concept of *X*. An empirical analysis of the metaphysics of causation demands that we investigate empirical phenomena that motivate our belief that causation is temporally directed.

To initiate the exploration, I will describe an experiment that I believe adequately captures at least one core part of the empirical content of the directionality present in our ordinary conception of causation. Then, I will use the theoretical machinery developed earlier in this book to formulate an explanation of the experimental results. Next, I will spell out the consequences of the explanation for pseudo-backtracking prob-influence in order to complete the discussion from the previous chapter concerning the possibility of influence being routed through a common-cause pattern. Then, I will relate my account of the causal asymmetry to the famous asymmetry of entropy and to the fork asymmetry. Finally, I will address whether it is best to explain the direction of causation by positing a fundamental direction of influence.

Remember that if events only contribute in one direction of time, that direction constitutes the fundamental future. Because it is trivial to explain the direction of causation when there is such a fundamental future, we can turn our attention to the more difficult task of exploring whether the apparently asymmetrical character of causation can be explained even when there is no relevant temporal asymmetry in fundamental reality. Although some people believe that the past is not at all susceptible to influence, I believe that past-directed influence not only makes sense but also that the widespread existence of past-directed influence in our everyday lives is a plausible hypothesis. So, in this chapter, I will continue to operate under the assumption that past-directed influence, promotion, bringing about, etc., are just as legitimate as their future-directed versions. I caution

---

[55] See Hausman (1998) for an extensive discussion of eleven causal asymmetries.

readers to avoid importing the asymmetrical temporal connotations of ordinary language into the discussion.

7.1 The Empirical Content of the Causal Asymmetry

The empirical analysis of the metaphysics of causation discussed in this book is targeted at the cluster of phenomena related to the existence and character of effective strategies. The asymmetry most straightforwardly relevant to 'effective strategies' appears to be the following. Some strategies are effective for bringing about future events, but no strategies are effective for bringing about past events. In this section, I will attempt to clarify this portion of the empirical content of the causal asymmetry, which I call the '*advancement asymmetry*'. The advancement asymmetry is the apparent fact that no matter what past event *E* you care to consider, agents who attempt to have *E* occur do not generally get *E* to occur any more or any less often than similarly placed agents whose goal is to have *E* not occur. If this advancement asymmetry can be adequately explained, that will account for why it is reasonable for us to construe the past as settled, even if the past is susceptible to influence from the present.

Consider the following schematic experiment called the '*asymmetry experiment*'. The asymmetry experiment is a slightly modified form of the promotion experiment from §5.1. Like various versions of the promotion experiment, it involves comparing the statistics from two different sets of experimental runs, each of which contains a zillion runs.

- For each experimental run, there must be at least one agent in the experimental run who counts as *the* agent of the experimental run. Ideally, each of these runs would be completely isolated from one another in the sense of having no physical interactions with each other, but in practice we make do by just preventing the agents from interacting with each other in any way that could plausibly make a difference to their ability to advance their goals.
- For the experiment as a whole, there exists a single description of a coarse-grained event, *E*, which is intended to represent an event that an agent might be asked to bring about or prevent. *E*'s location in the arena is specified relative to a starting time, START, that occurs during the experimental run. For simplicity, let *E* be constrained to lie either entirely to the future of START or entirely to the past. (If we can successfully account for this special case, we can account for events that are partly to the past, partly simultaneous, and partly to the future by breaking them up into past, present, and future subevents, and handling each separately.)

- Each agent is aware at START of the description of *E*. (The agents are allowed to be aware of the description of *E* throughout the experimental run.)
- Each agent is randomly assigned as a member of one of two teams: DO or DONT. Each team has a zillion members. The allowable methods by which the assignment is randomly generated can include many different kinds of processes already familiar to us. The experimenters can roll dice, or flip coins, or use a pseudo-random number generator from a computer, or extract a random bit using the microwave background radiation left over from the early universe. It is part of my prediction that the results of the experiment will be insensitive to the precise mechanism used to assign agents randomly to their teams.
- The agents *become* aware at START of which team they have been assigned to. The agents are not provided any information about which team they are on until START. They learn of their team by being sent some sort of message that arrives at the START time. The transit of the message does not significantly affect what happens in its environment as it travels to the agent.
- Each agent on team DO will be personally greatly rewarded iff *E* occurs in his experimental run.
- Each agent on team DONT will be personally greatly rewarded iff *E* does not occur in his experimental run.
- A team's score is the total number of times that *E* occurs in the experimental run.
- There are no rewards for how much the team scores or for one team scoring more than the other. The agents are not motivated to act for the sake of the team score. They are just trying to get their own reward.
- The agents are aware of all these rules to the extent they can be aware of them.

Summarizing informally, each agent knows of a possible event *E* and knows that he will be rewarded either for *E*'s occurrence or for *E*'s non-occurrence, but the agent does not yet know for which outcome he will be rewarded. Some time goes by, and then at a certain START time for each agent, the agent becomes aware of which team he is on, which indicates the outcome for which he will be rewarded. Then, the agent is free to try to make *E* occur, which he will want to do if he is on team DO, and is free to try to make *E* not occur, which he will want to do if he is on team DONT. Note that even though the members of team DONT want to stop *E* from occurring, the score of team DONT goes up when *E* occurs, so one can think of DONT's members as wanting to act in a way that has the effect of keeping the team's score low.

My bold prediction for the outcome of any asymmetry experiment is this:

- When *E* is later than START, team DO will sometimes outscore team DONT (depending on *E*).
- When *E* precedes START, teams DO and DONT will always have the same score regardless of *E* except for discrepancies attributable to the ordinary sources of error that occur in scientific experiments.

Here are some examples to motivate the prediction. Let *E* be, "The agent eats a slice of pizza sometime during the next three hours." The agents are just ordinary people on Earth. You randomly select a zillion of them and tell each of them (truthfully) that you will give him or her a sack of gold for eating a slice of pizza within the prescribed time, and you tell the other zillion to avoid eating pizza during that time period for a reward of a sack of gold. You monitor them with video cameras to keep track of whether they eat any pizza. Presumably, lots of people on the DO team will eat a slice, and many fewer people on the DONT team will eat a slice. So, team DO will outscore team DONT.

Now, repeat the experiment with *E* being, "The agent eats a slice of pizza sometime during the previous three hours." You monitor the behavior of two zillion people for four hours by video camera, you randomly assign them into DOs and DONTs, and then you inform them of their team at time START. Members of the DO team are informed that a sack of gold will be handed over next week to any of them who has eaten a slice of pizza during the three hours preceding START. Members of the DONT team are informed that a sack of gold will be given next week to any member who has not eaten a slice of pizza during the three hours preceding START. Then you wait a week or so in order to give them plenty of time to fiddle with the past as much as they like. My prediction is that there will be virtually identical numbers of DOs who ate a slice as DONTs who ate a slice. And the intuitive reason will be just that—even if they knew their task was going to involve either eating or not eating a slice of pizza—they were assigned the tasks randomly and did not know which task they would be rewarded for until after the opportunity to eat or to forego a slice had passed. That, at least, is how someone might explain the result, but I will offer a superior explanation in the next section.

Someone might think that there are some past goals one can have that are easy to achieve. For example, it is easy to succeed in one's goal of having been born in the past. However, that does not imply that one's present 'attempt at having been born' raises the probability of having been born in the past in the sense relevant to the asymmetry experiment. Let *E* be, "The agent is born sometime in the past." In that case, we get a zillion agents who are assigned the task of trying to be born previously and a zillion agents who are assigned the task of trying to avoid having been born. Presumably, when you conduct the experiment on human agents, all of them will turn out to have been born previously, so the score will be a zillion to zillion tie. That is evidence for the claim that people are unable to advance the goal of being born or the goal of not being born.

Note that the conception of agency employed in the description of the experimental design is extremely liberal and flexible. It can even include primitive devices such as thermostats. Imagine that the zillion thermostats on the DO team are trying to get the room to be at least 30 degrees and that the zillion on the DONT team are trying to get the room to be below 30 degrees. What instantiates a thermostat's "being informed that it is on team DO" is simply that its temperature setting is adjusted to be at say 35 degrees, and what instantiates its "trying" to reach its goal is merely the ordinary mechanical operation of a thermostat and whatever heating and refrigerating devices are connected to it. One of the desiderata for an adequate empirical analysis is that its concepts exhibit graceful degradation. Specifically, the operative concept of agency needs to be such that as an object continuously becomes less and less like an agent, its role in the explanation becomes less and less important. One can see this feature in the experimental design of the asymmetry experiment. For example, as we take the limit going from agents to non-agents, the results of the experiment will be such that there is never any past-directed advancement, but toward the future the variety of target event-kinds for which evidence of advancement accrues will decrease as one considers ever less intelligent or less capable agents.

I am fairly confident that my bold prediction concerning the asymmetry experiment is correct as stated even though I have never actually conducted any version of the asymmetry experiment.[56] However, it is slightly bolder than it needs to be in order to explain the reasonability of the dictum that the past cannot be influenced. If there were extremely subtle, hard-to-implement ways in which we could usefully influence the past, say by exploiting heretofore unknown properties of dark matter, the mere difficulty of applying such processes to ordinary events would still ground the utility of our thinking of the past as settled. Also, it is possible that the fundamental laws permit the kind of time travel scenarios where a macroscopic process, such as the world tube of a human being, occurs along a closed time-like curve. If such a case actually occurred, it would be compatible with the existence of an advancement asymmetry that held locally at each temporal stage of the time traveling process, even though it would not hold globally (by hypothesis). It is possible that the kind of consistency constraints imposed by closed time-like curves will make violations of my prediction for the advancement asymmetry possible, but my prediction is only intended for normal circumstances where there is no time traveling to the past. Although these considerations suggest that a prediction slightly weaker than the one I formulated is sufficient to vindicate our treatment of causation as temporally directed, I think these potential counterexamples to my prediction are remote enough that it will be convenient for the sake of discussion just to maintain the simpler working hypothesis that the

[56] I am currently seeking the funding needed for such experiments.

advancement asymmetry holds across the board. At least, I am confident that if anyone on Earth conducts the asymmetry experiment with the kind of technology we will have within the next millennium, the bold prediction will be confirmed in spades.

Presuming my prediction holds true, the results one gets from running various incarnations of the asymmetry experiment constitute the empirical content of the advancement asymmetry. It tangibly demonstrates that we are sometimes able to affect the future in a way that advances our goals but that whatever influence we exert on the past is unexploitable for the purpose of advancing our goals. That in turn suffices to explain why it is understandable for humans to conceive of influence as existing only toward the future.

7.2 Causation and Advancement

On the one hand, it is potentially misleading to talk of *the* asymmetry of causation as if there were a single asymmetric feature of reality that by itself vindicated the totality of platitudes associated with causal asymmetry. On the other hand, I do not think there is an unmanageably large and disparate set of unrelated causal asymmetries. It proves convenient to segregate the numerous platitudes that constitute our conception of causal asymmetry into two roughly defined sets. First, there are platitudes associated with agency or influence or the settledness of the past or (one might say) causal dependence. These include the principle that it is impossible to affect the past and the principle that effects do not make their causes happen. Second, there are platitudes associated with various material asymmetries, exemplified in the deterioration of living organisms, the diffusion of gasses, the flow of heat from hotter objects to colder objects, and the radiation of electromagnetic fields away from accelerated charges. One can even include more esoteric asymmetries related to quantum-mechanical measurement processes and the decay rates of neutral kaons. I will refer to these two collections of platitudes as the influence-based set and the pattern-based set.

My bifurcation of the platitudes concerning causal asymmetry is meant to parallel Michael Dummett's (1964) suggestion that intelligent plants would still have use for an asymmetric notion of cause despite their postulated inability to advance goals. It is perhaps inaccurate to conceive of intelligent plants as non-agents because in order to reason, they presumably need to have some ability to shift attention from one topic to the next, but quibbles aside, I agree that the traditional language of cause and effect would be useful for characterizing the many material asymmetries even if we set aside all issues related to the influence-based set of platitudes. For example, correlations holding among mundane events, such as rainstorms and barometer readings, often fail to hold after one has conditionalized on the existence or non-existence of an appropriate event from the past, such as the arrival of a low-pressure front. It is often much harder to find an appropriate mundane event from the future which will similarly screen the correlation. This asymmetry was formalized by Hans Reichenbach (1956) in terms of *conjunctive forks*, and the resulting asymmetry is known as the *fork asymmetry*. Explaining the fork asymmetry, so far as I can tell, does not require any commitment to the principle that events cannot influence the past or that the past is fixed. It merely involves correlations among instances of designated event-kinds and their relative locations in space-time, and thus should be associated with the pattern-based set. Whether the fork asymmetry should be construed as *causal* depends on how liberally one wants to define the term 'causal'. Reichenbach himself spoke of the "principle of the common *cause*" when discussing this very issue (Reichenbach 1956, §19).

Although it is a terminological quibble, my own inclination is to say that principles that concern only the pattern-based set of platitudes are not fully causal because they are not related closely enough to any reasonable conception of influence, whether in the sense of contribution or in some sense of difference-making. The motivation behind my choice of conceptual organization here is that the standard conception of probability used to formulate the common cause principle is one where probabilistic relations between localized events hold (or at least we are justified in believing they hold) because (like Dummett's intelligent plant) we observe certain patterns in the universe, not because of any commitment to underlying fundamental dynamical laws that might connect them in some productive or process sense of causation. Without some component of fundamental reality linking the events in such a way, there is no clear reason (that I can see) why counterfactually altering one localized event should have implications for what happens elsewhere.[57] Nothing I am claiming implies that principles like the fork asymmetry are unimportant or wholly unrelated to influence. I am merely judging that such principles are better categorized together with other material asymmetries that can be adequately characterized without recourse to notions of influence.

There are also examples of platitudes that arguably do not fit neatly into exactly one of the two sets I identified. Consider the maxim that effects never precede their culpable causes. One explanation for the reasonability of this maxim is that because of the general lack of any useful past-directed influence, we incorporate an implicit rule in our conception of culpable causation that excludes any event C from counting as one of the causes of E whenever E is located to the past of C. However, other explanations for the asymmetry of culpable causation are available that do not depend on the advancement asymmetry. Dummett's intelligent plant might come to appreciate the fork asymmetry and to think of events that

[57] Admittedly, counterfactual theories of causation like Lewis (1973b) have been designed to be compatible with a Humean approach toward causation that is fundamentally pattern-based, but such theories have been constructed to reproduce the kind of counterfactual relations one gets by using the dynamical laws of physics, in effect by mimicking forking accounts of counterfactuals. Readers may want to consult the supplementary readings I have provided to follow up on this topic.

screen off probabilistic correlations as explanatory. Because causes play a role in explaining how nature evolves, it is not crazy to call these explanatory events 'causes', and to assign the label 'common cause' to events whose occurrence and non-occurrence screen correlations, and to think of the direction of causation in terms of the fork asymmetry. The larger point here is merely that the array of principles implicit in our intuitive conception of causation are unsophisticated enough for there to be no clear fact of the matter about whether the principle "Effects do not precede their causes" stems primarily from the advancement asymmetry, or from some material asymmetry like the fork asymmetry, or from some mixture of the two.

Any decent explanation of the results of all asymmetry experiments will automatically explain why the advancement asymmetry holds. That in turn will explain why it is handy for just about any kind of creature to conceive of the past as not at all influenceable. And from there, one can leverage this conception of the past in order to formulate explanations of other asymmetries related to the influence-based set of platitudes. I believe a successful explanation of the advancement asymmetry is sufficient to account for the asymmetry of causation insofar as we are merely concerned with the influence-based set, though it will not suffice to explain the pattern-based set. Unfortunately, I cannot provide a rigorous argument in defense of my conjecture that the asymmetry of advancement suffices to account for the influence-based set of platitudes because there is no rigorous way to identify the platitudes constitutive of our notion of causation, nor is there a principled rule to separate them into two wholly distinct sets. As I just noted, the distinction between the two sets of platitudes is at best rough. My belief that the advancement asymmetry is all we need to explain in order to account for the influence-based set of platitudes is partly a consequence of (1) my having adopting empirical analysis as my method of inquiry so that there is no need to construct a metaphysics where the past is settled and immune to influence and partly a consequence of (2) my failure to imagine any empirical phenomenon that could bear on principles like "the past is fixed," "the past cannot be influenced," and "events cannot *make* previous events come about," other than through its bearing on the results of the asymmetry experiment. I do believe, however, that material asymmetries bear on a proper explanation of the advancement asymmetry, as I will demonstrate in the next section.

7.3 An Explanation of the Advancement Asymmetry

The explanation for why an agent is able to promote or inhibit certain *future* E's after learning which team it is on is just a straightforward application of everything discussed in chapter 5. So, in this chapter, we only need to concern ourselves with the case where the agent is motivated either to promote or to inhibit a *past* event E.

In this section, I will attempt to explain why agents are generally unsuccessful at affecting the past in ways that advance their randomly assigned goals.

Before presenting the detailed explanation, a summary of the overall explanatory scheme is in order. A crucial but easily forgotten lesson, emphasized by Price (1996), is that an asymmetric conclusion can never be derived from purely symmetric assumptions. Although it is difficult to ascertain the ultimate source of asymmetry in the asymmetry experiment, I believe that whatever ultimately accounts for the asymmetry does so through the principle of future-typicality, which explains through two different routes. Future-typicality helps to explain (1) the material asymmetries that are needed to get the asymmetry experiment up and running and (2) why agents cannot achieve their randomly assigned goals directly toward the past.

First, the paucity of bizarre coincidences as nature evolves toward the future arguably has some role in explaining several temporally aligned material patterns that are essential to the possibility of setting up an asymmetry experiment. To implement any asymmetry experiment, there needs to be an experimenter who initiates the experiment setup in a future-directed process. The selection of the agents, their random assignment to teams, and the delivery of the messages to them must all take place through processes that evolve toward the future. It is crucial to the experimental outcome that all the agents being tested in the experiment are temporally oriented in the same direction as each other and in the same direction as the experimenter so that the actions issuing from the agents are all to the future of the agents' processing of their goals and information. Also, the agents are presumably integrated into their environments in the ordinary way; they are not constituted by highly localized anti-thermodynamic fluctuations that instantiate a past-directed agent embedded in a normal future-directed environment. As noted in §5.8, the lack of bizarre evolutions toward the future plays some role in explaining the existence of these matter asymmetries. A proper explanation of why there are agents at all and why they are able to understand rewards and goals and strategies certainly requires citing much more than the non-existence of bizarre evolutions toward the future. Several factors were discussed in §5.8, and they do not need to be repeated here.

Second, future-typicality also bears on the results of the asymmetry experiment at the stage where the experimenter randomly assigns teams and sends information about the team assignment to the agent. Because the selection is effectively random, the lack of bizarre coincidences going toward the future ensures that on the whole, there will not be any correlations between which team is assigned to an agent and what is happening at the beginning of each agent's experimental run. I will attempt to spell out the explanatory role of this lack of correlations in the following subsections.

I will first address the possibility that we can advance our goals for the past by influencing what happens in the future. Then, I will address the possibility that we can advance our goals for the past in a manner that is entirely past-directed.

7.3.1 PROB-INFLUENCE THROUGH BACKTRACKING

Insofar as we understand the action of an agent as a future-directed decision and action process, it cannot prob-influence the past non-redundantly because that would flout the principle of causal directness.

Imagine the experimenter has just completed randomly assigning teams and is sending a signal to inform each agent of his or her team. Let us consider one of the agents being sent the message that she is on team DO, as depicted in Fig. 7.1. A signal comes in from the left and is just about to arrive at the agent at the START time. We can contrast this situation with what would have happened if the signal indicated team DONT. Insofar as we represent these occurrences as a pair of contextualized events, they should be virtually the same everywhere before START except in the tube representing the path of the signal. There can be slight differences owing to any subtle interactions between the signal and its environment, for example gravitational tidal forces and any chaotic ramifications thereof in the microstates.

For later convenience, let us define a prominent foreground and prominent background of any contrastive event $\tilde{C} \equiv (\overline{C_1}, \overline{C_2})$. The *prominent foreground* of $\tilde{C}$ is where $\overline{C_1}$ and $\overline{C_2}$ differ beyond a negligible amount. (Judgments of negligibility are imprecise, non-technical, and somewhat flexible.) The *prominent background* of $\tilde{C}$ is $\tilde{C}$'s region excluding its prominent foreground. The purpose of distinguishing the prominent foreground is to set aside any insignificant differences outside the path of the signal resulting from the signal being DO rather than DONT.

The relevant class of contrastive events that can serve as legitimate representations of the experiment all have a prominent foreground occupying the region depicted in Fig. 7.1 as C_{DO}. Our task will be to consider whether some reasonably chosen contrastive event $\tilde{C} \equiv (\overline{C_{DO}}, \overline{C_{DONT}})$ can promote E in a manner that becomes empirically evident when the corresponding asymmetry experiment is conducted. The contextualized event $\overline{C_{DO}}$ represents how things would likely be

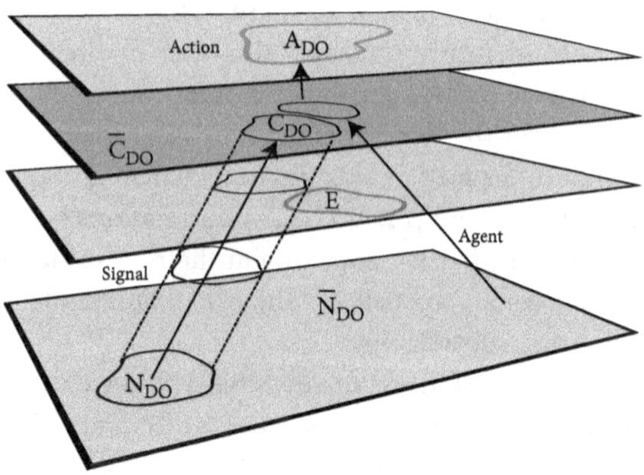

FIGURE 7.1 $\overline{C_{DO}}$ *shields the agent's actions from affecting E.*

at the START time if the agent were selected for team DO. Then, $\overline{C_{DO}}$ fixes everything after START, which represents what the agent will probably do in response to learning she is on team DO. In the figure, the agent is depicted as performing action A_{DO}. This can be compared to what $\overline{C_{DONT}}$ fixes for the future, which might be quite different because the agent could behave macroscopically differently in response to being on team DONT rather than team DO.

There also might be additional constraints imposed by nature on the states instantiating the experiment. For example, many contextualized events can represent the reception of DO rather than DONT, but only states that are compatible with the low entropy of the early universe can be actual. Instead of considering which particular constraints are appropriate, we can just leave that question open and define $\overline{C_{DO}}$ to be any contextualized event at START that fits the description of the experiment (with the agent being aware of the event-kind E and knowing the rules) and implements the arrival of a message that the agent has been assigned to the DO team. It can incorporate any additional constraints that are appropriate. Because the team assignments are selected by way of some localized random process, the appropriate $\overline{C_{DONT}}$ should be very nearly the same except that the message says the agent has been assigned to team DONT. In particular, it should incorporate the same constraints that were applied to $\overline{C_{DO}}$.

In order to facilitate my presentation of a reductio ad absurdum in the next subsection, let us now imagine the existence of some set of agents—call them 'the clever'—who know how to exploit their influence over the past in order to affect the probability of at least some target events. For their existence to count as a falsification of my prediction (that teams DO and DONT will be tied when E precedes START) there needs to be some feature they possess whose existence is verifiable independently of the results of the asymmetry experiment. It does not have to be some skill they can recognize or teach to others, but we cannot identify them retrospectively as clever merely on the grounds that they received their reward on numerous occasions. That would wrongly credit them with know-how if they were just lucky. They at least have to be able to demonstrate their ability reliably in numerous repeated runs of the same asymmetry experiment. If these clever agents do exist, it is appropriate to choose a contrastive event $\tilde{C}$ that is constrained to implement clever agents. This will serve to magnify the observed size of their alleged ability to advance goals for the past instead of having their abilities remain unrecognized amid all the statistical noise generated by us less talented agents.

The most straightforward way to think about an agent's action bringing about an effect E is to construe it as a future-directed process. Before the action takes place, the agent is not even thinking about E, and then at some point the agent turns her attention toward E, and then reasons about how best to bring about E, and then initiates some motion in order to raise the probability of E. The extent to which such a process can be said to promote E depends on this process promoting (toward the future) the agent's full action which then leads toward the hoped-for event E. But if we interpret actions in this way, it immediately follows from causal directness that the initial stages of the action going from $\overline{C_{DO}}$ forward in time through A_{DO} cannot do anything that is not already being done directly toward the past by $\overline{C_{DO}}$, which doesn't even instantiate the agent's knowing whether she should be trying to bring about E or prevent E.

7.3.2 DIRECTLY PAST-DIRECTED PROB-INFLUENCE

The previous section demonstrated that an agent cannot exert goal-advancing prob-influence on the past by virtue of some future-directed nomic connection that continues on to affect some previous time. However, we could instead interpret the agent's action as taking place directly toward the past. Consider any time slab instantiating an ordinary future-directed action, stretching far out into space so as to permit prob-influence over reasonable time scales. But then, re-conceive of the evolution of that action as proceeding from the future toward the past. Can such an action ever raise the probability of some past E? According to my account, it can. However, the kind of past-directed promotion that exists is worthless for advancing the agent's assigned goal in the sense that it will not be revealed by any asymmetry experiment.

As justified in the previous subsection, the relevant class of contrastive events that can serve as a legitimate representation, $\tilde{C}$, of the state of the experiment at START all have a prominent foreground occupying the region depicted in Fig. 7.1 as C_{DO}.

I will allow that an appropriately chosen $\tilde{C}$ might promote the previous E, but I will conclude that any promotion cannot be revealed in the actual world because such a result would violate future-typicality.

Suppose for the sake of reductio that we have conducted the asymmetry experiment a couple zillion times using the clever as agents, and the result is that, for any of the agents, there is a strong correlation between the random team choice and the instantiation of some target event-kind E that precedes START.

In order for the asymmetry experiment to have been conducted properly, there needs to be a process leading from the device that randomly chooses DO or DONT to the agent at START as well as to outside observers so that they can monitor the results. Likewise, there needs to be a process leading from the occurrence or non-occurrence of E in its prescribed location to some sort of recording of whether E occurred. Furthermore, neither of these processes are allowed to interact in any way that would alter what information they carry, for example by having the agent's accomplice observe whether E occurs and then change the record of the team assignment or selectively interfere with the experimental run. So let us select, from each experimental run, some full and sufficiently large state that occurs before START and after there is a record of the randomly selected team and a record of whether E occurred. Let S be the complete collection of these actual states from all the experimental runs. The members of S will each have a prominent foreground with at least two components occupying non-overlapping regions. One of the components instantiates some record of the random selection, DO or DONT. The other component instantiates a record of whether *E* has occurred.

The assumed demonstrated success of the clever implies that, among the members of *S*, what is instantiated in these two disconnected regions will be strongly correlated. *E*'s will occur more often in states with DO than in states with DONT. Because the device that selected the team assignment is required to be a paradigmatic random device, its output and hence the recording of that output should not be correlated with some independently chosen target event. Instead, the random selection should so sensitively depend on a vast number of microscopic variables that can only be correlated with the independently chosen *E* if fundamental reality correlates them by way of a conspiratorial development of matter. This is one sort of bizarre evolution that is definitely ruled out by future-typicality.

The situation here is analogous to the example from §5.3, where each bird reliably lands on the one perch out of a million that happens to have been randomly selected. Future-typicality by construction must rule out any correlations between the birds and the random device, and similarly any correlations between *E* and the random device. Ultimately, whether our world is one where future-typicality holds is a matter of empirical investigation, not stipulation, but it is an extremely uncontroversial hypothesis. As John Bell (1981) noted, a world with such "correlations would be even more mind boggling than one in which causal chains go faster than light. Apparently separate parts of the world would be deeply and conspiratorially entangled."[58]

In order to relate my account of past-directed influence back to the discussion of broad and narrow prob-influence from §5.6, I will now offer a simplistic gloss concerning our ability to influence the past. Many people think that the reason we cannot usefully influence the past is that we cannot influence the past at all. On my account, however, we do influence the past whenever the fundamental laws bestow past-directed relations of terminance. In fact, our influence over the past is much more extreme in magnitude and scope than our influence over the future when the dynamical laws are deterministic in both temporal directions. In such cases, virtually everything that happened in the past depends much more sensitively than future events on our present (fine-grained) behavior. For any actual (fine-grained) action we undertake, if it were hypothetically altered to instantiate a reasonable alternative action, the fundamental laws would virtually ensure a vast expanse of anti-thermodynamic behavior in the past. Although all of us possess this ability to make the past behave anti-thermodynamically, we cannot prove that we have this ability by demonstrating it because the actual history of the world is pro-thermodynamic. That is, we can exert broad promotion over the past to

[58] That Bell was discussing *quantum-mechanical* correlations when he made this statement is irrelevant to the plausibility of future-bizarre evolutions. The crucial factor is that our random devices are coupled extremely sensitively to a vast number of particles.

virtually ensure it behaves bizarrely, but we are unable to exploit this promotion because of the future-typical environment in which we are trapped.

7.3.3 SUMMARY

The two main principles I invoked to explain the results of the asymmetry experiment are causal directness and the future-typicality of actual states. Causal directness was critical because my argument would not have worked if the agent's action A_{DO} could hop back in time to E via some fundamental law that violates shielding. Future-typicality played a role too. My prediction for the asymmetry experiment can fail when the material content of the universe is laid out in a special way that defies what would be expected from the rules of chance. Continual repetition of the asymmetry experiment could result in agents on team DO reliably following certain kinds of past E's more frequently than agents on team DONT. Yet, the empirical phenomena needed to disconfirm my prediction for asymmetry experiments are in a class that future-typicality rules out. It is helpful to recognize that future-typicality is logically stronger than what is needed to explain the results of asymmetry experiments. Future-typicality rules out all sorts of bizarre evolutions toward the future, not just bizarrely coincidental correlations between the random generation of DO or DONT and the occurrence or non-occurrence of the target event-kind E.

The setup of the asymmetry experiment incorporates an essential temporal asymmetry. The experimenter informs the agents of their team assignments in a *future-directed* manner. The presupposed temporal asymmetry is evident in my use of the temporally directed word 'becomes' in the rule that agents *become* aware of their team assignment only at START. Our ability to identify one direction of time as the future in turn depends on material asymmetries, as discussed in §6.1. So the required temporal asymmetry is built into the behavior of the human beings (or other agents) who are conducting, or are subjects in, an asymmetry experiment as well as the broader environment needed to run the experiment.

Nevertheless, this temporal asymmetry does not beg the question about the possibility of past-directed advancement. That the past-edge of the fragment of history constituting the experimental run is constrained to include an incoming signal does not *by itself* imply that an agent cannot influence the past so as to further the randomly chosen goal sent to it. Testing my prediction does require the universe to avoid behaving so bizarrely that the asymmetry experiment cannot even be conducted, but it does not require the stronger principle of future-typicality to hold universally. The design dictates that the team assignments be made randomly in all the familiar ways: coins, dice, computers, etc. Whether that eliminates correlations between the team assignment and the target event-kind in the experimental run that occur before START is an empirical question, not a precondition of the experiment. For all we know, there may yet be some event-kind whose instances in asymmetry experiments are reliably and strongly correlated with the output of whatever random-number generators or coin flips or die rolls are used. That such a possibility is strikingly implausible is a good reason for thinking my prediction is correct but not a good reason for thinking the experimental setup begs the question.

The purpose of setting each agent's goal randomly and externally was to simulate a fundamentally indeterministic choice, supplying the agent with a goal whose existence was not the result of her history up until then. If we were judging agents without externally fixing their goals, the possibility would be open that an agent's having a goal for a past E would be correlated with the past existence of E without any interesting past-directed advancement. Readers who are familiar with the concept of an intervention from the literature on causal modeling approaches (Spirtes, Glymour, and Scheines 2000; Pearl 2000; Woodward 2003) should recognize this external fixing of the goal as an intervention into the agent's behavior. So, one way to think about the explanation I have provided is that it explains why the kinds of processes we ordinarily recognize as interventions—all of which are future-directed—are useless for affecting the past to advance our goals. Furthermore, this result is compatible with the hypothesis that the past depends counterfactually and causally on the present. For reasons I will discuss in §10.2, I am not providing a formal definition of an intervention in terms of the machinery of my account.

The argument in this section has demonstrated that the asymmetry experiment can be explained without any fundamental asymmetry. The asymmetry of bizarre coincidences is not a fundamental asymmetry, and conjoining it with the (temporally symmetric) principle of causal directness suffices for the conclusion that agents who try to advance randomly assigned goals concerning past events reliably do no better and no worse at advancing those goals regardless of what they try to do.

7.4 Pseudo-backtracking Prob-influence

In §6.4, I defended the conclusion that there is a bona fide form of influence that exists between some E_1 and some E_2 by virtue of a common-cause pattern (where they are intuitively two effects of a common cause C). This influence is pseudo-backtracking prob-influence. The previous section implies two notable consequences for such influence.

First, one cannot exploit a common-cause pattern for achieving goals "from the outside" by meddling with E_1 in order to advance one's goals for whether E_2 occurs. Recall the example of §6.1, where an infection promotes an itch and a fever. If we try to exploit this common-cause pattern by applying skin creme to make the itch less likely so that it will thereby make the infection less likely and thus the fever less likely, the actual world is guaranteed to reveal no evidence that such a strategy works. That is because it essentially incorporates past-directed advancement.

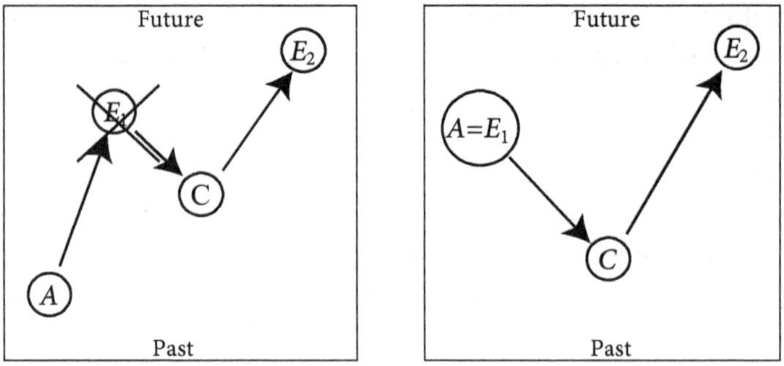

FIGURE 7.2 *The potential route of influence on the left cannot be exploited due to its past-directed part. The pseudo-backtracking influence on the right allows an agent to affect the probability of E_2 when E_1 is the agent's action, but it does not allow the agent to exploit that influence.*

Second, when the agent's action A is the effect E_1, the world does in one sense reveal the pseudo-backtracking prob-influence that links E_1 with E_2. E_1's are positively correlated with E_2's, and such correlations hold by virtue of the same kinds of nomic connections exhibited in all paradigmatic cases of causation. However, pseudo-backtracking prob-influence is not *exploitable* in the sense that if we randomly assign a goal to the agent, E_2 or $\neg E_2$, the actual world is virtually guaranteed to reveal that the agent's effort to achieve the goal does not correlate with success.

To investigate these consequences in more detail, let us return to the project of replacing the backtracking experiment we initially explored in §6.3 with three distinct experimental schemas that illuminate the structure of reality better. Remember that in §6.4, we considered an example where an agent interacts with a brain correlator and then chooses either green or yellow. If the correlator is sophisticated enough, the color of the raised flag will reliably match the color the agent selects. I claimed that there are two ways of construing whether E_1, the agent's choice of yellow, affects the probability of E_2, the raising of the yellow flag. According to the first, E_1 makes a difference; according to the second, it doesn't. I will encode these two interpretations in experiments B1 and B2, and then I will add a third experiment, B3, which represents attempts to *exploit* pseudo-backtracking influence to advance a randomly assigned goal.

Experiment B1: The goal of this experiment is to test the prob-influence exerted by $\tilde{E}_1 \equiv (\overline{E_1}, \overline{\neg E_1})$, where $\tilde{E}_1$ is a regular contrastivization of E_1 occurring at a single time (extending across the arena in a purely space-like way) where all the runs involving $\overline{E_1}$ instantiate agents choosing yellow; all the runs involving $\overline{\neg E_1}$ instantiate agents choosing green. In all other respects, $\overline{E_1}$ and $\overline{\neg E_1}$ are the same. In particular, we want to measure the degree to which $\tilde{E}_1$ prob-influences E_2, the existence of a raised yellow flag shortly after $\tilde{E}_1$.

My prediction is that the measured degree of prob-influence will always be (very nearly) zero.

Assuming this prediction is correct, my explanation is that causal directness prevents any non-redundant prob-influence from going first to the past and then to E_2. So the only way $\tilde{E}_1$ could prob-influence E_2 is in a purely future-directed way, say by having someone listen to the announced choice and meddle with the flag accordingly. This possibility is ruled out simply because the experiment is designed not to allow manipulation of the flag after the choice is made.

It is important when conducting this experiment that the instantiated backgrounds constitute a representative sample drawn from $\tilde{E}_1$. If you use a correctly functioning brain correlator to establish a correlation between the color of the raised flag and the agent's choice, and then use the resulting states to instantiate $\overline{E_1}$ and $\overline{\neg E_1}$, the initial conditions will be unrepresentative. It is a requirement of this experiment that $\overline{E_1}$ and $\overline{\neg E_1}$ agree on the condition of the flags at the time of the agent's choice. If a correlation between what $\overline{E_1}$ instantiates in the flag's region and what $\overline{\neg E_1}$ instantiates in the flag's region, the experiment counts as having been incorrectly conducted, and its results do not disconfirm my prediction.

Experiment B2: The goal of this experiment is to test the prob-influence exerted by the irregular contrastivization $(E_1, \neg E_1, \overline{S})$ at getting E_2 to occur. In this experiment, $\overline{S}$ is a condition shortly before the brain correlator interacts with the agent. The brain correlator is allowed to do its work, one of the flags is raised, and then some of the agents choose yellow while the others choose green.

My prediction for this experiment is that there might be non-zero correlations between the choice of yellow and the raised flag being yellow, the strength of the correlation depending on the quality of the brain correlator and the degree to which there are any interfering factors in the background environment that the brain correlator does not account for. To the extent that the agent's choice arises from a localized fundamentally chancy process after the agent has interacted with the brain correlator, there should be a corresponding weakening of the strength of the correlation, and in the limit where the agent's choice is determined by the result of such fundamental chanciness, there should be no correlation at all.

Assuming this prediction is correct, it can be explained in terms of the prob-influence exerted on E_2 by the irregular contrastivization $(E_1, \neg E_1, \overline{S})$. Owing to the fundamental laws, this contrastivization promotes the raising of the yellow flag to the extent that the brain correlator works as advertised.

Experiment B3: The goal of this experiment is to test the exploitability of the prob-influence exerted by the irregular contrastivization $(E_1, \neg E_1, \overline{S}')$ for getting E_2 to occur. We set up the experiment as in Experiment B2, but this time we include in the initial conditions, $\overline{S}'$, a device that will flip a fair coin that is green on one side and yellow on the other. The coin flip will occur after the brain correlator has completed its operation, and the coin flip and brain correlator are as isolated from each other as possible. Also, the agents have been informed (correctly) that they are free to observe the outcome of the coin flip before choosing and that they will receive a sack of gold for choosing the color that was the outcome of the coin flip. There is no additional reward or penalty for having a match between the chosen color and the color of the raised flag.

My prediction for this experiment is that no matter how effectively a brain correlator performs in Experiment B2, there will be (very nearly) no correlation between the agents' choices and the flag colors.

Assuming this prediction is correct, it can be explained by appealing to the general desire of humans to take easy money when they can get it, and to future-typicality, which rules out correlations between the outcomes of coin flips and the outputs of unrelated brain correlators. Because the brain correlator cannot be reliably correlated with the coin flip outcome without violating future-typicality or violating the requirement that the flip be generated by a chancy process, the brain correlator will fail to establish a correlation between the agent's choice and the flag color.

The presumed results of experiments B2 and B3 illustrate an empirical difference between pseudo-backtracking prob-influence and its exploitability. If the agency that is nomically linked with the brain correlator is left alone in the experiment, a robust correlation between the agent's choice and the flag outcome will be exhibited, and that correlation can be interpreted as the result of pseudo-backtracking prob-influence. To the extent that some process interferes with the usual nomic link, the correlation will be disrupted. This experimentally revealed difference makes evident the sense in which common-cause patterns cannot be exploited. The principle that we cannot advance any independently assigned goals through the mediation of a common-cause pattern is my replacement for the initially plausible but insufficiently accurate principle that two effects in a common-cause pattern cannot influence each other by virtue of their connection to a common cause.

For a final illustration that adds more complexity to the character of the agent, let us consider Ronald Fisher's (1959) hypothesis that there is a smoking gene that promotes both a craving for cigarettes as well as lung cancer, and that smoking itself is not a promoter of lung cancer if we hold fixed either the existence or the absence of the smoking gene. In the scenario under consideration, people with the gene and people without the gene can smoke without raising the probability of acquiring lung cancer. We can now ask whether, in such a case, abstaining from smoking is an effective strategy for reducing the probability of lung cancer.

The answer according to my theory is that, in full generality, there is no unique correct answer because it depends on the cognitive details. First of all, it is a given that any contrastivization of a person's choice to abstain that extends across space at one moment is guaranteed not to promote or inhibit cancer. The ambiguous case occurs only when the contrastivization holds fixed what happens at some earlier time and varies the choice while not specifying everything that happens in between.

Let us consider an agent with two cognitive components. One is a craving component that promotes smoking and is also the component whose existence is promoted by the smoking gene. The other is a rational component that is not affected in any relevant way by the smoking gene. Suppose for the sake of having a simple example that the condition of each component is not prob-influenced by the other. Both components together jointly prob-influence the agent's choice between smoking and abstaining in the following way. When the rational component is set to smoke or set to abstain, then the agent obeys, but if the rational component is set to neutral, then the agent smokes or abstains based on the presence of the craving.

When we hold fixed a contextualized event at some earlier time and consider a contrastivization of whether to abstain or smoke, we can do so by counterfactually twiddling the rational component or the craving component or both. Implementing the choice by a process that holds the rational component at neutral while only adjusting the craving will reveal that the choice promotes cancer because that part of the agent was positively correlated with the smoking gene. That case works just like the brain correlator example where the agent's choice of color partially influences which flag is raised. Implementing the choice by a process that maintains the craving while adjusting the rational component will reveal that the choice of smoking neither promotes nor inhibits cancer.

One can ramp this example up to far more complicated interactions that do not permit one to isolate the components of the decision process that are prob-influenced by the smoking gene from the components that are not. In such cases, it may easily become less clear what the appropriate contrastivizations are, hence, less clear how to abstract away from the fundamental details using contrastive events. Given any contrastivization, the fundamental laws will suffice for a determinate degree of prob-influence over the chosen effect-kind, but in complicated interacting cases, there may be no reasonable way (or too many acceptable ways) to decompose agency into components. This is not a metaphysical muddle because the fundamental material contents will develop according to fundamental laws regardless of how they are contrastivized, but it does indicate the possibility of cases where our heuristics for interpreting the relations among events is insufficient to settle whether one event raises the probability of another.

Let me emphasize that nothing about my non-standard analysis of common-cause patterns has any practical scientific consequences, nor does it take any sides in the debate between causal and evidential decision theory. One can, if desired, interpret the reliable correlation between a common cause and its two effects, E_1 and E_2, as exhibiting a genuine probability-raising influence from E_1 to E_2 and vice versa. The influence of E_1 on E_2 is of a kind most people would interpret as the influence of the past state together with an epistemic supplement, a conditionalization on the presence or absence of E_1. I agree with this interpretation as well, but for the reasons mentioned in §6.4, I think the causal interpretation is acceptable as well, even though it may seem initially odd to think of it that way.

7.5 The Entropy Asymmetry and Causal Directionality

A long-standing conjecture among physicists is that the asymmetry of causation is somehow explained by the entropy gradient exhibited by our universe, and my explanation is, broadly speaking, part of that tradition. In this section, I will attempt to contrast my account of the asymmetry with standard entropy-based approaches. Note that this section is only concerned with explanations of causal asymmetry that attempt to exploit the entropic asymmetry existing in the region where the cause and effect are located. In §7.6, I will discuss the so-called AKL approach, which is based on the idea that the low entropy of the early universe provides the crucial temporal asymmetry needed to explain the causal asymmetry. Because the AKL approach does not explain causal asymmetry by invoking the entropic asymmetry in the neighborhood of the cause and effect, it is not a subject of discussion in this section.

There are several good reasons to think that the direction of causation should not literally be identified with the entropic asymmetry. First, entropy is a well-defined quantity only for a suitably isolated physical system. There are two possible kinds of such systems: the universe as a whole, which is arguably perfectly isolated,[59] and so-called branch systems, which are more localized and are only approximately isolated.

Concerning the universe as a whole, there are significant technical problems adjudicating whether a suitable concept of entropy applies in the context of cosmology. Because general relativity governs the large-scale behavior of the universe and (in its standard interpretation) does not posit a fixed background geometry, it is a non-trivial task to define the relevant phase space appropriately. There is even a problem in defining appropriate concepts of energy and velocity, which are needed for an adequate conception of entropy. Even if a suitable concept of global entropy exists, identifying the direction of causation as the direction of time in which the global entropy increases makes the direction of causation depend on what occurs arbitrarily far away. This is problematic because our epistemic grasp of the direction of causation is based on spatio-temporally local evidence. Suppose our region of the universe, say a state ten billion light years wide, has all the traits associated with increasing entropy, but also imagine that space extends infinitely and that in all other regions of the universe, there are enough of the localized traits

[59] The universe is normally presumed to be physically isolated just by definition, but it is important that the universe not have any problematic boundary conditions, for example, edges or singularities where energy can enter or exit willy-nilly, or conditions that would violate the future-typicality of the known universe. Thus, the proposition that the universe is thermally isolated is in this sense an a posteriori claim, albeit one that would strike the average physicist as plausible.

of decreasing entropy such that the global entropy is decreasing. If the direction of causation goes by the direction of global entropy increase, then the true direction of causation will be divorced from what we see locally. Thus, it would not explain the phenomena motivating our belief that causation is directed toward the future.

Concerning branch systems, there are at least five obstacles to the identification of the causal asymmetry with the entropic asymmetry of branch systems. First, it is unclear how the entropic asymmetry would explain the advancement asymmetry. How would the lower *entropy* of the recent past ensure that we are unable to advance goals we might have for the past or at least make it difficult for us to do so? I am not aware of any attempt to address this question.

Second, if the direction of causation is set by the direction in which entropy of a branch system increases, then there ought to be no definite direction of causation when there is no suitably isolated branch system. But it is easy to imagine paradigmatic, systematic, widespread, asymmetric causation even in systems that are not even approximately isolated in the correct way. In such a case, the local physical conditions could either be considered part of a small non-isolated system in which case there would be no direction of causation, or as part of a very large isolated system, in which case the direction of causation would not necessarily match the behavior of the local matter.

Third, a related problem for identifying the causal asymmetry with the entropic asymmetry of branch systems is that in moving (figuratively) from cases of perfect isolation to cases where the system is not even approximately isolated in the right ways, the direction of causation needs to degrade gracefully. There would need to be some account of what it would mean for the direction of causation to steadily disappear. For example, a box of gas at equilibrium will necessarily make occasional slight deviations toward lower entropy. Does that mean that the direction of causation reverses for a brief moment, or is there some threshold fixing how much of an entropy decrease is needed to reverse causation, or is the direction of causation somehow a matter of degree? I do not want to conclude that there is no possible explanation available to resolve these questions, only that details would need to be spelled out.

Fourth, to the extent that some aspects of the environment embody a decrease in entropy, this need not conflict with the existence of what we think of as a widespread pattern of ordinary causal asymmetries. In Barry Dainton's (2001) example, the ocean begins spontaneously to transfer so much of its energy from a cold section to a warm section that the overall entropy of the Earth decreases. But this thermal phenomenon in itself does not preclude ordinary causal processes from taking place as usual. Fish can still swim normally, sailors can drop anchor, and so on. Thus there is a conceptual disconnection between the hypothesized reversal of the direction of causation and our practices that indicate to us the direction in which causation operates.

Fifth, one can imagine devices that systematically (as a result of heretofore undiscovered consequences of the fundamental dynamical laws) behave contrary to the usual thermodynamic regularities. For example, a sealed box of gas with a button could behave so that whenever the button is pressed, any gas in the box spontaneously collects temporarily at the center of the box, with no other significant thermal effects and no violation of energy conservation. The natural way to think of such cases is that pressing the button *causes* the gas to contract, and that the causal direction exists in spite of its anti-thermodynamic evolution. What makes the direction of causation future-directed in such a case is that it is a device that (1) is able to bring about an expected kind of event in the future, (2) actually does so whenever someone employs it, and (3) is integrated in a broader environment with all the usual material asymmetries.

Before discussing causal asymmetry any further, there is a lesson pertaining to the methodology of empirical analysis that deserves reinforcement at this stage. Some of the arguments above appeal to common-sense intuitions about the direction of causation in imagined test cases, yet in the introductory chapter I mentioned that the quality of an empirical analysis does not depend at all on whether such intuitions are rendered explicitly true. The above arguments, I must emphasize, do not violate the methodology of empirical analysis because in the current context, I am using the intuitions merely to flag conceptual disconnections that need to be bridged by any adequate empirical analysis. In order to defend the hypothesis that the direction of causation is equivalent to the direction in which entropy increases, one needs to know how entropy increase is connected to the empirical phenomena that give us a good reason to believe in the asymmetry of causation. Because the results of asymmetry experiments are included among these empirical phenomena, we have a right to demand that an entropy-based theory explain why the direction of *entropy increase* matters as opposed to alternatives like the direction of non-bizarre evolutions. These two directions both happen to point the same way in all circumstances that we know about. Thus, in order to see which principle better explains the empirical phenomena, we need to examine cases like Dainton's anti-thermodynamic ocean where they point in opposite directions. Ideally, we should run asymmetry experiments in environments that are spontaneously decreasing in entropy to see whether the entropy decrease by itself would allow agents to advance their goals for the past. Because such experiments are not feasible, we resort to our intuitions about the relative plausibility of competing hypotheses.

I am unfortunately unable to make a proper comparison between my own explanation and entropy-based alternatives because I am unaware of any existing explanation of the asymmetry experiment that appeals to the entropy gradient localized in the experiment's environment. The best I can do is to cast doubt on the relative plausibility of such an approach. I will do so merely by reporting that I am unable to discern any plausible reason why a decrease of entropy per se would make likely the existence of persistent correlations between the randomly chosen team assignments and the (somewhat arbitrarily chosen) E, a correlation that must hold if there is evidence in the actual world that some agents can reliably advance their goals for the past. The role of intuitions in this context is not to highlight truths that must be rendered explicitly true by the empirical analysis. Their role is to emphasize deficiencies of rival hypotheses that a potential opponent could offer to explain the experimental results.

Another legitimate role that intuitions can play in a proper empirical analysis is to highlight connections among concepts that can serve as data for comparing alternative ways to organize the overall conceptual scheme. This activity is common in mathematical physics, where choices about which mathematical structures best represent a physical quantity are often based on how well they apply to paradigmatic situations and how they help to generalize our theories. Empirical analysis is intended to be friendly to the use of intuitions about which hypothesized structures are simpler and more conducive to future development. To connect this observation to the study of causation, I will just note that I have attempted throughout this volume to design the two main concepts—terminance and prob-influence—to be maximally comprehensive and to be as simple as possible given the role that they play in linking general causal claims in the special sciences to claims about how the world evolves fundamentally. I am eagerly awaiting the development of competing empirical analyses of the metaphysics of causation so that my account can be compared to these alternatives on the basis of simplicity, comprehensiveness, and other recognized scientific virtues.

Let us now return to the discussion of causal asymmetry.

Even though it is unwise to equate the direction in which entropy increases with the direction of advancement, according to my account it is not surprising that these two directions happen to align. This is because the asymmetry of bizarre coincidences plays a large role in explaining both. In §5.4, I noted that future-typicality ensures that anti-thermodynamic behavior will be highly unlikely toward the future and that future-typicality implies past-bizarreness once we set aside two possibilities: (1) that we inhabit an equilibrium condition, which evidently does not apply to our environment, and (2) that we are at the extreme end of a thermodynamic fluctuation, which was dismissed earlier because we can arrive at that conclusion only by a self-undermining argument. In §7.3, I explained the role future-typicality plays in explaining the results of the asymmetry experiment. Although the asymmetry of bizarre coincidences is part of the explanation of an entropic asymmetry and an advancement asymmetry, the example of the anti-thermodynamic ocean and the gas-collapsing device demonstrate that violations of future-typicality can be strong enough to generate local entropy decrease while not being so severe as to vitiate the asymmetry of advancement. And the example of the birds landing on the random perches demonstrated that future-typicality can be violated without affecting thermodynamic regularities.

Cosmological evidence indicates that the asymmetry of bizarre coincidences present in our solar system is part of a broader asymmetry of bizarre coincidences that extends across the known universe. By virtue of the arguments in §5.4, the ubiquity of the asymmetry of bizarre coincidences can be explained by the fundamental laws together with an assumption about the layout of the material contents of the universe not too long after the big bang. Given the general character of the matter and fields in the early universe, and given the dynamical laws governing the local development of matter, it should not be surprising that the history of the universe is future-typical and past-bizarre, so long as there is no future structure—a low entropy big crunch, for example—that would impose a strong countervailing constraint on the world's future evolution. We currently have no evidence suggesting that a reversal of the asymmetry of bizarre coincidences lies in our future.

Although our universe appears to have this ubiquitous asymmetry of bizarre coincidences, it is useful to explore how the concept of advancement applies to the possibility of other regions of the universe being structurally very similar to our region but temporally reversed. It is conceivable that not every kind of physical process can be temporally reversed and still obey the laws of nature. But where the complete material content of the region E containing Earth can be temporally ordered in reverse in some region R without conflicting with the fundamental laws, we can ask whether the direction of causation is thereby reversed in R.

Among many possible models, the Gold (1962) universe depicted in Fig. 7.3 is one of the simplest. A Gold universe has boundary conditions that are nearly symmetrical so that it has a big bang and a big crunch where the material content at each temporal end closely resembles the temporal inverse of the material content at the other end. It also incorporates no fundamental temporal asymmetries. If one interprets the dynamical development of either temporal end by picturing it going toward the temporal middle of the space-time, both sides exhibit the same kinds of galaxies, the same kinds of chemical regularities, the same kinds of organisms, and the same kinds of intelligent creatures. The direction in which things evolve bizarrely would run toward the two temporal ends of the universe, and somewhere near the middle there would be a region without a unique asymmetry of bizarre coincidences. It might be that there are bizarre evolutions in both directions or in neither direction (insofar as near term physical evolutions go), or that these

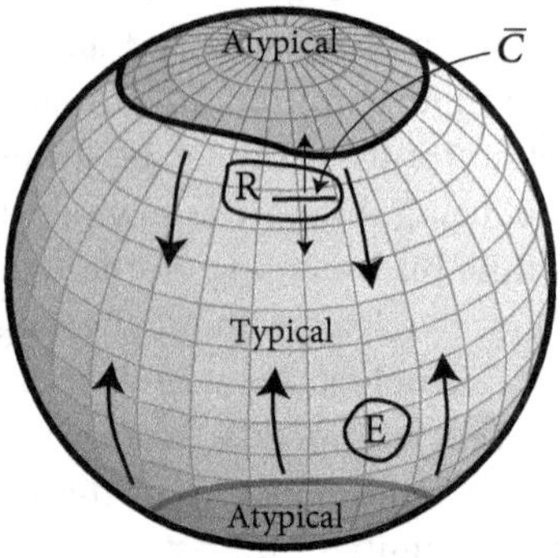

FIGURE 7.3 *The Gold universe has entropy low at both temporal ends and high in the middle. A contextualized event* $\overline{\text{C}}$ *located in R will likely fix an entropic increase in both temporal directions.*

two options are interspersed. Regardless of the details concerning the middle, the more important issue concerns how we should engineer our concepts to best make sense of how creatures on one side should construe the causation-like interactions on the other side.

Inhabitants of E and R should be in agreement about fixing relations, but they should disagree about which developments will count as bizarre because the direction of time that each group of inhabitants will designate as the future—what ends up being toward the temporal center of the universe—will also be the direction in which (locally) matter behaves non-bizarrely. Imagine a contextualized event $\overline{C}$ occupying part of a time slice in R, using a probability distribution that is reasonable in light of statistical-mechanical principles (without conditionalizing on what is happening at either of the two temporal boundaries). $\overline{C}$ fixes events in (what we Earthlings in E think of as) the future that develop non-bizarrely in the normal pro-thermodynamic way, and that development will radically disagree with the actual evolution that occurs in R. Inhabitants of R, whose derivative future coincides with our derivative past, will think of this same development as bizarre. Also, $\overline{C}$ fixes events toward (what we think of as) the past that instantiate (what looks from our future-directed perspective like) bizarre anti-thermodynamic behavior that approximates what actually happens. Such happenings would be judged by agents in R as non-bizarre.

So, is the direction of causation in a Gold universe in some sense relative? I think it is relative in the same sense as 'up' and 'down'. It is reasonable for both sets of inhabitants to associate a direction of causation with the direction in which their local matter behaves non-bizarrely, and they may find it convenient to extend this temporal orientation to other regions that do not exhibit the same orientation of matter. But informed inhabitants should also be able to recognize that it would be reasonable for an inhabitant of the distant region to think of causation as pointed in the opposite way. Furthermore, like the distinction between up and down, the direction of causation should be considered objective in two senses. No inhabitant can actually (in the narrow sense of 'ability') reverse the direction of causation. And one can say of the Gold universe that the future is objectively toward the temporal middle of the universe.

7.6 Recent Alternative Explanations of Causal Asymmetry

There are many explanatory schemes that have tried to explain the asymmetry of causation in terms of the seemingly special character of the initial conditions of the big bang. What I think is novel about my own explanation of the advancement asymmetry is that it provides an adequate account of how the special character of the big bang vindicates the platitude that we are unable to influence the past by explaining the results of the asymmetry experiment. It does so while easily evading the five previously listed obstacles for entropy-based explanations.

I do not know of any previous attempts to explain the results of the asymmetry experiment, but I can offer a few thoughts on three alternative approaches to the problem of understanding causal asymmetry: the Albert-Kutach-Loewer (AKL) approach, the Price-Weslake approach, and the fork asymmetry approach.

7.6.1 THE ALBERT-KUTACH-LOEWER APPROACH

In Albert (2000), Kutach (2001, 2002, 2007), and Loewer (2007, 2012), there are attempts to explain the asymmetry of causation by imposing a restriction on the possible worlds relevant to counterfactual dependence. One only considers those with the same kind of boundary conditions as the actual universe, a low entropy big bang. The motivation for postulating this restriction is based on the hope that it effects a near-settledness or quasi-fixity of the past, where most macroscopic happenings in the past depend at most microscopically on highly localized present events.

In my work cited above, I have detailed several obstacles for the AKL program, some of which have also been flagged (in the context of attacking Albert's and Loewer's versions) by Frisch (2005b, 2007, 2010) and Price and Weslake (2009). Although I cannot say much about Albert's account because it lacks sufficient detail, readers may find a comparison between my account with that of Loewer (2007, 2012) informative.

Loewer and I agree that the goal is to explain why we cannot exploit our influence over the past, though Loewer speaks of 'control' where I speak of 'advancement'. We also both model the relevant kind of influence as a form of counterfactual dependence. However, Loewer's (2007) statistical-mechanical (or SM-) conditional differs from my nomic conditional in several respects. (1) Loewer's theory employs the standard conception of probability that applies to human-scale causes. (2) Loewer's theory postulates a single "correct" probability distribution (sometimes applied to the early universe, sometimes applied to the universe's present macroscopic condition and then conditionalized on the macroscopic condition of the early universe). (3) Loewer's theory requires there to be a well-defined macrostate of the universe for any time we want to consider. (4) Loewer's SM-conditional applies paradigmatically to highly localized regions where people make decisions. (He intends his SM-conditional to apply to more general situations but provides only sketchy suggestions for how to extend its range.) For three reasons, I do not want to argue that Loewer's account is faulty. First, in the (2012) revision of his account, he has adopted a measure of our "degree of control" over the world that turns out to be a special case of my prob-influence. Second, Loewer only briefly mentions a couple of explanations for why certain kinds of attempts to influence over the past cannot be exploited and does not contrast his explanations with existing alternatives like (Kutach 2007, 2011c). Third, his (2012) modification to the SM-conditional appears to make it equivalent to my earlier (2001, 2002) "Entropy Theory" of counterfactuals. Because he has not yet addressed the problems I previously (2001, 2002, 2007) raised for the use of such counterfactuals, I will forego further comment here.

The major points of contrast between my present account and Loewer's can be summarized by noting that my explanation of the advancement asymmetry does not require assumptions nearly as strong as Loewer's. If my explanation of the presumed results of asymmetry experiments is adequate, the kinds of restrictions built into Loewer's SM-conditional are unnecessary for explaining the empirical phenomena relevant to the asymmetry of causation and may be safely dropped. In particular, we do not need the distant past to be held very nearly fixed under a counterfactual alteration of the present, and we do not need the counterfactual alterations to the present to be small, and we do not need a single "correct" probability distribution. My nomic conditional was constructed to hold for any hypothesized event with any probability distribution one chooses. Nomic conditionals are generally more helpful to a quality explanation when they incorporate probability distributions that are close enough to the kind invoked in statistical mechanics, but they make sense without a unique "correct" probability distribution.

Unlike Loewer's (2007) SM-conditional, my nomic conditional makes no reference to macrostates, which are unacceptably unclear in many important circumstances. It makes sense to speak of the macrostate of a box of gas because gasses have discernible properties like temperature and pressure that make sense over a reasonably wide range of distance scales, but what is the macrostate of a human being? If it does not include the microscopic genetic information and the detailed microstructure of my cells, how can we be sure it implies appropriate probabilities for the human's future behavior?

A further disadvantage of the AKL approach is that it does not exhibit graceful degradation by properly extending to a wider class of possibilities. An instructive case to consider is the kind of universe imagined by Ludwig Boltzmann (1895) where an infinite space-time is filled with atoms bouncing around in a seemingly random fashion constrained by classical laws that conserve momentum and energy. With enough randomness and enough time or space, any type of event that maintains a non-vanishing probability of occurring will eventually happen sometime and somewhere (with probability one). So there will certainly be large regions in a Boltzmannian universe with an asymmetry of bizarre coincidences and with agents and whatever material asymmetry is needed to instantiate agents and their asymmetric environments. Any creatures living nearly fourteen-billion-years along an increasing entropy-gradient in a suitable Boltzmannian universe would presumably be correct to think that their environment exhibits a causal asymmetry just as reliable as the one we are familiar with.[60] Yet, Loewer's account has

[60] I am not implying that they would be correct to infer from their available evidence that they live on an entropy gradient in a universe of thermodynamic fluctuations.

insufficient resources to judge the region as having a single direction of causation because there are no physically special conditions to warrant the status of a law. There is only a heterogeneous multiplicity of localized thermodynamic fluctuations. His justifications for the law-like status of the low entropy big bang are not available for this Boltzmannian universe.

My account presented in this volume, by contrast, is able to explain the asymmetry in terms of the physical properties of regions that are much smaller in space and time than the entirety of space for the first fourteen billion years, extending without amendment to appropriately time-asymmetric regions of a Boltzmannian universe. I agree with Loewer that our inability to advance goals for the past is explained in great part by the early state of the universe, but not by its restricting the malleability of the past or by its directly preventing us from exploiting our influence over the past. Instead, the explanatory role of the low entropy of the early universe comes from its generating (1) a temporally asymmetric pattern of material asymmetries that permit asymmetry experiments, and (2) a lack of bizarre evolutions toward the future. Regardless of the advantages of my account, defenders of Loewer's approach can take advantage of the fact that his two versions of the SM-conditional are both special cases of my nomic conditional and his definition of a person's "degree of control" is a special case of my prob-influence.

7.6.2 THE PRICE-WESLAKE APPROACH

Another theory that might provide a useful contrast is the (2009) account by Huw Price and Brad Weslake. For them, the goal is to explain what they call the "temporal asymmetry of disjunctive deliberation (TADD)." TADD can be defined in terms of an agent's robustly believed disjunctions of the form $\neg A \vee O$, where A is a possible action of the agent (and recognized as such by the agent) and O is any outcome.[61] For a person to believe a disjunction robustly is for that person to believe the disjunction and to be disposed to continue to believe the disjunction if the person were to learn that exactly one of the disjuncts is false.[62] Robustly believed disjunctions are believed more than is warranted by one's credences of the individual disjuncts. TADD is the (empirically verifiable) fact that people rarely believe disjunctions of the form $\neg A \vee O$ robustly when O temporally precedes A but believe many of them robustly when O temporally follows A.

Although I think the primary empirical phenomena to be explained are the results of the asymmetry experiment, I believe just about any explanation of the asymmetry experiment will automatically explain TADD rather straightforwardly. The main component of that explanation is that people recognize the pointlessness of deliberately trying to meddle with the past, and they interpret this lack of

[61] As in §3.3, each variable here represents a token event coarse-grained as some event type, and it also stands for the proposition that that event occurred.

[62] See Jackson (1979) for additional information on robustness.

exploitable influence as a lack of any causal connection between whether or not *A* and whether or not *O*.[63] Because they also conceive of *A* as under their control, they recognize any past *O* as not subject to their influence. Hence, disjunctions of the form $\neg A \vee O$ will be believed only to the extent that their disjuncts are believed on independent grounds.

By contrast, I am unable to understand the explanation for TADD being offered by Price and Weslake. They appeal to the temporal directness of agency, as do I, but they seem to think it can be given a "subjectivist" interpretation. What I find puzzling is their distinction between subjective accounts and the more objective "third arrow"[64] accounts, which derive the causal asymmetry from statistical-mechanical considerations—for example, using entropy or bizarreness or forks or some other non-fundamental asymmetries not essentially tied to agency. They primarily worry that objective accounts will not be able to explain TADD and the special role for agency in a theory of causal asymmetry. A subjective account, they suggest, is "in a better position to succeed" because of its explicit incorporation of the temporally directed character of deliberation.

First of all, consideration of which approach is in a better position to succeed is irrelevant once an explicit proposal exists, such as the one I provided above.

Second, regardless of how the "subjectivist" explanation is spelled out in detail, it is hard for me to see how the temporal orientation of deliberating agents can count as subjective except in an extremely attenuated sense. Subjective attributes paradigmatically vary among different subjects, as in people's taste in music or food, whereas paradigmatically objective attributes such as the solidity of a granite block are considered objective because they are disposed to interact with any person in pretty much the same way. Because all actual agents we know of are temporally aligned the same way, one cannot point to any evidence about temporally reversed agents to further the case for subjectivism. So, one must turn to imagined scenarios.

When conceiving of what these temporally reversed agents are like, there are a range of possible scenarios. One can imagine a reasonably large patch of space-time with the temporal reverse of agents and environments like ours, as in the Gold universe. That would vindicate the claim that agents vary regarding the direction in which they advance their goals. Still, this sort of variation is best described in terms of the causal direction implied by the environments and agents together as a whole, not a variation merely among agents. Nothing Price and Weslake have argued suffices to rule out the plausibility of a "third arrow" explanation of the reversed causal direction in such environments. After all, I offered a "third arrow" explanation above that incorporates agency essentially.

[63] I believe that the epistemic asymmetry probably also plays some role in our thinking of the past as settled.

[64] The first two arrows are the "causal arrow" and "temporal arrow" from (Price 1996).

If we instead postulate a small patch of space-time where a single decision is temporally flipped and the background environment is otherwise held fixed, there would be no regular pattern of advancement of agents' goals. Such cases are properly described as anti-thermodynamic fluctuations, not regions of reversed causation. There are other nomologically possible ways to implement a temporally reversed agent. Yet, worlds with reverse-agents embedded in something very much like our existing thermodynamically asymmetric environment would count as bizarre because they would continually instantiate fantastically conspiratorial fluctuations of matter. I think such cases count as scenarios so remote from any actual patterns of causation that it undercuts the applicability of the label 'subjective'. After all, many paradigmatically objective qualities can be made to seem subjective by framing them within a broader space of possibilities. For example, if we postulate the existence of creatures made of neutrinos, we can rightly say that these creatures would pass through the granite block just as easily as they pass through empty space. By accepting such possibilities as within the proper scope for assessing objectivity, we should judge the solidity of the granite to be subjective because it would be solid for humans but not for neutrino creatures. Adopting such liberal standards would render many paradigmatically objective attributes subjective. The imagined highly localized, temporally reversed agents are exactly the kind of far-fetched possibilities that are rightly ignored when we attribute subjectivity. In the end, it would be more perspicuous to describe the temporal orientation of agents as derivative rather than subjective.

7.6.3 THE FORK ASYMMETRY APPROACH

In this section, I will relate my explanation of causal asymmetry to attempts to explain the causal asymmetry in terms of the fork asymmetry mentioned in §7.2.

Hans Reichenbach (1956) proposed that for (presumably coarse-grained) events A, B, and C, when the following probabilistic relations hold, C is by definition a *conjunctive fork*.

1. $0 < P(C) < 1$
2. $P(A/C) > P(A/\neg C)$
3. $P(B/C) > P(B/\neg C)$
4. $P(A\&B/C) = P(A/C)P(B/C)$
5. $P(A\&B/\neg C) = P(A/\neg C)P(B/\neg C)$

These conditions state that C is positively correlated with both A and B and that both C and the absence of C each screen the probabilistic correlation between A and B. It is reasonable to generalize this definition to accommodate probability-lowering and also multiple common causes, but these details can be bracketed. When C is to the past of A and B and there is no other event to the future of A and B that forms a conjunctive fork with A and B, we say that the fork is *open* to the future. If all (or perhaps almost all) open conjunctive forks are open in the same temporal direction, there is a fork asymmetry. Reichenbach defined his famous common cause principle to ensure that if it holds true, all open conjunctive forks are temporally aligned so that they are all open toward the same temporal direction. Call it 'the OF-direction'. Then, Reichenbach proposed that the future be "defined" as the OF-direction, though I cannot claim to understand precisely what he means by that.

A direct comparison between my account of the time asymmetry of causation and explanations like Reichenbach's that are framed in terms of conjunctive forks is not possible without a thorough investigation of their contrasting approaches to probability. Remember that my account gets its probabilities by having them either stipulated in contextualized events (taking into account the insensitivity considerations mentioned in §5.2) or by having them arise from stochastic fundamental laws. Accounts based on Reichenbach's approach employ the alternative (and more common) conception of probability that applies to relations among spatio-temporally localized events or event-kinds, especially among mundane events as discussed in §2.2. According to my account, if we assume that fundamental reality obeys laws resembling the kind postulated in paradigm theories of fundamental physics, then fundamental relations do not hold between instances of these mundane events and thus any probabilistic relations among them must count as derivative. Because it is unclear to me how advocates of such probabilistic relations would react to my imposition of the distinction between fundamental and derivative and its application to events, I must forgo a thorough comparison of these competing conceptions of probability and just offer just two brief observations.

First, insofar as we are merely concerned with explaining the pattern-based set of platitudes regarding causal asymmetry, I accept that something very roughly resembling the fork asymmetry is a component of a full scientific explanation of causation. However, in doing so, I would be very reluctant to use the term 'causal' for any of the standard probabilistic relations, which I take to be operationally just statistical correlations that have had most of their noise filtered out. This includes the fork asymmetry as well. These probabilistic relations are too divorced from the fundamental laws and difference-making, in my opinion, to be appropriately labeled 'causal'. Perhaps a new term is needed to convey that these correlations (filtered of noise and reified as probabilistic relations) do not themselves engage in affecting events or bringing them about or making them happen, but just encode patterns that Dummett's intelligent plant could invoke for prediction or explanation. If I were pressed to speculate about where the fork asymmetry fits into the larger explanatory scheme, I would agree with Horwich (1987) that the fork asymmetry is explained in part by way of features of how the material content of the universe was organized shortly after the big bang and in part by way of the dynamical laws. I would then expand on this explanation by relating it to the asymmetry of bizarre coincidences, but because I do not believe this would lead to any great disagreement between me and those who propose to explain the asymmetry of causation in terms of the fork asymmetry, I will not discuss this topic any further here.

Second, insofar as we are concerned with explaining the influence-based set of platitudes regarding causal asymmetry—those regarding the seemingly settled character of the past or our lack of influence over it—there is room for disagreement between my account and those based on the fork asymmetry. It is hard for me to see how the fork asymmetry can explain the advancement asymmetry or anything else about the influence-based set of platitudes without introducing more theoretical machinery. A mere lack of conjunctive forks open to the past does not by itself make the past "settled" or make events immune from influence coming from the future or otherwise explain the advancement asymmetry.

I see my own account as providing an interpretation of counterfactual dependence and influence where the principle of causal directness and future-typicality play the starring roles and where some principle in the neighborhood of the fork asymmetry could play a supporting role. As I mentioned in §7.2, some account is needed of the existence of the agents, the messages, the sacks of gold, and other materials needed to set up asymmetry experiments. Something akin to the fork asymmetry might serve well for the purpose of explaining the required material asymmetries.

I also used future-typicality to explain why there are no correlations between the randomly selected team assignment and the presence or absence of the target event *E*. If there were such a bizarre correlation, it might count (together with some action of the agent) as something akin to a conjunctive fork open to the past, and if so there would be some connection between future-typicality and a lack of conjunctive forks open to the past, as suggested in §7.3.2. However, my explanation of the asymmetry experiment does not depend on a general lack of conjunctive forks open to the past, only that there are no conjunctive forks open to the past that involve the random team assignment as one tine and the occurrence or non-occurrence of the target event as the other tine and some action of the agent as the common cause. That is, my account depends on a much weaker assumption than the general non-existence of conjunctive forks open to the past. Assuming my account can be extended somehow to make sense of something like a conjunctive fork, my account only requires that they not be open to the past when they take the following form: the receipt of the team assignment DO by an agent in an asymmetry experiment counts as the common cause *C*; the seemingly random selection of DO counts as one of the effect, say *A*; and the occurrence of *E* counts as the other effect, *B*. (A similar condition holds for DONT.) If such open forks existed, that would be a truly remarkable feature of the universe, and I suspect that most readers will concur with my skepticism about their existence.

Explanations based on the fork asymmetry are inferior because it is plausible that there are other, less specific conjunctive forks open to the past. A good example of such a case was constructed by Weslake (2006), and I think others can be found by examining alleged counterexamples to Reichenbach's common cause principle. In any case, according to my account, at best only a very small subset of conjunctive forks open to the past could be potentially exploitable for advancing one's goals for the past, and so the focus on a general fork asymmetry appears to be unnecessarily general.

To summarize, the problem with using the fork asymmetry to explain the advancement asymmetry is twofold. In one sense, the fork asymmetry is too weak because the mere lack of conjunctive forks open to the past does not ensure the lack of exploitable influence over the past. In another sense, the fork asymmetry is too strong because the lack of exploitable influence over the past is compatible with many kinds of conjunctive forks open to the past.

7.7 Fundamental Influence Asymmetry

If there existed a fundamental temporal direction that ensured the non-existence of all past-directed influence, this would provide a simple explanation for the results of the asymmetry experiment. Such an explanation would also explain causal directness because all backtracking nomic connections must include a past-directed component. Furthermore, such an explanation of the advancement asymmetry would automatically be entirely insensitive to how we choose to characterize agents, tasks, and the advancement of goals. It would explain our inability to advance goals for the past as a trivial corollary of the much stronger thesis that nothing can influence the past.

There are two prominent ways one could arrive at the conclusion that fundamental reality disallows past-directed influence. The first is to conduct a thorough investigation of fundamental reality while trying to remain agnostic on whether there is past-directed influence. That would presumably involve just letting one's choice of fundamental causation-like relations be dictated by standard scientific criteria. If the investigation were to result in a convincing conclusion that fundamental reality includes a fundamental future ensuring that non-trivial pure contribution relations exist only toward the future, this would provide some motivation to infer that all forms of influence are future-directed. The second is to postulate by fiat such a fundamental future regardless of any past-directed pure contribution in the fundamental laws, as would exist when the laws are deterministic in both temporal directions.

The fundamental future can presumably be represented mathematically (without loss of important content) as a temporal orientation—a selection at each arena point of one direction of time to count as the direction in which events can purely contribute. In all four paradigm theories, the arena is structured so that for any of its points *p*, there are two disconnected components corresponding to "what happens in one direction of time from *p* (either before or after)" and "what happens in the other direction of time from *p*." A temporal orientation selects one of these components at each point in a continuous manner so that there are no discontinuities in which component is selected as one hypothetically travels from one point to another along any continuous path in the arena. By representing the direction of fundamental influence in this way, we are able to represent a global distinction between past and future even if space-time is relativistic (so that not all pairs of events are temporally ordered) and even if there are closed time-like curves (which prevent a linear ordering of all events in time).

7.7.1 FUNDAMENTAL INFLUENCE ASYMMETRY BY FIAT

I will now consider two ways of imposing a fundamental direction of influence by fiat within the context of theories that treat influence in terms of counterfactual dependence. One way is to restrict the kind of counterfactual dependence that represents influence to the future direction. Under this restriction, one would evaluate *C*'s influence on some chosen *E* by counterfactually altering *C* but not drawing any inferences whatsoever about what would have happened before *C*. Such a system would not keep the past fixed under counterfactual alterations to the present but would just remain silent about the past, ensuring that the past does not causally depend on the present. (Such an account could be supplemented with a non-causal variety of counterfactual dependence in order to support our practice of holding many past facts fixed when imagining counterfactual scenarios.)

A second way to impose a fundamental influence asymmetry would be to deploy a model of counterfactual dependence where the past is held entirely fixed under counterfactual alterations to a given time. The most prominent version of such an account is the forking model, which interprets a mundane contrary-to-fact conditional of the form $A \mathbin{\Box\!\!\rightarrow} C$ by having the counterfactual past match the actual past up until some time t not too soon before the obtaining of *A*. Then, at t, the indeterministic evolution of the counterfactual history departs from actuality and leads lawfully to *A*'s obtaining and later to the rest of the law-abiding counterfactual history.

Although positing a fundamental direction of influence by fiat provides a straightforward explanation of causal asymmetry and the results of the advancement experiment, I recommend against it on the grounds that leaving it out of one's account of fundamental metaphysics improves the overall merit of the explanation of the advancement experiment. The three most important factors are as follows. (1) By the lights of empirical analysis, it is of absolutely no value that the 'asymmetry by fiat' explanation corresponds better with our naïve conception of influence. (2) One can dispense with the fundamental direction of influence and still explain the asymmetry experiment without any controversial resources. (3) The presence of the fundamental direction of influence raises an unanswered question as to why it is aligned with the one temporal direction in which the universe does not evolve bizarrely.

Regarding (1), although imposing a fundamental direction of influence by fiat does greatly increase the match between what the theory says about influence and what people ordinarily think about influence, according to empirical analysis that counts for nothing. In an empirical analysis of *X*, so long as one can make sense of why it is understandable for people to hold their folk opinions of *X*, there is zero additional credit to be earned by having the folk opinions rendered explicitly true. For anyone who rejects the utility of empirical analysis in favor of orthodox analysis, I have no persuasive argument against a fundamental direction of influence. The existence of a fundamental influence asymmetry, so far as I can tell, does not interfere with or conflict with any of the other structures needed to provide an adequate analysis of causation. For example, if one adopts the position that a fundamental passage of time is needed to account for the consciousness of time, having that same fundamental direction also ground the causal asymmetry would not generate any problems for my account of causation.

If one accepts empirical analysis, though, there is a decent argument for the conclusion that the fundamental direction of influence is dispensable. A standard appeal can be made to ontological parsimony as one finds in arguments for the elimination of preferred rest frames in classical and relativistic physics. For illustration, let us just consider classical mechanics. A key consequence of Newton's tenet that the true motion of bodies is absolute is that there is some structure in fundamental reality corresponding to (or defining) the absolute speed of each corpuscle. One might incorporate such a structure into one's model of fundamental reality by defining world lines for the points of space or perhaps a constant time-like vector field in Galilean space-time.[65] Then, the absolute speed of any corpuscle is its speed relative to this fundamental standard for 'being at rest.' For all we knew, the force laws could have made use of absolute speeds, but no empirical phenomena demanding such laws have ever been found, nor is there anything in later physics to suggest that absolute speeds (for massive corpuscles) play any interesting role in physics. Simply by formulating our account of the arena structure to avoid postulation of absolute speeds, we can eliminate them from our account of fundamental metaphysics without any predictive or explanatory cost. Indeed, we benefit by moving from a Newtonian space-time to the more structurally impoverished Galilean space-time because the non-existence of absolute speeds provides a good explanation for why no forces refer to them.

Similarly, assuming that my explanation in §7.3 was sufficient to account for all the empirical phenomena relevant to the asymmetry of influence, the postulated fundamental future counts as a redundant component that can be excised without cost from the account of fundamental reality, leaving us with a derivative future. This argument leaves open the possibility that some other empirical phenomenon justifies belief in a fundamental future.

---

[65] Galilean space-time was introduced in §2.4.1. For details, consult (Friedman 1983, pp. 87–92).

Regarding (3), I do not see how postulating a fundamental direction of influence would explain the existence of an asymmetry of bizarre coincidences or an entropic asymmetry. Nor does it appear to augment or improve our account of the many material asymmetries that are already explained in terms of the fundamental laws of physics and boundary conditions. Because my explanation of the results of asymmetry experiments relies on future-typicality and the fundamental laws, there is automatically a tight connection between the asymmetry of advancement and the asymmetry of bizarre coincidences. A theory that postulates a fundamental direction of influence cannot appeal to this tight connection, which raises a worry that it is not integrated enough with the physics to substantiate its government of influence. The worry is that the fundamental direction of influence is tacked on to avoid the appearance of past-directed influence but that it does not play any substantive role in explaining why there are effective strategies with regard to future-directed influence or why the past-directed evolution of the actual world is bizarre. Again, because it does no great harm to tack on a fundamental influence asymmetry to one's theory of fundamental reality, this deficiency is far from decisive.

7.7.2 FUNDAMENTAL INFLUENCE ASYMMETRY BY HAPPENSTANCE

The case for a fundamental direction of influence would be significantly strengthened if there were some reason to think that fundamental physics required the non-existence of past-directed terminance besides a pre-theoretical disbelief in past-directed influence. One way this could occur is if we came to believe that the fundamental laws required a kind of chanciness that severely restricted the extent of past-directed terminance. For a long time, it was mistakenly believed that quantum mechanics required fundamental future-directed chanciness.[66] If that had turned out to be correct, one might argue that physical states do not imply enough about the past for there to be any significant form of past-directed influence.[67] One can even imagine laws where ubiquitous determination holds toward the future but no determination or probability-fixing toward the past. Concerning the possibility of a fundamental direction of influence being implied by the fundamental laws of physics, I will just make two observations.

If there is a fundamental future, we get a simple explanation of asymmetry experiments. Events never termine toward the past; thus, agents cannot promote past events. But in some cases, our theories of fundamental reality also (1) obey

---

[66] This dogma has since been refuted by adequate models of the signature quantum effects that posit ubiquitous determination, for example Bohmian mechanics, which was described in §2.13.2.

[67] Recall that the kind of fundamental chanciness existing in some interpretations of quantum mechanics permits a limited past-directed nomic connection which I labeled 'parterminance' in §2.2. A parterminant of *e* is just like a terminant of *e* except that it does not necessarily establish a probability distribution over *all* the events that could happen at *e*. Global states in the traditional GRW interpretation of quantum mechanics exert past-directed parterminance.

causal directness, (2) are consistent with a widespread asymmetry of bizarre coincidences, and (3) can be altered to remove the fundamental direction of influence. For example, some of the indeterministic interpretations of quantum mechanics described in §2.13.3 obey causal directness regardless of the temporal asymmetry in their fundamental laws, and they can be reformulated so as to be temporally symmetric by tacking on the appropriate past-directed transition probabilities.[68] This would provide enough structure to explain the asymmetry of advancement while ignoring the fundamental influence asymmetry. Because the fundamental theory would have enough resources to explain the advancement asymmetry from §7.3, the fundamental direction of influence could be seen as superfluous.

Such a theory would provide a redundant explanation of the asymmetry experiment and thus of the asymmetry of advancement. It would be able to explain the advancement asymmetry both in terms of the fundamental direction of influence but also along the lines of my explanation in §7.3. This explanatory redundancy is exhibited in many other circumstances. Recall John Locke's famous example where a man is sleeping and is secretly transported to an unfamiliar room where he is locked inside with a friend. Upon waking, the man stays in the room voluntarily in order to converse with the friend. What explains why the man remains in the room? On the one hand, his voluntary decision led him to stay in the room independently of whether the room is locked. On the other hand, regardless of whether he chose to stay in the room, he was locked in. In such cases, it is fair to say both that his voluntary decision and his being locked in the room are explanatory by themselves even though each factor is dispensable given the existence of the other. Similarly, the fundamental future can explain the advancement asymmetry while also being redundant in the sense that causal directness and the typicality asymmetry can account for the advancement asymmetry without invoking any fundamental direction of influence.

Also, I think that even if we accept the existence of a fundamental direction of influence, it still makes sense to accept the past-directed partial influence that exists by virtue of a common-cause pattern. These forms of influence are not fundamentally past-directed because they exist by virtue of the future-directed terminance of a suitably large background state together with some plain coarse-grained event to its future (contrasted with another plain coarse-grained event). As I noted in §4.12, partial influence counts as one reasonable way to make precise

[68] Reformulating theories to be temporally symmetric in such a way would result in the past-directed transition probabilities radically mismatching patterns of actual outcomes toward the past. As a practical matter, this would be an unreasonable alteration to the theory. However, because we know that our environment is past-bizarre, we already know that insofar as we think of the present state as evolving toward the past, it is a bizarre evolution that does not respect the rules of chance that hold toward the future. Thus, we have good reason to believe that if there were past-directed probabilities resembling future-directed chances, we could not see them in action because of the ubiquitous past-bizarreness. So, it would do no harm to accept a symmetric version of the fundamentally chancy theory as a legitimate candidate theory.

our intuitive conception of influence using the machinery of my account. Just on the basis of theoretical unity and the fact that past-directed partial influence does no harm, it makes sense not to clutter the metaphysics of causation by trying to rule out counterintuitive versions of partial influence just because of an instinctive disbelief in past-directed influence.

7.8 Summary

In this chapter, I have put forth several controversial theses. First, I think that what we ordinarily think of as the settledness of the past should really be understood as the advancement asymmetry. The advancement asymmetry is what makes it understandable for us to conceive of the past as settled and immune to influence even if it is influenceable and even if it is affected much more sensitively by the precise condition of the present state than the future is. Second, the empirical content of the advancement asymmetry is captured by the results of the asymmetry experiment I laid out in §7.1. Third, this empirical content can be explained without using any fundamental temporal direction. Fourth, the advancement asymmetry also explains why we cannot exploit common-cause patterns as a route for backtracking influence, while allowing that there is genuine probabilistic influence between effects of a common cause that obtains by virtue of the common-cause pattern. The unexploitable nature of this influence suffices for a principled distinction between effective and ineffective strategies, enough to make sense of why we are justified in thinking naïvely of the probabilistic relationship between two effects of a common cause as non-causal. Finally, I discussed whether we should believe in a fundamental direction of influence, and I suggested that because we do not need it in order to account for the empirical phenomena relevant to the asymmetry of causation, we should just keep an open mind and let our opinions be guided by our best theories of fundamental reality, which include fundamental physics. If it turns out that our best guess at the fundamental laws involves no past-directed terminance, so be it. If not, the principle of causal directness together with future-typicality suffices to explain the advancement asymmetry. Because it is so easy to add a fundamental direction of influence without significant harm and because we tend to like theories with temporal symmetries, our standards for assessing fundamental theories may well be too slack for us to be able to ascertain decisively whether there is a fundamental future. If my assessment of causal asymmetry has been roughly on track, though, it might not be so important in the end whether a fundamental future exists because a derivative future is likely in position to ensure that we cannot manipulate the past.

{ PART III }

The Top Conceptual Layer of Causation

{8}

Culpable Causation

The presentation of my account of the metaphysics of causation is now complete. The account of singular causation was given entirely in terms of fundamental relations among fundamental events, with the regimentations of full and partial singular causes being terminant and contributor respectively. The account of general causation was provided in terms of derivative relations among derivative events, where claims of the form "*C*'s cause *E*'s" were rendered as relations of the form '$\tilde{C}$ promotes *E*'. The bottom and middle conceptual layers of causation, I have argued, are sufficient for an adequate account of both singular and general causation in the sense of providing a general structure such that if any singular or general claim is made precise enough for there to be a corresponding relation in my metaphysics—a relation of contribution, or terminance, or prob-influence, or partial influence, etc.—then fundamental reality will suffice for a definitive answer as to whether the causation-like relation obtains. In this chapter and the next, I will attend to the top conceptual layer in order to address how the kind of singular causation that humans tend to employ in everyday causal reasoning and in the sciences relates to the metaphysical conception of causation I have laid out.

There are numerous respects in which the kind of singular (partial) cause that appears in my metaphysics of causation—contributor—does not match the kind of singular cause that is implicit in claims like, "One of the causes of the fire was the fuel spillage." These differences are significant enough to provoke an accusation that my account of singular causation is deficient. To review, here are some notable differences:

- The ordinary usage of 'cause' is far more discriminating than 'contributor' because contributors include absolutely everything that plays a fundamental role in the effect's coming about, whereas 'cause' neglects almost all such contributors.
- Unless fundamental metaphysics takes a special form, events from the future contribute to present events, whereas ordinary usage typically only counts events from the past as causes.
- Necessarily, every event is a contributor to itself, but people tend to think that events rarely if ever cause themselves.

TABLE 8.1 The Three Conceptual Layers of Causation

Layer	Subject	Metaphysical status	Standards of adequacy
Top	Non-metaphysical aspects of causation	Derivative	RELAXED
Middle	Derivative metaphysics of causation	Derivative	STRICT
Bottom	Fundamental metaphysics of causation	Fundamental	STRICT

- Events that play the role of singular cause in most metaphysical theories can be coarse-grained, whereas contributors are always as fine-grained as fundamental reality allows.

In order to discourage readers from conflating different interpretations of the word 'cause', I introduced[69] the expression 'culpable cause' in §1.4 to designate the kind of cause implicit in philosophers' regimentation of our folk attributions of singular causation. Although 'culpable cause' is terminology I introduced, the concept of a culpable cause is not my invention, nor does it have any technical meaning in my account. 'Culpable cause' is merely a bookkeeping device to designate the kind of cause routinely discussed in philosophical discussions of singular causation. I suspect this kind of cause has been sometimes called an 'actual cause', for example in (Glymour and Wimberly 2007), but it would be misleading to apply the word 'actual' to a cause when 'the actual world' refers to fundamental reality, and 'actual events' refer to instantiated fundamental events. Culpable causes have also been called "egalitarian" (Hall 2004) causes, but I prefer not to use that expression because the so-called egalitarian causes are far from egalitarian. They exclude the vast majority of events that play an essential role in the fundamental development of nature leading to the occurrence of the effect. Like Orwell's pigs, some egalitarian causes are apparently more equal than others.

My treatment of culpable causation has been deferred until this chapter because it plays no essential role in the metaphysics of causation. Culpable causes are relevant to our understanding of causation because they play a role in the special sciences including our explanatory practices. In brief, culpable causes exist because our concept of causal culpability is a useful cognitive device by virtue of the causation-like relations from the middle and bottom conceptual layers of causation. This chapter is dedicated to making good on my promise in §1.10 to demonstrate that once we have adopted the structures from the bottom and middle layers as part of the empirical analysis of causation, culpable causes become inessential to the *metaphysics* of causation and can thus be safely reassigned to other disciplines that can evaluate theories of causation in terms of the more appropriate RELAXED standards of adequacy.

[69] Recall that 'culpable cause' was invoked by Alicke (1992) for other purposes. Recall also the possible ambiguity discussed in §1.10.

Despite my best attempts to convey what I have in mind, the claims expressed in this chapter have been frequently misinterpreted, so, as a precautionary measure, I will issue a few preliminary disclaimers that I hope will inhibit at least some of the miscommunication. First, I will not be arguing that culpable causes do not exist. They do exist. Second, I will not be arguing that culpable causes are unimportant to science. They are important to science. Third, I will not be arguing that our talk of culpable causes should be eliminated in favor of talk of contributors and promoters. It is acceptable to refer to culpable causes. Fourth, I will not be arguing that culpable causes should be left out of a philosophical account of causation. If I thought they should be left out, I would not be discussing them throughout this chapter and the next. Fifth, I will not be arguing that the roles culpable causes play in our explanatory practices should be ignored. Quite to the contrary, I think their explanatory role is so important that we should not hinder our understanding of it by mistakenly thinking that culpable causes need to be integrated with the *metaphysics* of causation. Instead, culpability can be shown proper respect by understanding it as suitable for an empirical analysis of the non-metaphysical aspects of causation, which is the fitting home for any special sciences that invoke causal culpability and scientific approaches toward causation in the special sciences such as the philosophical and scientific work based on causal modeling.

The conclusion I will advance is that causal culpability should not be considered part of the *metaphysics* of causation. This rhetorical maneuver is not a verbal trick carried out by redefining 'metaphysics' in some highly idiosyncratic way to exclude causal culpability by fiat and thus evade the many problems it generates for theories of causation. Instead, it is based on a slightly idiosyncratic precisification of 'metaphysics' that enforces a necessary condition on the adequacy of any metaphysical theory, namely the condition I defended in §1.10, which says a satisfactory metaphysical theory needs to obey STRICT standards of adequacy.

The underlying idea from §1.6 and §1.8 was that fundamental reality is consistent,[70] and that any apparent conflicts in a metaphysical theory must be explicitly ameliorated because there is no other discipline to which metaphysicians can delegate an apparent conflict for amelioration. Because metaphysics, as I understand it, primarily concerns fundamental reality and secondarily anything in derivative reality that is so closely connected to fundamental reality that no other identifiable discipline can be reasonably expected to ameliorate its apparent conflicts, metaphysics needs to avoid having any genuine conflicts. The many special sciences employing causal notions, on the other hand, can get by just fine using the RELAXED standards that allow for more flexibility and imprecision in one's theories. There is always a place for them to delegate apparent conflicts because the metaphysics of causation developed in this volume provides a comprehensive

[70] By this, I mean that it obeys the metaphysical correlate of the law of non-contradiction. Thus, any theory of fundamental reality with apparent conflicts must be either inaccurate or incomplete.

explanatory backstop to fill in the details for any special science whose treatment of causation is not STRICT.

Chris Hitchcock (2007) advances a thesis that is somewhat similar to mine:

> The folk make causal attributions when they assign praise and blame for various salutary and untoward outcomes. As philosophers, we naturally seek an objective basis for these folk causal attributions. In fact, the scientific concept of causation captures all there is to this objective core; beyond that, there are only our value-infected attributions of causal responsibility. As we have seen, the causal structure described by causal models does not yield causal relations with the grammatical form '*C* causes *E*', where *C* and *E* report events, facts, or states of affairs. The objective causal structure of the world has a more complex grammar. However, in searching for the objective basis of our folk causal attributions, philosophers have mistakenly assumed that this objective basis will have the same logical structure as those attributions, that of a binary relation between events. This mistake has led philosophers to seek the objective basis of folk causal attributions in a metaphysical concept of causation, rather than in the causal structure of the scientific conception. This picture, if correct, would help to explain the failure of philosophers to capture the metaphysical concept of causation in objective terms. Metaphysical causation,[71] it turns out, is an unstable compromise between the scientific and folk attributive concepts of causation: it seeks to retain the logical structure of the folk attributive concept while retaining the objectivity of the scientific concept.

Hitchcock defers to causal modeling approaches for the scientific concept of causation whereas I have formulated my own scientifically-based metaphysics of causation, but we have roughly the same idea in mind: to the extent that the notion of culpable cause tries to satisfy two opposing masters—science and common-sense judgment—it serves neither master well.

It is the purpose of this chapter to provide an argument that is in the same spirit as Hitchcock's. I will conclude that if one has already adopted the metaphysical machinery of my account, culpable causes do not need to be included as part of this metaphysical machinery but can instead be categorized as non-metaphysical, specifically as part of an empirical analysis of the non-metaphysical aspects of causation. The advantage of such a categorization is that any conflicts in a theory of culpability can then be delegated to my metaphysics of causation, freeing us from having to find a conflict-free set of rules for when an event counts as a culpable cause. This permits us to agree to many common-sense rules of thumb governing

[71] Note that by 'metaphysical causation', Hitchcock is referring to causation as understood in the literature on the metaphysics of causation, not causation insofar as it is related to fundamental reality. I believe that I have captured a "metaphysical concept of causation in objective terms" in a scientific conception.

causal culpability without having to ensure that they can all be consistently applied to every realistic scenario.

The argument is structured as follows. First, I assume the acceptance of empirical analysis as the operative method of investigation. I briefly defended empirical analysis in chapter 1, and I have made additional commentary available so that no more comment is needed here.

Second, given the discussion in chapter 1, I assume that all the empirical phenomena bearing on causation can be sorted (at least roughly) into two groups: empirical phenomena regarding causation insofar as it is "out there in reality" and psychologically-oriented empirical phenomena. The first category subsumes all phenomena we uncover by investigating how the world itself behaves in its causation-like way. This includes almost all the roles causation plays in digestion, combustion, gravitation, cogitation, and communication. The second category subsumes the empirical phenomena revealed when we examine creatures to ascertain how they think about causation. This includes people's judgments about what caused what, the amount of time infants look at unnatural causal behavior, and any evidence concerning what people think should bear on a causal explanation of any given event.

The core step of the argument—to be presented in the next section—concerns the role of culpability in accounting for empirical phenomena of the first category. The core step concludes that culpability *per se* has no role. Claims about causal culpability are undoubtedly partially about the external world. My contention is that the extent to which claims about culpability concern the external world exactly matches the extent to which such claims are parasitic on facts about terminance and prob-influence. Once all the relations of terminance and prob-influence have been taken into account and held fixed, any further details that bear on whether a given event is culpable for the chosen effect do not concern the external world.

Because metaphysical theories of culpable causation are typically offered in the hope that they will match philosopher's intuitions about which happenings are singular causes (of some chosen effect) in any reasonably realistic scenario, the sought-after notion of cause—culpability—is relevant to the second category of empirical phenomena and thus is relevant to the psychology of causation and theories about explanation. Furthermore, as I discussed in §1.10, there is no apparent reason why we need a concept of culpability that is simultaneously optimized for metaphysical purposes (by being related to other metaphysical concepts in a principled system that obeys STRICT standards) and psychological purposes (by matching strongly held pre-theoretical intuitions about culpability) if we already possess a principled STRICT account of the metaphysics of causation and a related RELAXED account of causal explanation and our psychology of causation. Lacking any such reason, it is best to think of culpability as inhabiting the top conceptual layer of causation, which includes all aspects of causation that do not need to be part of the STRICT portion of the theory of causation.

In the next section, my goal is to spell out the core step of the argument. After that, I will identify three useful roles that causal culpability can play despite its lack of metaphysical significance. First, our intuitions about culpability are handy for learning about promotion. Second, culpability is relevant to certain kinds of causal explanation. Third, disputes over the culpable causes of some chosen effect can be about causation "out there in reality" insofar as claims about culpable causes serve as proxies for claims about terminance and promotion. In the next chapter, I will illustrate how the greater flexibility allowed by the RELAXED standards applies in practice by presenting a simplistic theory of culpable causation that coheres with the metaphysical scheme I have advocated.

8.1 The Empirical Insignificance of Culpability

The core step of the argument is presented here in two parts. The first part attempts to demonstrate that facts about culpability—that some event *C* was causally culpable for a given event *E*—do not count as empirical phenomena in any single region, R, of space-time in a way that goes beyond the more detailed empirical phenomena in R that culpability facts presuppose. The second part shows that culpability facts do not improve one's account of general empirical phenomena: predictions about what will happen in locations other than R. Thus, culpability *per se* plays no role in an optimal account of any empirical phenomena relevant to causation qua some structure of the external world. That is, it plays no role once we already have in place the more detailed structure provided by the comprehensive metaphysical relations of my account.

My argument concerning culpability in this chapter does not rely on the details of my particular account. Specifically, it does not require that the empirical phenomena relevant to causation be related to the effectiveness of strategies nor does it presuppose that the fundamental causation-like relations are precisely the relations I have identified as terminance and contribution. Although this volume has mostly relied on the auxiliary hypothesis that fundamental reality resembles the paradigm theories of fundamental physics, it is an open question whether fundamental reality more closely resembles our folk attributions by instantiating fundamental cause-effect relations holding among mundane events. My argument in this section is not meant to dismiss this possibility. Instead, my conclusion should be understood as claiming that once we have a complete and accurate account of all fundamental causation-like relations, possibly including these cause-effect relations among mundane events, any further discrimination among all such causes would not be needed to explain any external-world-oriented empirical phenomena. I personally do not think the incorporation of cause-effect relations among mundane events would help any empirical analysis of the metaphysics of causation, but if people are interested in constructing a model of how the kinds of interactions in physics relate fundamentally to cause-effect relations among mundane events, I wish them the best of luck.

One final note deserves emphasis. The sort of explanation that is relevant to my argument in this section is a "complete story" causal explanation. Although many kinds of explanation depend essentially on abstracting away from the details of a completely comprehensive account of why some effect occurred, the purpose of my metaphysics of causation is to provide a completely general explanatory resource to which the special sciences can delegate any apparent conflicts concerning singular or general causation. Because a "complete story" causal explanation can always have its information watered down to recover other kinds of explanation, we do not need to consider them here.

8.1.1 PART I: SINGULAR CAUSATION

In the first part of the core step, the topic under discussion is whether culpability *per se* constitutes some sort of empirical phenomenon in a particular fragment of history. A fragment of history is a region of space-time together with all the material contents in that region. If the arena—defined in §2.1 as the fundamental container of all the fundamental material contents of the universe—is just space-time, then a fragment of history can be identified with some full fundamental event, *f*, located in some space-time region R. If the arena is not space-time, a fragment of history can still be adequately defined in terms of an abstraction from the arena and its material contents. We just stipulate *f* to be some fundamental event that includes all the fundamental materials needed for a complete, maximally detailed account of the metaphysics of causation in the space-time region R.[72]

Remember also that an event is never a culpable cause simpliciter but only a culpable cause of some chosen event. So when identifying culpable causes within a fragment of history, one must first choose a coarse-grained event, *E*, to play the role of the coarse-grained effect.[73] To summarize, *E* is any happening in space-time that we choose, R is any space-time region that encloses *E*, and *f* is some fundamental event that specifies the shape and contents of R as much as fundamental reality allows.

---

[72] As discussed in §2.13.1, some models of quantum mechanics posit a quantum configuration space as one component of the arena in addition to the space-time component of the arena. If the arena contains more than space-time, *f* can be identified with some full event large enough so that making it any larger has no consequences for what exists (derivatively or fundamentally) in R. For example, *f* might be defined to include the entire universe's quantum state throughout the duration of time corresponding to the space-time region R. It is also possible that space-time is not even part of the arena but is metaphysically derivative. For example, it is possible that fundamental reality is constituted by superstrings vibrating in an eleven-dimensional arena. In such cases, *f* may again be identified with some fundamental event that suffices for everything that happens in R when supplemented with appropriate fundamentally arbitrary parameters.

[73] As mentioned in §2.1.1, the discussion throughout this book construes causation in terms of events, but none of the important conclusions require an event ontology. In this case, I am using 'fragment of history' to highlight the possibility that the arena is not space-time and that the happenings in space-time are therefore not necessarily coarse-grainings of fundamental events.

One of the constitutive features of the concept 'causal culpability' is that it is much more selective and discriminate than notions like contribution and partial influence. Many events do not count as culpable causes of *E* even though they play a fundamental role in how the instance of *E* comes about or make a difference to the probability of *E*. In order to serve its purpose, the concept of culpable cause needs to exclude events whose influence on *E* is negligible and needs to focus instead on a limited class of events that are somehow important to the occurrence of *E*. And the issue to be addressed now is whether, for any event *C*, *C*'s culpability for *E* is a fact that we have empirical access to, beyond our access to the less discriminate causation-like relations.

Some aspects of *f* count uncontroversially as empirical phenomena in the ordinary sense of 'empirical'. For example, *f* might instantiate a hammer sinking a nail into a wooden board. In that case, the existence of the hammer is empirically accessible, the motion of the nail is empirically accessible, and the later presence of the nail in the board is empirically accessible. Certainly, as a general matter, it can be quite unclear or vague or indeterminate precisely which aspects of the fragment of history should count as empirical phenomena. However, regardless of any limitations on our grasp of what is empirical, we can safely infer that 'being a culpable cause' *per se* is not an empirical phenomenon in the sense that once we hold fixed the full array of fundamental details in *f*, there is no further fact of the matter that can be checked to assess whether *C* is genuinely culpable for *E*. There are no experiments one can run to confirm or disconfirm whether *C* is a culpable cause of *E* rather than a non-culpable contributor. This sense in which culpability *per se* fails to count as empirical is extremely weak, but I am not attempting to defend a strong thesis here. The conclusion is merely that relations of culpability do not constitute some empirically accessible metaphysical or ontological addition to the fundamental event, *f*. This conclusion is on a par with the claim that we have no *empirical* access to whether I have two pencils in my hand rather than two pencils plus their mereological fusion.

It is also worth recognizing that causal culpability does not play any role in how nature evolves. An event *C* is culpable for some effect *E* not because *C* instantiates some culpable-for-*E* attribute that the laws of nature incorporate in their rules for how things evolve. Rather, if *E* eventually occurs with the right kind of relation—whatever that may be—to *C* and possibly to other events, then the predicate "culpable for *E*" is applicable to *C*. *C* is a culpable cause of *E* not by virtue of what *C* instantiates but by virtue of the richer array of components in *f*. The fact that culpability is to be *inferred* from the details of *f* (rather than supplement *f*) implies that facts about culpability do not require an empirically discernible addition to *f*.

8.1.2 PART II: GENERAL CAUSATION

For the second part of the core step, the topic under discussion is whether culpability relations among events in R are useful for characterizing testable claims about what happens in regions other than R. The conclusion I defend in this subsection is that culpability is not a concept that helps to optimize our predictions about what happens elsewhere so long as we already have general dynamical laws and an inventory of the kinds of attributes that are instantiated in the actual world. Nor does culpability improve any "complete story" explanation (in terms of dynamical laws) of why some happening elsewhere evolved the way it did. Whenever we are trying to evaluate whether some instance of C (contextualized as $\overline{C}$) will bring about a certain effect E, our best estimate of the laws alone provides our best guide for what will happen (or what has happened) in situations we have not yet investigated. The guidance provided by the laws is not bolstered by information about whether an event, C', was culpable for some event, E', in a particular fragment of history. One reason for this is that facts about culpability can take into account the outcomes of fundamentally chancy processes that occur in f, and these outcomes need to be ignored in our predictions of what will happen elsewhere.

For example, imagine we know that events of type $\overline{C}$ are followed by events of type E eighty percent of the time by virtue of what follows nomologically from $\overline{C}$. Then, suppose we learn that an event of type $\overline{C}$ was instantiated in some fragment of history without learning any information that should lead us to adjust our belief that $\overline{C}$'s probability distribution is our best guide to which of its members was instantiated. If we were then asked to make a prediction about whether an instance of E occurred after $\overline{C}$, the rational response would be to predict that there is an eighty percent chance of E. Now imagine we are given further information that in some remote fragment of history, there was an instance of $\overline{C}$ followed by an instance of E and that C was culpable for E. If this information does not motivate altering our credences about the operative laws or our opinions about how $\overline{C}$ is likely to be instantiated in our nearby fragment of history, it should not affect our prediction that E is eighty percent likely. After all, the defining characteristic of a dynamical law is that it provides a general rule for how things evolve through time, and the defining characteristic of a fundamentally chancy outcome is that it cannot be reliably predicted beyond what can be predicted about its chances. More generally, facts about culpability do not help in formulating testable claims about what happens in other regions that improve on what follows from the laws and the likely layout of the material content.

Combining the two parts of the core argument, culpability facts do not augment or improve predictions or "complete story" explanations for what happens in R or out of R. Thus, the concept of culpable cause is not optimized for any empirical phenomena relevant to the *metaphysics* of causation. This conclusion implies that if we have a STRICT account of singular causation in the form of terminance and contribution relations and a STRICT account of general causation in the form of prob-influence relations, then there are no further empirical phenomena that could serve as a target for a STRICT account of culpable causation. Hence, we are free to treat culpable causation not as part of the metaphysics of causation, but as a subject for the special sciences, broadly construed. For example, we can regiment our conception of culpability to fit data from the psychology of causation or adapt it to fit a desired account of causal explanation.

Recall that in §1.10, I mentioned that one of the tasks central to orthodox investigations of the metaphysics of causation was to provide a STRICT account of culpable causation. One can see that this activity is prevalent in the professional philosophical literature on causation just by examining the many instances in which common-sense judgments of singular causation are employed as counterexamples to theories of causation. For example, certain cases of preëmption are routinely (Menzies 1989, 1996; Hitchcock 2004; Schaffer 2008) understood as at least prima facie counterexamples to the claim that probability-raising is necessary for singular causation. This orthodox demand of an adequate account of causation is unnecessary according to my empirical analysis of the metaphysics of causation.[74]

To summarize, my conclusion is simply that the kinds of intuitions about culpability that philosophers standardly attempt to regiment do not need to be included in the *metaphysics* of causation. Instead, we are free to formulate theories of culpability safe in the knowledge that any potential ambiguities in one's rules for identifying the culpable causes of some effect can always be cleared up by citing all of its contributors and how they fit together as terminants. Stated this way, my conclusion that causal culpability does not constitute an empirically discernible addition to the fundamental metaphysics ought to sound trivial because the fundamental events and their fundamental causation-like relations were deliberately constructed to include *all* singular happenings that play any role in bringing about events. I designed the conceptual scheme proposed in this volume to make this conclusion a truism. I am certainly not trying to defend a contentious claim,

[74] That culpable causes are not to be left out of an adequate metaphysics of causation is a proposition largely taken for granted among philosophers. However, Nancy Cartwright, in her (1994) *Nature's Capacities and their Measurement*, explicitly defends the view that culpable causes are needed in order to provide an adequate account of effective strategies. Her argument presupposes a conception of probability that my theory does not employ, and that alone ensures that her argument cannot by itself rule out the possibility that my account succeeds at adequately addressing effective strategies without using culpable causes. As I noted in §2.2, her conception of a probabilistic relation, in the tradition of Reichenbach (1956), Suppes (1970), and many others, is the kind one infers from statistical data, typically concerning mundane events. (Recall the definition of mundane event from §1.10.) An example of such a probabilistic relation is the conditional probability of a person having liver cancer, given the person's history of heavy alcohol consumption. For all that I have argued, Cartwright may be correct that if one starts from such probabilistic relationships, one cannot get to causal regularities without facts concerning causal culpability. However, I did not employ that particular notion of probability but instead a notion of probability derived from either chanciness in the fundamental laws or from a stipulation of the contextualized events used to abstract away from the fundamental material content. (The utility of the stipulated probability distributions is in turn justified in terms of the insensitivity considerations discussed in the literature on the foundations of statistical mechanics.) It was just my starting point to assume that the conception of probability my account employs is cogent enough to serve as an adequate basis for an explanation of effective strategies. Someone could certainly attempt to challenge the viability of the notion of probability I relied on, but that would go well beyond the considerations offered by Cartwright in her defense of the need for culpable causes.

say, that talk of culpable causes is meaningless, or fails to refer to events in the external world, or does not illuminate features of fundamental reality that are important components of reality. The reason I have dedicated so much verbiage to defending what should be uncontroversial was not to establish any claim about reality or fundamental reality, but rather to clarify the boundary of metaphysics and thereby make apparent that the same RELAXED standards that are acceptable in the special sciences are also acceptable for theories of causal culpability.

8.2 Culpability as a Heuristic for Learning about Promotion

It might be puzzling—if culpability is metaphysically superfluous—why people would ever come to have a significant number of shared intuitions about it. A sizable component of the answer is that our intuitions concerning culpable causes are useful for simplifying the process of identifying promoters, which in turn can help us manipulate the world to our advantage. The identification of regularities of the form '$\check{C}$ promotes *E*' does not *require* the notion of culpable cause because one could in principle just analyze lots of data about which initial states lead to which final states in order to find prob-influence relations. But with resources for collecting such data being limited, it is useful for creatures to have heuristics that identify promoters and inhibitors quickly, and the culpability concept implicitly incorporates rules of thumb for just such a purpose.

For illustration, consider an experimental setup with a single window and two mechanical throwers, *A* and *B*, that are each capable of tossing a single rock at the window. On each experimental run, both throwers simultaneously toss their own rocks at the window. For simplicity, assume that (1) each thrower has an accuracy that remains constant over time, (2) there are no other potential causes of window breakage, (3) any rock either definitely goes through the window or definitely misses, (4) when both rocks strike the window, they do so at the same time in a symmetrical manner with no borderline cases, and (5) when a rock strikes the window either alone or with the other rock, its momentum is sufficient to break the window and carry the rock through the opening. The data collected from each experimental run includes tracking the complete path of each rock from its thrower. The experiment is repeated so many times that there is every reason to believe that the statistical outcomes reflect any underlying chances, fundamental or derivative.

From such an experiment, we can learn the probability of window breakage that is fixed when *A* and *B* are operating together simultaneously. But suppose we want to know about *B*'s role in the causation. There are two distinct questions one could ask. One is, "When *A* and *B* are operating together simultaneously, what fraction of the experimental runs result in breakage that is caused (at least in part) by *B*?" and the other is, "What is the chance that the window will break if *B* operates alone?"

Notice that the second question has an answer we can ascertain by conducting numerous experimental runs with *A* being shut off or removed. Direct observation of such runs would reveal the probability of breakage when *B* works alone. We do not need to know how often *B* was culpable of window breakage in trials of *A* and *B* together.

Yet, it is much easier to answer the first question and simply repeat that answer for the second question. The pattern of culpability exhibited in *B*'s tosses when *A* and *B* are running simultaneously is an excellent guide to *B*'s likelihood of breaking the window when operating alone. In the simplified environment of this example, the culpable cause is easy to identify. A thrower's toss is a culpable cause of the breakage if and only if its rock makes a continuous flight to the window. We can just observe the throwers in action, and from any broken window, we can trace back the path of any rocks to the thrower(s) responsible for the breakage. This rule adequately satisfies all relevant pre-theoretical intuitions about which thrower to blame for the breakage. One might argue that our intuitions are not perfectly clear about how to interpret culpability when both rocks break the window simultaneously, but owing to the symmetry, they have to be equally culpable. It proves more convenient to assume that when the rocks strike the window at the same time, both throwers are each fully blameworthy for the breakage, so I will make this assumption in order to more accurately estimate the effectiveness of the tosses.

B's culpability in a fraction, *x*, of the trials where *A* and *B* operate together simultaneously is strong evidence that *B* alone (when contextualized using roughly the same background conditions) will fix the probability of breakage at *x*. Thus, the rules of how to attribute culpability allow us to draw useful inferences to other situations and avoid the need for gathering additional experimental evidence. The crucial simplification that allows culpability to track promotion in this example is the lack of any significant interaction between the flights of the rocks. If *A* were constructed to periodically interfere with *B*'s accuracy, the fraction of times that *B* was culpable would not equal *B*'s accuracy in cases where *B* operates alone. When we are in a position to know that the interaction between *A* and *B* is slight, we can make use of our instinctive grasp of culpable causes to infer the degree to which the *B* system alone promotes the breakage of windows.

Although this is just one example, there is nothing blocking its generalization to many other situations. The inference from patterns of culpability to relations of prob-influence is a defeasible one, but much of our causal reasoning involves macroscopic objects that interact with one another in limited ways. In such cases, our practice of identifying culpable causes by tracing back in time from the effect to other prominent events by way of some recognizable process can be expected to be useful for drawing quick inferences about promotion.

By justifying the utility of causal culpability in terms of learning about promotion, I do not mean to exclude other justifications. Culpable causation is arguably implicit in many other practices, such as moral culpability. I have focused on learning about promoters because it explains why a notion of causal culpability would be useful for virtually any reasonably intelligent creature that is capable of perception and basic action.

8.3 Culpability as an Explanatory Device

Culpability also plays a role in causal explanation. Although explanation in general is difficult to characterize adequately, there are some kinds of explanation that we grasp well enough. One kind of explanation for why some event *E* occurs is to cite certain previous promoters. Citing promoters helps to explain *E*'s occurrence because the existence of the promoters makes the effect more likely than it otherwise would have been, and one role of explanation is to place occurrences in a context where they are less surprising.

We often have epistemic access to more facts than just which promoters were present. We might have evidence that a particular promoter did not *successfully deliver* its promotion to the effect: that it was preëmpted or that its connection to the effect somehow fizzled out. I will discuss such cases in more detail in §9.5 to unpack the metaphor of successfully delivering promotion, but the basic idea should be evident from examples where a rocket's fuse burns out before it can initiate a launch. Acquiring information about the fizzled connection undermines the usual inference one draws between the presence of a promoter and the likelihood of the effect. Thus, a more thorough explanation of why *E* occurred in some particular scenario would involve citing only the promoters whose probability-raising was effectively communicated to the effect, the events that raise the probability of *E* and do not later fizzle out or succumb to preëmption. It is easy to see why one should avoid citing fizzled promoters as causes of *E*. Knowing of the existence of *C*'s fizzled connection puts people in an epistemic position where the information that *C* occurred no longer makes *E* less surprising. Nevertheless, there is no need to require that our models of fundamental reality must avoid granting unexplanatory events the same ontological status as explanatory events. Explanation is an epistemological notion that can be more flexible and context- and interest-dependent than the metaphysics itself, and nothing about the explanatory role of culpability requires that it has a prominent role in metaphysics, as I precisified the term 'metaphysics' in §1.10.

8.4 Culpability as a Proxy for Terminance and Promotion

On the one hand, I contend that culpable causation has no role to play in an empirical analysis of the metaphysics of causation, given the comprehensive applicability of the causation-like relations from the bottom and middle conceptual layers of causation. On the other hand, scientists and ordinary folk routinely make claims about culpable causes and engage in debates over whose claims of culpability are better confirmed by the available data. In this section, I will illustrate how the practice of citing culpable causes can be legitimate within the bounds of the metaphysics of causation I have endorsed. The resolution of the tension is that our practices for settling questions about causal culpability are vindicated by the existence of singular causation-like relations (explained by concepts from the bottom conceptual layer of causation) and the existence of general causation-like relations (explained by concepts from the middle conceptual layer of causation). If a debate about culpability proceeds far enough to where the disputants agree about all the relations handled by the bottom and middle layers, there is nothing left but a dispute about which precisification of culpability one prefers or about which events deserve more emphasis in a causal explanation.

I will first present the reasoning with regard to the food analogy from §1.1 to illustrate what I think should be recognized as an uncontroversial illustration of when a dispute becomes merely a semantic quibble. Then, I will present a structurally identical exposition of how to resolve the debate over the causes of the mass extinction at the boundary of the Permian and Triassic periods, the P-T extinction.[75] The net result is an explanation of how scientific debate over causal culpability can be legitimate by virtue of the existence of terminance and promotion relations, without requiring that there ultimately be some *further* fact of the matter as to which events (culpably) caused the P-T extinction.

Suppose we have a substance *S* in front of us, and we are interested in answering the question, "Is *S* food?" We can think about resolving this debate in stages. In the first stage, there is a precise enough fact of the matter about the detailed character of *S* which perhaps can be ascertained by conducting chemistry experiments and other related investigations. If the fully detailed characterization of *S* is such that it is not even a remotely reasonable candidate for being identified as food, say by being composed entirely of neutrinos or molten lava, we can rule definitively that *S* is not food. Likewise, if *S* is paradigmatically food according to all the reasonable heuristics we could use for judging whether something is food, then the debate should be settled in favor of its being food.

But suppose we have a case that is more controversial, such as a beetle or an aspirin. In stage two, in a further attempt to resolve whether *S* is food, we can consider whether *S* is nutritional because nutrition is the primary reason we have a food concept. Whether *S* is nutritional is a testable fact in the sense that, in principle, duplicates of *S* can be fed to various people whose later health outcomes can be monitored. Such facts also incorporate a lot of implicit relativization; what counts as a nutrient for one population may not count as a nutrient for another. The same substance might be nutritional along one dimension, for example providing glucose, but non-nutritional along another dimension, for example

[75] This particular controversy was mentioned by Jim Woodard at the 2008 meeting of the American Philosophical Association, but any example will do.

by aggravating an allergy. All of this relativization is easily understandable, and insofar as disputes about food are debates about nutrition, these debates are to be settled by relativizing claims to the relevant parameters. If the disputants can be satisfied that their disagreement was merely the result of a disagreement over which parameters they were using to judge the nutritional character of *S*, then we have a resolution to the debate.

Finally, in stage three, we can consider the possibility that the debate continues over *S*'s status as food even after all of its relations concerning nutrition have been settled to everyone's satisfaction. Because the disputants agree on all of the nutritional facts (including any facts about the effects *S* would have on any person who ate it), the only remaining source of contention is due to the mismatch in extension between 'nutrient' and 'food'. However, if that is all the debate consists of, it is no longer an interesting scientific debate about *S* itself. For example, one person might say an iron crowbar is food because iron is a nutrient and an iron crowbar is nothing but a bunch of iron; another person might disagree by saying that a priori an iron crowbar is not food and that this conceptual truth is revealed just by asking ordinary people whether an iron crowbar is food. In general, one might view the third stage as involving disagreements over the relative value of various precisifications of the food concept, or one might view it as involving debates about human psychology—for example, whether people on the whole identify *S* as a food—but in any case, the substantive debate is not about *S* itself.

Now, let us apply these lessons to the debate over what caused the P-T extinction, as summarized by Benton (2005). One theory proposes a cataclysmic strike of meteors as a source of atmospheric debris, and another theory proposes the extensive volcanic activity that formed the Siberian Traps. There appears to be a substantive scientific debate about which theory is correct, and evidence can be amassed in favor of one theory over another in an uncontroversially scientific manner. The practice of citing either the volcanos or the meteors as culpable is legitimate to the extent that debates about whether meteors rather than volcanos were the cause of the extinction can be cashed out in terms of debates about contributors and promoters. To the extent that such citations go beyond what is justified purely in terms of terminance and promotion, they implicitly rely on heuristics whose justification does not lie in metaphysics or in sciences that concern causation "out there in reality." Rather, they rely on intuitions about culpability that exist in part because these intuitions are handy for creatures with limited epistemic capacity who need efficient means to gather information about promotion. According to the methodology of empirical analysis, debates that go beyond terminance and promotion are resolvable only to the extent that they are debates about evidence concerning people: what factors they employ when reasoning about causation, providing causal explanations, producing moral judgments, etc.

For illustration, let us now adjudicate the debate in stages. In the first stage, because there are presumably complete facts about what exists fundamentally, including every microphysical detail concerning the extinction, there is a fact of the matter about how many meteors existed and a fact of the matter about how much volcanic activity existed around the relevant time, t, some 251 million years ago. If the causal contributors to the extinction do not include significant volcanic activity, we can in principle come to know that the volcanic activity should not be counted among the culpable causes of the extinction, and so the debate would end with the volcano theory being eliminated. The same principle applies, mutatis mutandis, if the full set of causal contributors contains an insufficient number of meteors. If both meteors and volcanos are known to be present in enough quantity at t, we need to look at further details to settle the debate. For the sake of discussion, we can even imagine everyone having ideally precise information about all the microscopic details of history available to them. If the disputants still disagree as to what caused the P-T extinction in light of the full microscopic history, we need to move to the second stage.

In the second stage, the disputants can consider various promoters that are suggested by contextualizing the precise state of the world at t (and other nearby pre-extinction times) and comparing them with what follows from various hypothetical contrasts. One can do so by taking the actual state at t and coarse-graining over the volcanic activity and meteor activity to blur over the many microscopic differences in how those events are instantiated (adding any reasonable probability distribution). What results is a contextualized event, $\overline{C}$, representing the actual condition of the world at t. A reasonable contrast for evaluating the promotion exerted by the volcanos is a contextualized event, $\overline{\neg C_v}$, which is identical to $\overline{C}$ except that the erupting volcanos and those that are on the way toward eruption are replaced by dormant volcanos. Similarly, a natural contrast for evaluating the promotion exerted by the meteors is a contextualized event, $\overline{\neg C_m}$, which is identical to $\overline{C}$ except that the meteors heading toward Earth are replaced with empty space. Finally, we can consider a contextualized event, $\overline{\neg C_{v\&m}}$, that is missing both the meteors and the eruptions. One measure for the volcanos' promotion of the extinction, E, is $p_{\overline{C}}(E) - p_{\overline{\neg C_v}}(E)$. Similarly, $p_{\overline{C}}(E) - p_{\overline{\neg C_m}}(E)$ is one good measure of how much the meteors promoted the extinction. However, there are many other ways to contrastivize the actual world to highlight various causal factors. In a highly idealized sense, we can come to know about all the relevant prob-influence relations by knowing the actual state of the world and knowing what the laws dictate about suitable contrasts. Just as in the case of nutritional facts, it can turn out that on one way of abstracting away from the historical details, the meteors promoted the extinction more than the volcanos, and on another way, the volcanos promoted the extinction more. There might even be interactions between the effects of the meteors and the effects of the volcanos. To the extent that the dispute over the cause of the P-T extinction is a result of the employment of different assumptions about how to abstract away from the detailed microscopic history, one might be able to secure a resolution to the debate by making explicit all the various prob-influence relations. If so, we have settled the scientific debate. Otherwise we move to stage three.

For the third stage, we can suppose the disputants agree on the complete microscopic details of the actual history of the P-T extinction and the full range of probabilistic relations given the various ways of contrastivizing events. Is there anything of substance left to argue about? I cannot see how there could be. We already know everything about what *did* happen and everything about what *could have* happened. When the actual and the non-actual have both been comprehensively and adequately accounted for, there is nothing to argue about except anthropocentric factors: which simplifications count as more explanatory, whether the explanatorily relevant notion of causation should liberally permit chains of causation, etc.

I will make some additional comments about the explanatory role of singular causes at the end of chapter 10.

8.5 Commentary

Although one can make sense of the legitimacy of scientific debates about which events were the culpable causes of some chosen effect, the philosophical consequences of the metaphysical superfluousness of culpability should not be underestimated. A wide variety of concepts appear to rely on some notion of culpability, especially those concepts that require "the right kind of causal connection" between two events. For example, for a person *P* at time *t* to have *perceived* the object *O* arguably requires the right kind of causal connection between a temporal stage of *P*'s perceptual system and some temporal stage of *O*. What underlies our judgments of the right kind of causal connection appears to be approximately a proper subset of cases where the stage of *O* is a culpable cause of the stage of *P*. If we know that the stage of *O* was not one of the culpable causes of *P*'s perceptual condition at *t*, then we are usually justified in judging that *P* did not at time *t* perceive *O*. The important consequence of my account of culpable causation is the conclusion that insofar as the notion of perception is parasitic on the notion of culpable cause, it is parasitic on psychological heuristics that are geared toward learning about metaphysical aspects of causation. Because our concept of perception arguably incorporates a restricted version of our concept of culpability, a proper investigation of perception requires untangling the metaphysical connections that explain the utility of having some concept of perception from the common sense heuristics we use to identify cases of perception. Thus, an empirical analysis of perception will surely lead to a much different account than accounts based on orthodox conceptual analysis, both in its assessments of individual cases and in the nature of the connections between perception and other concepts. What goes for perception goes for many other concepts that incorporate notions of causal culpability. Empirical analyses of concepts like belief, meaning, reference, and knowledge, among others, will be substantially different from orthodox accounts.

{9}

The Psychology of Culpable Causation

Though causal culpability is metaphysically superfluous, it undoubtedly plays a prominent role in how we think about causation, including many of our explanatory practices. An adequate account of the metaphysics of causation ought to play a role in explaining why it is reasonable for humans to believe in culpable causes and why we have certain shared intuitions about culpability. Orthodox metaphysical accounts explain the reasonability of such beliefs by claiming in effect that these beliefs are true in the most literal sense. There are cause-effect relations out there in reality (in many cases holding between fairly localized singular events) as part of the world's metaphysical structure and people have a more or less accurate epistemic grasp of them. According to my account, belief in culpable causes is reasonable because there exist (metaphysically fundamental) terminance relations and (metaphysically derivative) prob-influence relations, and our intuitions about culpability serve as cognitive shortcuts for dealing with them.

In this chapter, I will construct a toy psychological theory whose primary purpose is to illustrate how my account of causation leads rather naturally to several heuristics for judging culpable causation. The toy theory shows how culpable causes help us learn about prob-influence along the lines of the discussion in §8.2. A secondary purpose of the toy theory is to complement my argument for locating culpable causation in the top conceptual layer of causation by demonstrating how many alleged problems in the metaphysics of causation dissolve once we acknowledge that a theory of culpable causation can be acceptable and informative and explanatory even if it has genuine conflicts and thus does not satisfy STRICT standards of adequacy. Once we reject that we should hold out for a complete and consistent systematization of cause-effect relations "out there in reality" that correspond to our folk conception of causation (or some moderately regimented version of it), many traditional puzzles about causation are easily resolved.

It is not my aim to provide anything remotely close to a full theory of the psychology of causation, nor even to provide a comprehensive theory of how people make judgments about culpable causes because that would be far too ambitious a topic. It would also distract from the main task of demonstrating that there is a reasonable link between my metaphysics of causation and the psychology of causation, broadly construed to include causal explanation. Furthermore, in order to keep this chapter as concise as possible, I have had to relegate some standardly discussed topics to an extended version of this chapter that I have made available.

Although I have attempted to construct the psychological theory in this chapter to accord with a wide range of stock intuitions about causation, it deserves to be called a toy theory for three reasons. First, it is a woefully simplistic theory that does not take into account the wide range of psychological data relevant to this topic and is only intended as a preliminary gesture.

Second, it does not produce any quantitative psychological predictions. For example, it does not provide enough structure to predict how much people's confidence in their judgments will change as they consider hypothetical situations that are ever more remote from ordinary experience. The toy theory does suggest some crude default predictions, but because I am unable to offer any principles that indicate where its predictions will be overridden by a more sophisticated treatment, there is no sure way to tell which failures of the default predictions are a result of its being based on an inherently defective scheme and which are merely the result of its being the toy theory it purports to be. So, whatever seeming success the toy theory has at explaining our common-sense intuitions about culpability should be weighed against the fact that it is not risking falsification with any bold predictions as a more serious theory would. (Also, I cannot address how the toy theory of culpability could be integrated with an account of the psychological mechanisms needed to implement assessments of culpability.)

Third, I am not pretending that the theory is free of counterexamples. On the contrary, one of my aims in discussing the toy theory is to illustrate a theory of causation that only meets RELAXED standards of adequacy. I will deliberately provide conflicting rules of thumb for identifying culpable causes in the technical sense of 'conflict' from §1.8. As foreshadowed in §1.10, my toy theory will not provide any formal rules sufficient to ameliorate these conflicts but will instead blithely delegate the conflict-resolution to my metaphysics of causation. In other words, whenever the rules of thumb I present for evaluating whether C is a culpable cause of E result in contradictory judgments in some realistic scenario, my theory declares that if you want consistency, you either (1) select one of the rules of thumb that is generating the consistency and stipulate that it is inapplicable to the scenario being considered, or (2) forgo talk of culpability in favor of contribution. You say it isn't clear whether C is a cause of E according to my theory? Fundamentally, all the contributors are partial causes of E, and there is always a definitive answer as to whether one fundamental event is a contributor to another. The more restrictive conception of singular cause that I have labeled 'culpable cause' is useful for epistemological purposes like causal explanation and discovering promotion relations, but these practices do not require STRICT consistency; a system of managed inconsistency is adequate.

Remember that because the purpose of the toy theory is to complement the metaphysics, its shortcomings do not undermine the metaphysical system provided in previous chapters. Psychological considerations could serve as evidence against a metaphysical account of causation only if the metaphysics were to make highly implausible the provision of a reasonable account of how humans could have the shared intuitions about causation that they have.

9.1 The Toy Theory of Culpable Causation

My metaphysics of causation says that (1) fundamentally, causation consists of terminants and contributors, which play the role of full and partial singular causes respectively;[76] and (2) we can abstract away from this kind of singular causation to get promotion relations, which adequately characterize general causation. If this is correct, our folk conception of singular causation among mundane events—culpable causation—is our imperfect way of grasping facts about terminance and promotion and the like.

Because one of the main reasons we have a notion of culpable cause is that it aids our discovery of promotion or prob-influence relations—a hypothesis I suggested in §8.2—we should expect this function to reveal itself in our judgments. It will turn out in §9.4 that there are discrepancies between what we would judge culpable if we cared only about whether that one particular instance of C affected the probability of that one particular instance of E and what we would judge culpable if we cared more about the *discovery* of prob-influence relations that apply to more general circumstances. When such discrepancies appear, according to my theory, we should expect our instinctive judgments concerning culpable causes to track the latter because such thinking would have greater practical utility.

A tension inherent in the idea of culpable causation is that it is a notion of *singular* causation that tries to incorporate features that essentially belong to *general* causation. On the one hand, it purports to apply to individual fragments of history, and, on the other hand, it privileges some contributors as more important to the occurrence of the effect than others. But the causal significance of each contributor in a single case ultimately derives from the fact that some kinds of events are *generally* good at bringing about other kinds of events. Culpability is what we get when we try to project onto individual fragments of history principles that govern general causation. Our implicit rules for assessing culpability are structured to mitigate the tension between the singular and general aspects of causation, but they do so imperfectly. Some of the implicit rules are easy to evaluate, but are less valuable as a guide to promotion relations. Others are harder to evaluate but provide a better guide to promotion relations. None of the rules carve nature at the joints. Our implicit conception of a culpable cause is a kludge that serves us well enough in practice, but whose implicit rules arguably do not systematize in a fully coherent way.

---

[76] Recall again that there are several important respects in which terminant relations do not match what we intuitively think of as causal, e.g., by being reflexive and not necessarily being asymmetric.

I think the core idea at the heart of culpability is this:

An event is a *culpable cause* of *E* iff it successfully induces *E*.[77]

To begin the investigation of this guiding principle, I will first impose a simplifying assumption, second comment on 'induces', and third comment on 'successfully'.

First, in assessing causal culpability, the starting materials include (1) a sufficiently filled-in scenario, which is a possible fragment of history with some sort of laws governing its temporal evolution, and (2) a chosen occurrence in that scenario, the effect. The goal is to identify any happenings in that fragment of history that deserve to count as "one of the causes" of the effect. If I were unconcerned with overly cluttering the discussion with technicalities, I would make explicit that my discussion of culpable causation is compatible with the hypothesis that space-time is metaphysically derivative. But because the required terminology might be confusing, I will present this chapter (without loss of generality) under the assumption that some sort of space-time is the (fundamental) arena.

Second, I have introduced the term 'induce' to serve as a rough and ready psychological surrogate for 'promotion'. Because our native conception of culpable causation does not take into account the vast background that is usually required for promotion, it is best to avoid defining culpability exclusively in terms of promotion. We have at least some grasp of the idea that one event *C* can help make *E* occur. One could say that *C*-events have a tendency to result in *E*-events, *C*-events lead toward *E*-events happening, or *C*-events have a causal power to bring about *E*. In this chapter, 'induce' should be interpreted liberally enough to accommodate this variety of ways in which a cause can help make an effect come about.[78]

---

[77] This guiding principle is one variant of the hypothesis that singular causation can be adequately understood in terms of probability-raising processes. This should not be surprising because such theories are motivated primarily by the goal of incorporating (1) some sort of production or process or mechanism with (2) some sort of counterfactual dependence or difference-making or probability-raising. Because the metaphysics of causation I have presented represents fundamental causation along the lines of (1) and derivative causation along the lines of (2), the theory of culpability that complements the metaphysics should incorporate both aspects. There are some existing proposals along these lines, like (Schaffer 2001), but I do not know of any account that resembles the version presented in this chapter.

[78] I invite readers to interpret 'induces' liberally enough to include models of causal tendencies expressed in terms of forces or hastening or intentions. For example, there is a sizable literature in psychology based on the suggestion of Talmy (1988) that many of our intuitions about causation can be effectively modeled in terms of our conception of force vectors. Wolff, and Zettergren (2002) report that a force-based approach successfully predicts a range of causal judgments regarding material objects. For example, if a motorboat is attempting to go away from a buoy but a strong wind blows it back until it hits the buoy, people will say the wind caused the boat to strike the buoy. Also see Wolff (2007). The pronouncements of this "force dynamic" model of causation, I believe, overlap enough with the pronouncements one gets from a well-designed model based on difference-making in order to justify the following claim. If it is useful for a creature to possess the psychological faculties described by one of these two models—the "force dynamic" model or the difference-making model—it is useful for a creature to possess the psychological faculties described by the other. Similar comments apply to cases where someone hastens the occurrence of an effect that would have happened later without the action and to cases where someone acts intending for a certain effect to occur. The scenarios where these

Nevertheless, in order for me to connect the toy theory of culpable causes to my formally defined relation of promotion, it will facilitate communication if "C induced E" is primarily understood as "C raises the probability of E" which in turn can be related to promotion insofar as talk of plain coarse-grained events like C can be translated into the language of contrastive events.

When sorting through various candidate causes of an effect, we normally think of each candidate, c_i, under some not-too-convoluted coarse-grained description, C_i. For brevity, I will use the expression "c (as C)" to refer to the fine-grained event c under the coarse-grained description C. In order for it to be connected to the metaphysics, though, the event also needs to be thought of as a contrastive event, $\tilde{C}$, which comports with the observations of §4.8 that we often tend to use implicit contrasts when thinking of culpable causes. In all cases that we need to consider, the contrastive event is intended to be a contrastivization of the coarse-grained event having its background conditions filled in with a reasonable contextualization of C's actual environment at the same time as C. For brevity, I will use the shorthand "c (as C qua $\tilde{C}$)" to signify that c has been coarse-grained as C and contrastivized as $\tilde{C}$.

The practice of switching between coarse-grained and contrastive events applies to the effect as well. In order to keep the discussion in this chapter manageable, I will initially treat effects as plain coarse-grained events. In §4.8, I described how my account can handle contrastive effects as well, illustrated by the statement, "Adding a dash of salt causes the dish to be tasty rather than bland." Such contrastive effects can be accommodated by considering fixing relations rather than prob-influence relations. For example, in seeking the culpable causes of the dish being tasty rather than bland, we would ignore events like the presence of working kitchen equipment and the presence of groceries. These are promoters of the dish being tasty rather than not existing at all, but they are not promoters of the dish being tasty rather than bland. So, throughout the rest of this chapter keep in mind that my talk of promoting the effect E is meant to extend to contrastive effects and the events that fix them.

Third, as we proceed through the following discussion, I will progressively spell out four candidate interpretations for 'successfully' in the definition of 'culpable cause'. This will result in four distinct formulations of culpability. Each successive version builds on the previous one in order to match our instinctive identification of culpable causes better. I will first lay out the simplest version of culpability, culpability$_1$, to establish a basis for (1) clarifying how the effect and its potential causes are individuated, (2) specifying some parameters people tend to employ when judging promotion, and (3) exploring a preliminary guess at what it means

models disagree are important for debates in psychology, but I will not be concerned with their differences because the toy theory is only intended to establish a fairly reliable link between promotion and our assessments of culpability, not to insist that people's reasoning about causal tendencies must closely match probabilistic relations.

for an instance of promotion to count as successful. Then, I will examine some deficiencies of culpability$_1$ in order to motivate an improved conception, culpability$_2$, which takes into account the contrastive character of causes and the fine-grained character of the effect. After explaining how culpability$_2$ addresses the problems with culpability$_1$, I will reveal some deficiencies culpability$_2$ has by virtue of its not taking into account anything that occurs temporally in between a candidate cause and the effect. Culpability$_3$ modifies culpability$_2$ by taking into account intermediate happenings, which allows it to be more discriminating by ruling out some candidate causes for failing to deliver their inducement successfully through an appropriate process. The final notion, culpability$_4$, extends culpability$_3$ by chaining together instances of culpability$_3$. I will then attempt to connect these last two technical notions to our intuitive conception of culpability, suggesting that we tend to vacillate between culpability$_3$ and culpability$_4$ depending on our explanatory purposes. Culpability$_1$ and culpability$_2$ merely serve as heuristic devices to help me communicate the content of the toy theory and to illustrate how it addresses standard examples in the philosophical literature on causation.

9.2 Culpability$_1$

Here is an initial refinement of the schematic definition of causal culpability:

> An actual event c (as C qua $\tilde{C}$) is *culpable*$_1$ for an actual event e (as E) iff $\tilde{C}$ is a salient, significant promoter of E.

Culpability$_1$ captures the idea that culpability is successful promotion in the most naïve way possible. The cause occurred; it promoted the effect; the effect occurred.

9.2.1 SALIENCE

A *salient promoter* is a promoter people tend not to ignore as part of the causal background. In the psychology literature, the expression '*focal set*' refers to the set of contextually salient events that serve as candidate causes. There is a sizable literature on principles that determine which events are part of the focal set, and a more sophisticated account of culpability would presumably benefit from being integrated with a general psychological theory of focal sets, but that is far beyond the scope of this discussion. I will just mention a few issues that are particular to my toy theory.

The striking of a match counts as a salient promoter of its flame whereas the presence of oxygen does not, even though either one alone would not promote the flame in the absence of the other. What makes the striking stand out more than the oxygen has little to do with its role in nature and a lot to do with how we think of it. Reasons for conceiving of a promoter as worthy of special consideration include that it is the action of an intentional agent, that it is an unusual event, or that it deviates from what should be happening either in a moral sense or in the sense of an object performing its perceived function or in the sense of an object's deviating from its inertial path.[79] The implicit contrasts we use to select promoters play a large role in the process of identifying salient events. When an event takes place that is commonplace and either unchanging or in accordance with how things are supposed to be, we tend not to notice a contrast and therefore tend not to flag the event for further consideration. Most of the reason the presence of oxygen does not count as salient is that oxygen is almost always present at the Earth's surface and so we tend not to think of its absence as worth considering. The striking of a match counts as salient largely because it is an intentional action, involves a noticeable change, and is much rarer than other promoters like the presence of oxygen or the dryness of the match.

Some evidence exists that moral categories play a role in our selection of which events potentially count as causes, for example, Alicke (1992), Knobe and Fraser (2008), and Driver (2008a, 2008b). This would be surprising in a model of moral judgments where step one is to ascertain which events count as causally relevant without any appeal to morality, and step two is to apply moral principles to assess those events for moral culpability. Although investigation of the role of morality in people's identification of culpable causes is in its infancy, the claim that our beliefs about morality play a role in whether some chosen event counts as a cause would not be surprising given my theory of causation. Because the concept of culpable cause is parasitic on the notion of promotion, culpable causes inherit the contrastivity of promotion. And, as noted in §4.8, the default contrasts people use in assessing causal promotion include what people believe is normal or what they believe should happen. We can think of "what should happen" as what typically happens, or as what will happen if things work as they are intended or designed to function, or as what the law or morality dictates. All these senses of 'what should happen' can play a role in identifying candidate causes. For example, when determining why a particular bridge collapsed, we tend to sift through events that differ from the norm in one of these senses. We might flag the existence of an unusually heavy load as a candidate cause just because it is atypical. Or we might flag the failure of a certain joint to maintain rigidity as a candidate cause because the purpose for which it was installed was to hold its beams rigidly together. Or we might flag the inspector's negligence because he was legally obligated to check the joints and morally obligated to make a good-faith effort. Actions people take in accordance with the law and morality are ceteris paribus less likely to be salient because routinely considering them would usually result in an unmanageably large number of candidate causes.

Another factor governing whether an event counts as salient is how broadly it is coarse-grained. The coarse-graining is often selected by some sort of default

[79] See Maudlin (2004) and the discussion of default and deviant states in Hall (2007) and Hitchcock (2009).

conception of an event, but we also have the ability to select a coarse-graining as salient in a more sophisticated manner. Imagine observing a person who is the subject of a psychological experiment. The subject attends to an unlit button on a panel; the button lights up with a green color; and the subject responds by pressing the button. It is natural to conceive of the situation as one where the lighting of the button caused the person to press it or where the lighting of the button as green caused the person to press it. One would not normally think of the cause as "the button lighting as either green or yellow" because there is no reason to suppose the button can light up as yellow or that a yellow light would induce the subject to press the button. However, if you are told the subject was instructed to press the button when and only when the light appeared as either green or yellow, and you see the button turn green and then the person pressing it, it would be reasonable for you to describe the cause as "the button lighting up as green or yellow." That description is appropriate because you know the most informative description of what is promoting the person to press the button is its lighting up as either green or yellow. It is reasonable to select this "green or yellow" contrastivization to inform one's selection of a salient candidate cause even though nothing in this particular case prevents one from accurately describing the cause more narrowly as "the button turning green." (Communication of the intended contrast also plays a role here.) This feature accords with my contention that our intuitions about culpability are often tuned in order to be useful for conveying information about promotion. Unlike Yablo's (1992, 1997) principle that "causes must be proportional to their effects," however, culpable causes in my account need not be coarse-grained in a maximally informative way.

9.2.2 IRREFLEXIVITY

Even though an event always determines or fixes itself, we generally judge that events do not cause themselves. This can be explained by noting that an event's self-determination or self-fixing is entirely trivial in the sense that it holds regardless of the laws and regardless of the character of the event. The triviality is pragmatically evident in the pointlessness of adopting the strategy to bring about *E* by bringing about (some contextualization of) *E*. Also, in presenting a causal explanation for *E*, it would be pointless to cite *E* as a cause because that would provide no new information. Because trivial fixing relations are always useless in practice, it makes sense for humans not to think of them as instances of causation at all. In general, to represent this pragmatic feature, we can simply declare that as a rule, no event is culpable for itself. It is conceivable that this rule might be overridden, perhaps to make a theological point, but it is reasonable to suppose it holds generally of mundane events.

9.2.3 ASYMMETRY

Because past-directed prob-influence is apparently always useless for the advancement of goals, it is reasonable for us to conceive of the past as settled and thus to think of events as not genuinely promoting past effects. If we instinctively think of events as not promoting past effects, it is reasonable for us not to count any events as culpable for previous happenings. This general rule can be overridden by prompting people with time travel stories or tales of magical past-affecting wands, and to the extent that people come to accept the possibility of such past-directed promotion—often because it is of a kind useful for advancement—they can come to override the default rule of thumb that events do not cause anything toward the past.

9.2.4 SIGNIFICANT PROMOTION

When an event *C* increases the probability of the effect from nearly zero to some appreciably large value and the effect occurs, we tend to think of *C* as a culpable cause, barring some reason to think otherwise. But in many cases, the promotion is not significant enough in magnitude to warrant our assigning it culpability for the effect. Judgments of significance are guided in part by the absolute amount by which the probability of *E* is increased, but there is an asymmetry in how we treat probability raising when it involves unlikely events compared to when it involves likely events. For example, if *C* and *E* both occur and *E* had a 99.9999% chance of occurring in the presence of *C* but would have had a 99% chance of occurring in the absence of *C*, then people will be less likely to classify *C* as a cause than they would if *C* raised the chance of *E* from 0.0001% to 1% despite the same increase in the absolute magnitude of probability. This difference in judgment is understandable in terms of either of the two following psychological rules. The first is that we reckon probability-raising at least partly in terms of ratios, not absolute increases. When the contrast probability is lower, the degree of promotion will be a higher factor; 1% is ten thousand times greater than 0.0001% whereas 99.9999% is barely greater than 99%. The second possible psychological rule is that we think of culpability as something which itself is susceptible to chance. The subject may know that *C* increased the probability of *E* from 99% to 99.9999% but recognize that *E* probably would have happened anyway and thus judge that *C* only had a relatively small chance, maybe around 1%, of being something that made a difference to *E*'s occurrence.

Another aspect of judging whether the promotion is significant enough occurs when the resulting chance of the promoted effect is still small. If *C* raises the probability of *E* from 10^{-100} to 0.01, and there are no other candidate promoters, and *E* occurs, then we tend to identify *C* as a cause of *E*. Other cases, though, are less clear. Suppose the causal background is such that the event *E* has a 10^{-29} chance of occurring without any salient cause. If some *C* raises the probability of *E* from 10^{-29} to 10^{-20}, it has increased the chance of *E* a billion-fold but only raised it to a minuscule level. In such cases where *C* occurs, followed by *E*, it can be unclear whether we should attribute *E*'s occurrence to *C*.

9.3 Shortcomings of Culpability$_1$

Culpability$_1$ measures the success of $\tilde{C}$'s significant promotion of E in the crudest way possible. The promotion is successful if and only if E occurs. In this section, I will examine some deficiencies of this measure of success by providing several examples where culpability$_1$ fails to match some pre-theoretical judgments concerning culpable causation. I will respond to these faults in the next section by defining an improved concept, culpability$_2$.

9.3.1 PRECISE CHARACTER OF THE EFFECT

Consider a fragment of history with two cannon-like machines that launch paint balls toward a single canvas mounted on a wall. The machine on the left is able to hit the canvas with 99% accuracy and selects its paint balls from a random assortment of one hundred different hues, not including periwinkle. The machine on the right is able to hit the canvas with 1% accuracy and only uses periwinkle-colored paint balls. The machines are fired simultaneously once and a single paint splat forms on the canvas, which happens to be periwinkle in color. Let the fine-grained effect e be the full state (five seconds after the machines are fired) of the canvas and its immediate environment, including any parts of the wall within a few meters of the canvas. Let C_l and C_r be the firing of the left and right machines respectively, and let E be the event of the canvas having paint on it five seconds after the firing. Which of the machines were culpable for E? Our intuitive judgment selects C_r and not C_l by virtue of the fact that the right machine is the only one capable of making a periwinkle splat. But C_l is culpable$_1$ for E because C_l and E occurred and C_l raised the probability of E significantly over what it would have been had the right machine fired alone. Thus, culpability$_1$ does not match our instinctive identification of the culpable causes.

9.3.2 OVERLAPPING CAUSATION

Now suppose the left machine is aimed slightly to the left of the canvas so that when it splatters paint onto the canvas, it also splatters paint to the left of the canvas and it never splatters to the right. Suppose the right machine is aimed so that it splatters to the right when it hits the canvas and never splatters to the left. Suppose that both use green paint balls and that e instantiates a splattering of paint onto the canvas and onto the wall to the right of the canvas. Which machine caused the canvas to acquire paint? We tend to select C_r and not C_l. One good reason is that the right machine is the only one capable of making a splat that spreads to the right of the canvas. But C_l is culpable$_1$ for E because C_l and E occurred and C_l raised the probability of E significantly over what it would have been had the right machine fired alone. Thus, culpability$_1$ does not match our instinctive identification of the culpable causes.[80]

9.3.3 PROBABILITY-LOWERING CAUSES

Suppose as before that the left machine firing alone is 99% accurate and the right machine is 1% accurate, but now introduce an interaction between C_l and C_r so that if they fire simultaneously, the accuracy of the left machine drops to 1%. Suppose the precise event e that occurs is a splattering of green paint onto the canvas and on the wall to the right of the canvas but not to the left. Which machine caused the canvas to acquire paint? We tend to select C_r and not C_l, again because of the paint to the right of the canvas. But C_r is not culpable$_1$ for E because C_r lowered the probability of E from 99% to approximately 2%. Thus, culpability$_1$ does not match our instinctive identification of the culpable causes.

9.4 Culpability$_2$

In an attempt to reduce the mismatch between culpability$_1$ and our instinctive judgments of culpability, we can define an improved successor concept, culpability$_2$. A metaphorical way to think about what makes culpability$_2$ essentially different from culpability$_1$ is that culpability$_1$ is the notion we get when we judge the success of a candidate cause merely by whether reality meets the goal we impose from the outside—our choice of how to coarse-grain the effect—whereas culpability$_2$ is the notion we get when we judge the success of a candidate cause in terms of whether the outcome it ended up inducing—an achievement defined in terms of a contrastive effect—also happens to promote the goal we have imposed from the outside.

To unpack this idea, let us first review how culpability$_1$ attempts to approximate our intuitive notion of culpability. The starting point for evaluating culpability is a fine-grained event, e, coarse-grained as E, which serves as the effect whose causes we seek. To find its causes, we look for events that induce E. An existing coarse-grained C induces E when some salient contrastivization $\tilde{C}$ of C significantly promotes E. Then, culpability$_1$ counts an instantiated event's inducement of E as successful if and only if E occurs.

The guiding idea behind culpability$_2$ is to measure successful inducement in terms of what results $\tilde{C}$ was "trying" to promote and how its attempt fared. Let us first define R to be the region occupied by our chosen effect E as well as a fair amount of its surrounding environment at the time t of E's occurrence. Second,

---

[80] See also, (Schaffer 2000a).

$\tilde{G} \equiv (\overline{G}, \overline{\neg G})$ is defined to be the contrastive event occupying R that is fixed by $\tilde{C}$.[81] Then, we can think of $\tilde{G}$ in terms of its prominent foreground and background.[82] What $\tilde{C}$ was "trying" to promote is $\overline{G}$ rather than $\overline{\neg G}$, localized in $\tilde{G}$'s prominent foreground.

For an illustration, consider the case of overlapping causation from §9.3.2 where the left machine is 99% accurate and can hit to the left of the canvas, and the right machine is 1% accurate and can hit to the right of the canvas, and the machines do not significantly interact with each other. The chosen actual effect *e* is a splattering of paint on the canvas and to the right of the canvas but not to the left. We stipulate our coarse-grained effect of interest, *E*, to be the existence of some paint on the canvas. Construed as a contrastive event, the firing of the right machine is $\tilde{C_r} \equiv (\overline{C_r}, \overline{\neg C_r})$, which fixes $\tilde{G} \equiv (\overline{G}, \overline{\neg G})$ for region R. In this example, $\overline{G}$ and $\overline{\neg G}$ are very nearly alike except that $\overline{G}$ is far more likely than $\overline{\neg G}$ to instantiate a paint splatter on the right side of the wall. Because $\overline{C_r}$ and $\overline{\neg C_r}$ very nearly agree on the likely motion of the left paint ball, everything that happens with the left paint ball is excluded from what is (trivially) promoted by (the prominent foreground of) $\tilde{G}$. What $\tilde{C_r}$ is "trying" to promote is not *E*, paint on the canvas, but paint being somewhere on the canvas and/or on the wall to its right rather than no paint in that region.

The next step in assessing culpability$_2$ is to characterize how $\tilde{C_r}$'s attempt at promotion "fared" in terms of a contrastive effect occupying R that is more finely-grained in order to account for what actually occurred (fundamentally) in R. Let us say that g_e is the unique full fundamental event occupying R. For concreteness let us suppose that g_e instantiates one splotch of paint on and to the right of the canvas, and another splotch of paint far to the left of the canvas. It also instantiates the wall, the lighting, the sounds, and other details in the room. We then slightly coarse-grain g_e as G_E to circumvent the problem that often all fundamental events equally have zero probability even though some are far more likely than others in the more intuitive sense captured by slightly coarse-graining them with some reasonable probability distribution.

We then construct a new event, $\overline{E_1}$, by using $\overline{G}$ as a starting point and removing all the members of $\overline{G}$ that are not members of G_E, renormalizing the resulting probability distribution to make $\overline{E_1}$ a well-defined contextualized event. The role of $\overline{E_1}$ is to represent in a slightly fuzzed way the portion of what $\overline{C_r}$ was "trying to make happen" that actually occurred at *t*. A reasonable choice of coarse-graining should fuzz $\overline{E_1}$ enough to eliminate stray traces of dust and similar fine details but

---

[81] This fixing of a contrastive event was defined in §3.7. For simplicity, I am assuming we are not dealing with cases of past-directed time travel or any other conditions that would allow $\tilde{C}$ to fix more than one contrastive event for R.

[82] Recall from §7.3.1 that the prominent foreground of $\tilde{G}$ is the region of the arena where $\overline{G}$ and $\overline{\neg G}$ differ significantly, and the prominent background is the complementary region where they do not differ significantly.

not enough to eliminate the detailed pattern of the paint splatter or the brightness of the light on the wall or any audible sounds in the room.

We are in the process of constructing a contrastive event to represent the effect, and $\overline{E_1}$ serves as the protrast, the first member of the ordered pair of contextualized events. We now want to construct another contextualized event, $\overline{E_2}$, to serve as the contrast. In doing so, there are two main considerations we should attend to. The first is that we should restrict $\overline{E_2}$'s members to those that are already in $\overline{\neg G}$ so that we properly respect what $\overline{\neg C_r}$ was "trying to make happen." The second is that we should eliminate members of $\overline{\neg G}$ that would result in spurious identifications of promotion. I will return shortly to the question of how this is to be accomplished, but once the appropriate members have been removed and the probability distribution renormalized, we will have our sought-after contrast, $\overline{E_2}$.

The final step is to pair these two contextualized events together to form $\tilde{E} \equiv (\overline{E_1}, \overline{E_2})$, which represents everything $\tilde{C_r}$ successfully promoted for region R. To evaluate whether C_r successfully induced some chosen E, we now only need to consider the degree to which $\tilde{E}$ (trivially) prob-influences E. We can always read off of $\tilde{E}$ the degree of prob-influence for any plain coarse-grained event E in R as follows. The degree to which $\tilde{E}$ prob-influences E is equal to the proportion of $\overline{E_1}$'s members that instantiate a member of E (in the correct region) minus the proportion of $\overline{E_2}$'s members that instantiate a member of E (in the correct region). The degree to which $\tilde{E}$ prob-influences E is by construction equal to the degree to which $\tilde{C_r}$ successfully promoted E and thus is equal to the degree to which C_r successfully induced E. If this degree of successful inducement is significantly positive, then C_r counts as a successful inducer of E.

In order to fill in the gap left in this procedure—removing appropriate members from $\overline{E_2}$—there are at least two guiding principles we should apply. The first principle involves removing aspects of $\overline{E_2}$ that are not of the right kind to be prob-influenced by $\tilde{C_r}$. The second principle involves removing aspects of $\overline{E_2}$ that can be attributed to other independent causes. We remove an aspect of $\overline{E_2}$ by stripping out members of $\overline{E_2}$ to equalize the probabilities that $\overline{E_1}$ and $\overline{E_2}$ fix for that aspect. I will illustrate both principles with examples.

The first way to tell whether some aspect of $\overline{E_2}$ should be removed is to examine what kinds[83] of effects $\tilde{C_r}$ promotes for region R. Suppose for example that a particular location on the canvas and wall is lit by several spotlights that flicker on and off every now and then with $\tilde{C_r}$ not prob-influencing anything about the lights. Because $\tilde{G}$ does not prob-influence anything regarding amount of light striking the wall or canvas, we should adjust E_2 (to match $\overline{E_1}$ with regard to its pattern of luminosity) so that $\tilde{E}$ does not prob-influence the amount of light striking the wall. In this way, we render C_r not culpable for the canvas being lit the amount that it is, for the wall being at room temperature, for the existence of a roach at a particular location on the floor, and so on.

[83] The relevant kinds here exclude any kinds that are too difficult for people to cognize.

A special case of this principle involves transferring the prominent background of $\tilde{G}$ to $\tilde{E}$, as can be seen in our current example of overlapping causation where $\overline{E_1}$ fixes a very high probability for the particular pattern of paint on the wall to the left side of the canvas (which came from the left machine). Unamended, $\overline{E_2}$ would fix a very low probability for any particular splotch of paint because $\overline{\neg C_r}$ leaves open the full range of possibilities for where the left machine's paint can land. Unamended, $\tilde{E}$ would thus count as successfully promoting the splotch of paint on the left, which disagrees with our judgment that the firing of the right machine was not one of the causes of the left machine missing the canvas. To resolve this problem, it is reasonable to strip out members of $\overline{E_2}$ to make the prominent background of $\tilde{E}$ match the prominent background of $\tilde{G}$. The prominent foreground of $\tilde{G}$ is the subregion of R that includes everywhere the right paint ball could have landed, and its prominent background is everywhere else, including the actual location of the paint splotch on the left. The technical implementation of the solution is to discard any members of $\overline{E_2}$ that disagree with G_E in the prominent background of $\tilde{G}$, and then renormalize its probability distribution. By doing so, the prominent background of $\tilde{E}$ will be the same region as the prominent background of $\tilde{G}$.

The second way to tell whether some aspect of $\overline{E_2}$ should be removed is to infer that this aspect is already attributable to some alternative cause that is independent (in the sense of not being significantly prob-influenced by) the candidate cause. When we have good grounds for attributing an aspect of the effect to another cause by virtue of some signature detail in what the alternative promotes, it should be removed from what $\tilde{E}$ promotes. When we do not have good enough grounds for attributing it to an alternative cause, then the culpability of C_r is not ruled out.[84]

For example, imagine a scenario where both green paint balls have landed on the canvas and overlap somewhat. $\overline{E_1}$ fixes a probability of one for the particular pair of paint splotches on the canvas, and without being further amended, $\overline{E_2}$ fixes a very low probability for paint being exactly at the location where the left machine's splotch of paint actually ended up. Consequently, $\overline{C_r}$ successfully promotes the left machine's paint hitting the target, which is the incorrect judgment. To remedy this situation, we should try to identify which aspects of $\overline{E_1}$ cannot be properly attributed to C_r (or are better thought of as attributable to causes other than C_r) and to conditionalize $\overline{E_2}$ accordingly to eliminate its promotion of those aspects. In our current example of the two overlapping splotches on the canvas, we can make a judgment as to which part of the paint pattern is attributable to the left machine's firing and discard from $\overline{E_2}$ any members that do not instantiate this part of the paint pattern. This alteration makes $\overline{E_2}$ agree with $\overline{E_1}$ as to the location

[84] One could at this stage incorporate additional considerations related to causal grouping—overdetermination and joint causation—but unfortunately I have had to abbreviate this presentation of the toy theory.

of the left paint splotch and thus ensures that $\tilde{E}$ prob-influences the existence of the splotch from the left machine to degree zero, rendering C_r as not successfully inducing the left machine's splotch of paint landing where it did.

We can summarize this sketched procedure in terms of an semi-formal definition for culpability$_2$.

> An actual event c (as C qua $\tilde{C}$) is *culpable*$_2$ for an actual event e (as E) iff a region R (including and surrounding e) has a contrastive effect $\tilde{E}$ imposed on it that significantly promotes E.

The imposed $\tilde{E}$ is generated by taking what $\tilde{C}$ fixes for R, conditionalizing its protrast with a slight coarse-graining of the full fundamental event occupying R, and adjusting its contrast in parallel in light of what $\tilde{C}$ and other independent salient events induce or promote for R.

Because this is a toy theory, I am forced to leave underspecified the precise implementation of the procedure for constructing the contrastive effect. For example, I cannot say whether some kinds of aspects are more salient for the purpose of stripping out aspects from $\tilde{E}$. In any case, one notable deficiency of the method described in this section is that it does not work nearly as well when the contrast in the candidate cause is likely to interact with stuff in the environment and leave traces in the fine-grained effect. Unfortunately, I will have to forego how to refine the method further.

It is a good exercise at this stage to consider other cited shortcomings of culpability$_1$ in order to see how culpability$_2$ helps secure better agreement with our pre-theoretical judgments of culpable causation. In the example from §9.3.1, the right machine fires periwinkle paint balls and the left fires some other color. The firing of the left machine will not count as a successful inducer of E. Although $\tilde{C}_r$ promoted paint on the canvas, it did not promote the existence of periwinkle paint on the canvas, and yet periwinkle paint is the only color of paint on the canvas and thus was the only color of paint that was represented in the constructed $\tilde{E}$.

In the example from §9.3.3, the firing of the right machine lowers the probability of paint on the canvas, yet we still think of it as culpable for the existence of paint on the canvas. We can now make sense of this judgment. Even though $\tilde{C}_r$ makes paint on the canvas much less likely, it significantly promoted the probability of the precise pattern that happened to appear. C_r was successful at inducing the more finely grained effect and so was culpable$_2$ for E.

Given my previous suggestion that the reason we have a notion of culpability is that it allows us to more quickly infer promotion relations, it is worth considering why it would benefit us to have intuitions that match culpability$_2$ rather than culpability$_1$. I certainly do not think our psychological mechanisms for attributing singular causation implement the precise form of my exegesis of culpability$_2$. However, it is a plausible hypothesis that we reckon culpable causes by scouring the evidence included in the environment of the fine-grained effect and piecing together which candidate causes were responsible for which aspects by attributing each chosen aspect to a candidate cause (or candidate group of causes) when it is the only candidate that could have promoted that aspect.

In many circumstances, culpability$_2$ is not too much harder to assess, and it is often much more responsive to the observed evidence than culpability$_1$. As discussed in §8.2, recognizing which of the two machines is culpable for the effect often allows one to make good estimates about how much each machine individually promotes the effect. In circumstances where the paint-ball-firing machines barely influence one another, one can easily gather statistics on how often the splotch of paint spreads significantly toward the right and thereby infer the fraction of times that a C_r event was culpable for the existence of paint on the canvas, E. That, in turn, provides a good estimate for how likely the machine on the right would place paint on the canvas when operated alone.

9.5 Shortcomings of Culpability$_2$

The primary shortcoming of culpability$_2$ is that we sometimes rule out a candidate cause because it does not successfully deliver its inducement to the effect through an appropriate process. In technical terms, culpability$_2$ does not properly account for 'fizzling', a term from (Schaffer 2000a).

Intuitively speaking, *fizzling* occurs when a process is "heading toward" bringing about E but reaches a stage where it is no longer bringing about E. Framed within the context of the toy theory, fizzling can be defined using the following procedure. First, assume that there is some actual event c (as C qua $\tilde{C}$) that promotes some E. There is no need to assume that E is instantiated. Second, we can consider any region R that is intermediate between $\tilde{C}$ and E, typically a region that lasts only for a moment. Second, let i be the actual full event occupying all of R. Third, construct a contrastive effect $\tilde{I}$ for the region R employing the same procedure used to evaluate culpability$_2$. Fourth, check whether $\tilde{I}$ significantly promotes E. If it does not, i counts as a fizzle with respect to E. If there is an actual event identified by this four-step process that counts as a fizzle, the process leading from c (as C qua $\tilde{C}$) to the promoted E counts as having fizzled.

A good example of fizzling occurs when a fuse is burning at time $t = 0$ and is "going to" launch a rocket at $t = 2$, and that nothing else of interest is going on. The default contrast built into $\tilde{C}$ at $t = 0$ is the fuse just lying there unlit with nothing in the background environment that would suggest that it could become lit in the near future. Suppose that shortly before time $t = 1$, the fuse burns out prematurely. In this case, the full event i at time $t = 1$ instantiates a burned out fuse, and the contrastive effect $\tilde{I}$ represents a short non-burning fuse rather than a long unlit fuse. Because this $\tilde{I}$ does not significantly promote the later rocket launch, i counts as a fizzle.

Let us now consider several examples where our knowledge of intermediate events motivates rejecting a candidate cause. These will illustrate how culpability$_2$ counts as a defective approximation of our instinctive concept of culpability.

9.5.1 SAVED FIZZLES

A *saved fizzle* is when there is some c (as C qua $\tilde{C}$) promoting some E, the process leading from c to E fizzles, and yet E occurs anyway. A simple case of a saved fizzle is when a lit fuse that is on its way toward launching a rocket spontaneously burns out for a while and then spontaneously becomes lit again and leads to the launching of the rocket. The spontaneous event here can be conceived as a highly improbable event that does not occur by virtue of any recognizable previous event but results from some fundamental or derivative chanciness. Intuitively, the initial lighting of the rocket was not one of the culpable causes of the rocket's launching, but it is culpable$_2$ for the launching because what actually occurred is very nearly what would have occurred had the fuse not burned out.

The next two subsections discuss cases of preëmption, that are also special cases of saved fizzles.

9.5.2 EARLY CUTTING PREËMPTION

Preëmption occurs when some event is culpable for a fizzle. *Early cutting preëmption* occurs when the caused fizzle precedes the induced effect. For illustration, consider the pair of machines that fire paint balls, but also introduce a shield that can spring into place and absorb one of the two paint balls without leaving any noticeable trace of which ball it absorbed. Suppose both machines are placed very close together and aimed so as to fire green paint balls in very nearly the same direction toward the middle of the canvas, so that the pattern of paint each would likely produce is the same. The machines are fired at the same time, but the left ball by chance happens to be absorbed by the shield, and the right paint ball lands on the canvas. Which machine caused the canvas to acquire paint? We tend to select C_r and not C_l by virtue of the fact that there is a continuous path that the right paint ball follows coming from the right machine all the way until it splatters on the canvas. But C_l is culpable$_2$ for E, intuitively because it significantly raised the probability of the fine-grained effect that happened to occur.

9.5.3 LATE CUTTING PREËMPTION

Late cutting preëmption is a special case of preëmption where the preëmption (or fizzling) is the occurrence of the effect. This kind of preëmption is illustrated by replacing the canvas in the previous example with a fragile window and adjusting the machines so that the paint balls are launched with random speeds. Suppose both shots are on target and that the ball from the machine on the right arrives at the window first, shattering it at time t. C_l is culpable$_2$ for E because the firing of the machine on the left significantly raised the probability of (a slightly coarse-grained version of) the actual effect, e. However, we instinctively judge that C_l is not culpable for E because when the window broke, the left paint ball had not yet reached the window. That event counts as a fizzling of C_l's process.

9.6 Culpability$_3$

Culpability$_2$ is deficient because it ignores everything that happens after the candidate cause and before the effect, which makes it unable to take into account the presence of fizzles. However, we do not need to modify the definition of culpability$_2$ much in order to take into account events that happen at other times. To construct a superior concept, culpability$_3$, we simply enlarge the region R in the definition of culpability$_2$ to include what happens at other times, including events located between the candidate cause and effect as well as events occuring after the effect. Often, the additional information acquired includes fizzles that allow us to rule out certain candidate causes.

A definition of culpability$_3$ can now be stated:

> An actual event c (as C qua $\tilde{C}$) is *culpable*$_3$ for an actual event e (as E) iff a region R (including and surrounding the process leading from c to e) has a contrastive effect $\tilde{E}$ imposed on it that significantly promotes E and includes no fizzling of this process.

This definition differs from the definition of culpability$_2$ primarily by (1) enlarging the region of consideration, R, to include the whole process from c to e and its environment, not just the time of the effect; and (2) forbidding the process heading toward E from fizzling. Presumably, the procedure for evaluating what contrastive event, $\tilde{E}$, is imposed on R needs to be made more sophisticated as well.

The three examples in the previous section included an event c that was judged culpable$_2$ for E but where its process leading to e fizzled. Such events cannot be culpable$_3$ for E because the definition of culpability$_3$ requires the non-existence of the fizzles that were previously cited. So, these examples provide evidence that culpability$_3$ extends culpability$_2$ to accommodate intuitions about causal mechanisms and continuous processes.

It benefits us to have intuitions that match culpability$_3$ rather than culpability$_2$ because culpability$_3$ is not appreciably harder to assess and because it provides more accurate information about prob-influence relations. As illustrated in the examples of preëmption, our intuitions about culpability are likely being driven by perceptions of the paths of the projectiles, so that we are tacitly employing the kind of information captured in culpability$_3$. Imagine we are trying to evaluate the accuracy of the left and right machines in conditions where they are aimed at the same target from very nearly the same location. If we were to try to evaluate their accuracies using culpability$_2$, we would fail because we would not be able to sort out which of the two splotches of paint came from which machine. By assessing culpability$_3$, which one can discern merely by observing the paths of the balls on repeated trials, the accuracy of each machine is equal to the fraction of the trials in which its ball strikes the canvas.

9.7 Culpability$_4$

Culpability$_3$ takes into account intermediate events that *rule out* certain candidate causes as unsuccessful inducers of the effect, but it is also reasonable for humans to have a notion of culpability that takes into account intermediate events that *rule in* additional candidate causes that would not otherwise count as culpable. This more inclusive notion is culpability$_4$. According to the toy theory, culpability$_4$ exists by virtue of an appropriately linked chain of causes. We can define it formally as follows:

> An actual event c (as C qua $\tilde{C}$) is *culpable$_4$* for an actual event e (as E) iff there is a chain of culpability$_3$ relations running from c to e.

Because there are no hard and fast rules about which intermediate events count as salient inducers or how they are to be rendered as contrastive events, culpability$_4$ is sensitive to our choices of how to abstract away from the fundamental material layout. By being extremely permissive about event salience, one can achieve an extremely long chain of very finely grained events that are only slightly apart in time. Being extremely permissive as a general policy would result in so many culpability$_4$ relations that culpability$_4$ would have little utility. So, our employment of culpability$_4$ needs to be restricted to a suitably limited class of salient events if we want it to do interesting psychological or explanatory work.

The culpability$_4$ notion should not be thought of as the toy theory's replacement for culpability$_3$ or a decisive improvement on culpability$_3$. Sometimes culpability$_3$ matches our intuitive conception of culpability better than culpability$_4$ and sometimes culpability$_4$ matches it better. The definition of culpability$_4$ ensures that whenever C is culpable$_3$ for E it is also culpable$_4$ for E, but there are often cases where C is culpable$_4$ without being culpable$_3$. These include cases that are widely recognized as counterexamples to the transitivity of causation. In §4.9, I described two scenarios where our intuitions match culpability$_3$ rather than culpability$_4$. But let us consider a simpler example here, adapted from (Hall 2000):

> A train is rolling along a track that bifurcates and then rejoins after one hundred meters. Suppose that all the relevant details about the background

> environment are the same on the left side of the track as they are on the right. As the train approaches the junction, the engineer flips a switch that makes the train take the left track. Then, after the train passes the section where the left track rejoins with the right track, the train crosses a road.

Let *e* (as *E*) be the event of the train crossing the road, and let *c* (as *C* qua $\tilde{C}$) be the activation of the switch for the left track rather than the right track. *C* is culpable$_4$ for *E* because $\tilde{C}$ significantly promotes the train moving along the left track, which in turn significantly promotes the train crossing the road, *E*. Note that the intermediate event does not use the contrast that is fixed by $\tilde{C}$ but is chosen by reckoning salient contrasts at the intermediate time. The switching event is arguably not culpable$_3$ for *E* because $\tilde{C}$ does not promote the slightly coarse-grained version of *e*. After all, the probability of the train's reaching the road is the same whether the switch is thrown or not. As with most judgments of culpability$_3$, it is possible in principle to argue that there is some just barely coarse-grained version, E', of the fine-grained effect, *e*, such that the switching, qua $\tilde{C}$, promotes E' and is thus culpable$_3$ for *E*. Such an E', though, would need to include the kind of fine details about the likely character of the train had it taken the left track versus the right. For example, there might be more flies near the left track, so that a train taking that route would tend to displace more flies. If such a finely grained construal of the effect were to be countenanced as part of our standards for judging culpability, there would be many more culpable causes than we actually judge. Thus, we can set aside (as too deviant) such a finely grained construal of the effect.

People who are fully aware of what happened in this scenario will be likely to say that the switching event was not one of the causes of *E*, largely because it is clear that the switching makes no significant difference to how *E* comes about. Their judgments match what is culpable$_3$. However, if the example is altered slightly, people's judgments will likely match what is culpable$_4$:

> A train is rolling along a track that bifurcates and then rejoins after one hundred meters. As the train approaches the junction, the engineer flips a switch that makes the train take the left track. Then, just after the train starts along the left track, a rare chancy event occurs: a tree standing between the two tracks topples. Given that the tree falls, it has a fifty percent chance of falling across the right track and a fifty percent chance of falling across the left track. The tree happens to fall across the right track, blocking any possible train traffic there. The train crosses the road with no trouble because the train is traveling on the left track.

Our intuitive judgment in such cases is that the switching event was one of the causes of the train successfully crossing the road. Again, the switching is culpable$_3$ for the train's traveling along the left track, which is later culpable$_3$ for the train's making it to the road crossing, *E*; thus, it is culpable$_4$ for *E*. And again, the switching would not be culpable$_3$ for *E* because at the time the switch is thrown, the chance of the train's eventually reaching the road crossing is the same whether it goes along the left track or the right track. Because the switching event does not prob-influence anything concerning the tree, the assessment of what $\tilde{C}$ successfully promoted is not supposed to change (according to the simplistic method for constructing contrastive effects discussed in §9.4). This pair of examples shows that sometimes our common-sense judgments of culpability match culpability_3 but not culpability_4 and that sometimes they match culpability_4 but not culpability_3.[85]

When people ask, "What are the causes of *E*?" they usually do not distinguish between these two different kinds of culpability. But once it is apparent that the toy theory posits these two distinct versions of culpability as guides to our (often presumed to be univocal) implicit notion of culpable cause, it follows that the toy theory has a conflict in the technical sense introduced in §1.8. The toy theory tells us that one good rule of thumb for assessing culpability is that an event is culpable for *E* iff it is culpable_3 for *E*. It also tells us that another good rule of thumb for assessing culpability is that an event is culpable for *E* iff it is culpable_4 for *E*. Because there are realistic circumstances where an event is culpable_4 without being culpable_3, the theory provides a conflicting account of which events are culpable. Furthermore, nothing in the toy theory ameliorates this conflict by specifying conditions that adjudicate which version of culpability should supersede the other. According to empirical analysis, these genuine conflicts do not imply that the toy theory is incoherent, nor do they imply that one of the two rules of thumb needs to be rejected as fatally flawed. On the contrary, both versions of culpability have limitations as guides to our cognition of culpable causation, and each offers different benefits. When investigating the psychology of culpable causation at a fairly high level of abstraction, as the toy theory does, it is acceptable to employ RELAXED standards where these kinds of conflicts do not need to be ameliorated with an explicit rule. Insofar as we are just sketching the outlines of a full psychological account, we do not need to specify in every possible instance whether culpability_3 or culpability_4 is the "correct" account of culpability. And insofar as we are investigating the metaphysics of causation, we do not need an account of culpability at all. The conflict in the toy theory that exists by virtue of its not privileging culpability_3 over culpability_4 or vice versa does not count as a reason to reject the toy theory qua toy theory.

---

[85] Because this chapter is only sketching a toy theory of the psychology of culpable causation, not every deficiency can be discussed, but I believe interested readers would benefit from exploring how the motivation I have suggested for distinguishing culpability_4 from culpability_3 by formulating a more sophisticated scheme for constructing the contrastive effects than the one that I assumed when extending the considerations in §9.4 to handle the temporally extended process leading from the cause to the effect. Specifically, the scheme I presented does not take into account that one's chosen contrast in the cause ought to interact with stuff in the background to help generate the proper contrast to use for representing the effect. If one does so, it may be possible to render the switching event culpable_3 for the train's making it to the road crossing, though I suspect people do not reason very clearly about sequences of merely hypothetical interactions beyond simple cases.

That point having been noted, nothing in empirical analysis forbids a special science theory from being conflict-free, nor does it discourage our favoring one theory over its rivals for being conflict-free, nor does it countenance scientists against seeking theories that meet STRICT standards of adequacy.

I have already discussed why it is reasonable for people to have intuitions about culpability that match culpability$_3$. Now I would like to cite a few reasons why it is reasonable for people to have intuitions about culpability that match culpability$_4$. For one, fixing plausibly obeys unidirectional transitivity and continuity, as discussed in §4.9 and §4.10, and to a great extent, relations of culpability serve as cognitive proxies for promotion relations. So, it is often convenient for us to think of culpability relations as being continuous and transitive just like the promotion relations they approximate. As Example 9.7 and the two examples from §4.9 demonstrate, it is not correct to think of our intuitive conception of culpability as transitive, but in a wide range of situations, it is convenient to think of *C* as successfully inducing *E* by virtue of successfully inducing an intermediate event, which successfully induces another intermediate event, and so on until *E* occurs. Because the metaphysics of promotion among contrastive events is too complicated for people to manage cognitively, it is understandable that people approximate the unidirectional transitivity of promotion by largely ignoring the background conditions and just thinking of causation as occurring by virtue of a localized chain of events or a localized continuous process.

For a second reason, thinking of culpability as existing by virtue of chains of culpable causation is useful in assembling the full set of events relevant to a causal explanation of an effect that arises through a complicated nexus of events. When there is a sizable set of salient events that play some role in the occurrence of an effect *E* and we are interested in providing a detailed account of why *E* occurred, we often not only want to know what events, C_i, were successful inducers of *E* but also a further explanation of why these C_i occurred, which often involves identifying and citing the events that successfully induced them, and then at a deeper level of explanation the events that successfully induced them. The totality of all such events are the ones that are culpable$_4$ for *E*. They count as causes of *E* in the sense of being events that played a significant role in how the total historical development brought about *E*.

For a third reason, thinking of culpability as existing by virtue of chains of culpable causation serves as a tool in learning about promotion. I will mention just two examples. First, in Example 9.7, the switching event does not promote the train's crossing the road because the chance of the train reaching the road is the same regardless of which track it takes. However, if we judge counterfactual dependence with hindsight,[86] by contrasting what actually happened with what

[86] For similar observations, see Edgington (2004), Kvart (2004), Northcott (2010), and my discussion of "infection by culpable causation" in the supplementary material I have provided concerning Morgenbesser's coin.

would have happened had the engineer guided the train down the right track, holding fixed the contingency that the tree fell across the right track, then the train's success should count as having counterfactually depended on the engineer directing the train to the left track. In the particular circumstances of this example, such counterfactual reasoning is a misleading guide to the promotion relations because the switching event did not improve the chances that the train would make it safely to the road. However, in a wide variety of cases, after-the-fact events such as the tree falling are indicative of the existence of some hidden condition of earlier states. When a tree falls toward the right in a seemingly spontaneous manner, that is often because there is some hard-to-identify-in-detail feature of the previous condition of the tree that induces its falling at that time to the right. If the tree's falling to the right were due to such a condition rather than brute chance, it would be correct to say the switching event successfully promoted the train's crossing the road. So, because we instinctively judge counterfactual dependence by presuming that the tree would still have fallen across the right track if the train had gone to the right, we often succeed at inferring the correct promotion relations, promotion relations that we would never be able to detect if we restricted our attention to what was happening at the time the switching event occurred.

For a second example of how culpability$_4$ serves as a heuristic for learning about promotion, consider causation that occurs via some enabling (or disabling) condition. Promotion stemming from enabling or disabling conditions are sometimes difficult to detect, and culpability$_4$ helps us filter through possible candidates more quickly. An enabling condition can be thought of as an event that is normally considered part of the background and induces an effect *E* in the presence of a more salient inducer of *E*, which counts as an activating condition. A disabling condition is similarly an inhibitor that lies in the background. For example, we might recognize that some migratory species, say the canvasback duck, annually visits a certain lake for mating. One year, the ducks do not successfully reproduce. That should lead us to suspect that there is some inhibitor of duck reproduction, perhaps in the water. Because there are many chemicals in the water, it might be difficult to identify what, if anything, inhibited the reproduction. However, if we can see that a factory is pouring some sort of liquid into the lake, then it is reasonable to suspect that a chemical from the factory is culpable$_3$ for the condition of the water. Because we have previously learned that waterborne chemicals are sometimes culpable$_3$ for reduced bird reproduction and because we know just by looking that the factory is plausibly culpable$_3$ for some sort of effect on the watershed, we are justified in inferring that there is a reasonable possibility that the factory is culpable$_4$ for the failure of the ducks to reproduce. This can justify restricting the testing to chemicals used in the factory instead of testing for the full array of epistemically possible chemicals in the lake. If it were illegitimate to identify potential causes by using what we know about chains of culpability, we might waste time testing other possible sources. For example, if we can be sure that the chemicals stored in some nearby warehouse never left the warehouse, we can be reasonably sure that they are not culpable for any effects on the water supply, and thus reasonably sure that they are not culpable for the canvasbacks' troubles. This indicates that it is likely unnecessary to test the water for these chemicals.

One of the consequences of having both culpability$_3$ and culpability$_4$ is that many questions about culpability that might initially seem straightforward become extremely messy. An exemplary complicated case is Hesslow's (1981) thrombosis example. Taking a birth control pill regularly is a promoter of thrombosis by virtue of its direct role as chemical in the body. But the birth control pill is also an inhibitor of pregnancy, which itself is a promoter of thrombosis. So, there are two routes by which thrombosis is probabilistically influenced. For the sake of discussion, we can modify the example to have them approximately cancel each other out over the course of time, so that taking the pills on the whole has no net probabilistic influence on thrombosis. Imagine that some woman takes the birth control pill, does not become pregnant, and is not afflicted by thrombosis. Is her consumption of the pills one of the causes of her being free of thrombosis? It might seem that the pills cannot be culpable$_3$ for her failure to contract thrombosis because they do not prob-influence thrombosis. It might also seem that the pills are culpable$_4$ for her not getting thrombosis because there are many chains of successful promotion that run from her taking a given pill to her lack of thrombosis at later times. So, the two notions conflict in their attribution of causal culpability. A univocal assessment is made even more difficult when we take into account that taking pills for a full year consists of many localized events: the daily occurrences where she ingests a single pill. It is plausible that many of these events exert different degrees of promotion and inhibition through different intermediate mundane events. Furthermore, whether an event is culpable$_4$ depends on which events are permissible for employment in chains of culpable$_3$ causation. Remember that if we identify salient events liberally, allowing all sorts of non-standard contrasts and coarse-grainings, just about any event will count as culpable$_4$ for her not having thrombosis. So, the relevant culpability$_4$ would have to be restricted to some appropriately salient events in order to match our psychological judgment that there do not exist a vast multitude of thrombosis preventers. But we do not have clear intuitions about how to break down vast causal networks (like those present in the daily operation of the human body) into relevant component events. In summary, it is safe to say that in complicated interactions like those exhibited by the thrombosis example, it is difficult to make unequivocal statements about culpable causation that are well grounded in our practices of attributing culpability.

9.8 Summary

The toy theory of our psychology of culpable causation that I presented in this chapter was an attempt to connect our folk intuitions about causal culpability to the metaphysics of causation, especially promotion. The toy theory is deliberately sketchy and vague in numerous respects and to the extent that it is precise enough to make predictions, it is surely in conflict with some psychological data. My goal was merely to provide an example of how to approach an empirical analysis of the non-metaphysical aspects of causation. Such an analysis should not try merely to systematize people's judgments about what events are singular causes or are relevant to causal explanations of singular effects but should try to connect this data to the metaphysics of causation. The pair of empirical analyses together ought to make sense of how our intuitions and reasoning about causation help us track metaphysical relations like promotion.

{10}

Causation in a Physical World

Hartry Field (2003) observed that "the problem of reconciling Cartwright's points about the need of causation in a theory of effective strategy with Russell's points about the limited role of causation in physics ... is probably the central problem in the metaphysics of causation." The theory presented in this volume is my attempt to solve this problem.[87] It provides a comprehensive theoretical framework for understanding how causal claims in the special sciences (and in everyday life) can be understood as rules of thumb that are useful by virtue of a fundamental reality that resembles paradigm theories of fundamental physics. It is helpful to conclude this volume by taking stock of what my theory has and has not accomplished and taking note of how the theory can be further developed and related to other theories of causation. Readers should keep in mind that this summary is overly simplistic and needs to be understood in light of the numerous qualifications offered in previous chapters.

10.1 Summary

Although this volume has introduced a non-standard methodology for evaluating theories of the metaphysics of causation and several new technical terms, the underlying theory is remarkably simple. Fundamentally, events are causally linked by determining (or by fixing probabilities for) each other. These relations amount to a comprehensive set of *singular* causal relations. Derivatively, events are causally linked by relations of probability-raising (or probability-lowering) understood as a form of difference-making or counterfactual dependence. We can say that contrastive events prob-influence coarse-grained events or that they fix other contrastive events. These prob-influence and fixing relations constitute a comprehensive set of *general* causal relations. Details aside, that amounts to a complete metaphysics of causation. My theory barely touches on the multitude of non-metaphysical issues regarding causation, including the practice of discovering

---

[87] Readers familiar with Field's article should note that I have emphasized some features of fundamental physics that Field has pointed out in his section one.

causes and formulating causal explanations, but those activities are not within the intended scope of my account.

To begin, we can consider how to apply traditional labels to my account. For example, philosophers have often attached the label 'realist' or 'anti-realist' to various components of a theory. The main purpose of my distinction between fundamental and derivative is for 'fundamental' to serve as a replacement for 'real' so that debates about realism can be reconfigured as debates about what is fundamental. A crude way to put it is that one should be realist about fundamental reality and anti-realist about derivative reality and any non-existents. A more refined way to think about the issue is that 'real' can be usefully clarified by distinguishing between two senses of 'real'. We can define real_1 to apply to a possible existent iff it is part of fundamental reality, and real_2 to apply to a possible existent iff it is part of reality, either fundamental or derivative. Realism_1 is the kind of realism important in metaphysics and ontology, whereas realism_2 is important for meaning and reference and truth. We can legitimately refer to derivative things like trees and poems, because they are real in the sense of not being non-existents, but we can do so without having to include them as components of the actual world (in addition to what is fundamental). Concerning the sense of 'realism' relevant to metaphysics, my theory is realist_1 only about singular causation (in the non-standard sense captured by relations of contribution and terminance) and is anti-realist_1 about general causation and causal culpability. Concerning the sense of 'realism' relevant to concerns about whether causal claims are true or refer to relations that would hold without there being any creatures around, my theory is realist_2 about all forms of singular and general causation, including contribution, prob-influence, and culpable causation.

Another distinction that talk of fundamental and derivative is intended to supplant is the traditional divide between objective and subjective. It is fair to apply the label 'objective' to the fundamental causation-like relations, unless it includes the kind of fundamentally perspectival stuff posited by solipcists and phenomenalists. However, 'subjective' and 'objective' are potentially too misleading when applied to derivative causation. For example, all prob-influence and fixing relations are as objective as the fundamental laws in the sense that they hold regardless of what anyone thinks about them and regardless of the existence of any observers. But the fundamentally arbitrary parameters that characterize how to coarse-grain various events are conventional artifacts like coordinate systems. We could similarly puzzle over the length of a coastline. Is it objective because its magnitude does not depend on what I believe about the coastline? Or is it subjective because I can make it have just about any magnitude I desire (within an appropriate range) merely by adopting different length scales to use for its evaluation? Owing to misunderstandings that can occur when people apply 'subjective' to relations that hold partially by virtue of fundamentally arbitrary parameters, I think it is more perspicuous to speak of fundamental and derivative instead.

As I see it, the two most noteworthy contributions of this volume are (1) the identification of a methodology for conducting philosophical investigations and (2) a new scientific explanation of why causation is seemingly directed only toward the future. The methodological advance involved a focus on formulating *experimental schemas* whose results constitute the target empirical phenomena. The primary distinguishing feature of the methodology is an insistence that honing a system of concepts for explaining the results of such experiments is all one needs to do in order to complete a satisfactory metaphysics. In the particular case of causation, I identified the *promotion experiment* as representing the primary empirical phenomena motivating our belief in causation, and I also identified a modification of it, the *asymmetry experiment*, to capture the empirical phenomena that motivate our thinking of causation and influence as directed toward the future. Tellingly, when I tried to isolate the empirical phenomena behind our belief that we cannot take an action E_1 that influences an event E_2 by way of their common causes, I found no single decent experiment. That indicated a deficiency in the idea that we cannot influence events through a common cause, and as a result I rejected this naïve doctrine. In its place, I substituted the principle that we cannot advance any independently assigned goals through a common cause, as embodied in the experiment B3 described in §7.4. I am confident that this principle can survive rigorous attempts to disconfirm it, even if we develop sophisticated brain correlators.

The secondary distinguishing feature of empirical analysis is that it places no value whatsoever on regimenting a concept to make explicitly true (rather than explicitly false but pragmatically reasonable as a crude non-technical approximation) the various platitudes that appear to govern the proper application of the concept. For example, many experts believe a successful account of the metaphysics of causation must "solve the problem of preëmption," which is the task of designing one's metaphysical scheme so that it counts as non-causes (in the metaphysics) those events that are uncontroversially recognized by experts (and presumably by the untutored public) as mere preëmpted would-be causes. When a theory identifies paradigmatic preëmpted non-causes as genuine causes, orthodox standards count this as a theoretical deficiency, which needs to be explained away or outweighed by other considerations.[88] According to empirical analysis, by contrast, we should not judge a metaphysics of causation as even prima facie deficient for implying that the preëmpted would-be cause is a bona fide cause and that its perceived status as a non-cause is a matter of pragmatics. Another way of thinking about this issue is that common sense intuitions are generally so easily explained away that we hardly need to bother spelling out the details.

---

[88] Think here of the eight strategies discussed in Collins, Hall, and Paul (2004) for accommodating such discrepancies.

A related feature of empirical analysis is the way it exploits the difference between metaphysics and non-metaphysics to segregate the philosopher's investigatory task into two reasonably distinct projects of conceptual regimentation. For example, preëmption is arguably not a metaphysical distinction at all.

One reason to categorize a distinction as not genuinely metaphysical is when it does not figure in the STRICT portion of how fundamental reality accounts for all empirical phenomena. If we have reason to believe that (1) all of the empirical phenomena associated with singular and general causation can be explained in terms of a model of fundamental reality that obeys STRICT standards of theoretical adequacy and (2) this model does not benefit from the distinction (between preëmpted and non-preëmpted events), then the distinction can be left out of the metaphysics and accommodated in a theory that merely satisfies RELAXED standards of adequacy.

A second reason for categorizing preëmption as non-metaphysical arises when we attempt to answer the question, "What empirical phenomena, if any, motivate the distinction between a preëmpted event and a non-preëmpted event?" The asymmetry experiment has empirically testable results that reveal the existence of effective strategies for advancing goals for the future and the non-existence of effective strategies for advancing goals for the past. No one has ever proposed an experiment concerning causation "out there in reality" whose results are best explained using a distinction between preëmpted would-be causes from genuine causes. I suspect no relevant experiment is forthcoming because identifications of preëmption are made *retrospectively* in light of the outcomes of all the relevant causal interactions, not prospectively in a way that can be checked empirically. If this hunch is correct, then preëmption is metaphysically a non-issue.

It is easy to construct an experiment to reveal the utility of being able to identify paradigmatic cases of preëmption. For example, a person who observes a fragment of history and judges whether some c (as C) was preëmpted by p (as P) from causing some particular E can make better predictions about whether contextualizations of P that incorporate C in their background inhibit E. Observers who have some ability to distinguish preëmption will likely be better at predicting E in situations where C and P occur than observers who only count *regularities* among the occurrences of C, P, and E. But such an empirically observable result only goes to support my conclusion that we have an instinctive preëmption-attribution capacity as part of our broader ability to designate certain events as culpable causes because these judgments are efficient means for learning about promotion. It is empirically verifiable that creatures (or computers or what have you) who employ heuristics for assessing causal culpability can learn causal regularities and apply them to novel situations faster than creatures who have to slog through repeated runs of the promotion experiment for every new circumstance that arises. What makes these empirical phenomena count as outside of the metaphysics of causation in the framework I have laid out is that these experiments are not testing causation "out there in reality" independently of how we think about causation but rather how various fragments of history are categorized by creatures (or computers or whatever).

By dividing the subject of causation into a metaphysical and non-metaphysical component, empirical analysis engages in two distinct projects of conceptual regimentation with the standards of theoretical adequacy appropriate to them.

I will now emphasize two previously discussed qualifications to this overly brief summary of empirical analysis. First, the focus on experiments is not intended to ignore the fact that singular happenings count as empirical phenomena too, and they need to be encompassed as part of the overall body of empirical data. For example, that a baker's windowsill was opened at midnight on Easter of 1798 in Kaiserslautern counts as a bona fide empirical phenomenon even though the "experiments" that reveal it are activities like (1) being there at the time and looking and (2) searching the library for a reputable report. Such singular empirical phenomena need to be included as part of the overall data that any account of causation must be consonant with. My account accommodates singular empirical phenomena by treating every occurrence as an instance of some fundamental event. Each of these fundamental events has full singular causes in the form of terminants and partial singular causes in the form of contributors. These concepts are adapted to a complete explanation of each singular happening because fundamentally these happenings are maximally determinate, and every slight influence is part of an absolutely comprehensive explanation of why that precise event occurred.

Second, adequate consideration must be taken of the broader set of empirical phenomena that do not fall under the scope of the motivating principle behind the analyzed concept. For the metaphysics of causation, after I adopted the preliminary guiding idea that we believe in causation because we know some strategies are reliably better than others, it was important that I strove to ensure that the resulting experimental schema respect that there are abundant causal happenings having nothing to do with agency or strategies. To that end, I formulated the promotion experiment so that it applies not only to circumstances where the initial conditions instantiate one strategy rather than another, but also to circumstances where the initial conditions instantiate one configuration of physical stuff rather than another. In that way, my preliminary grasp at causes as means by which certain distinctive effects are reliably brought about was successfully demonstrated to be merely a special case of the broader class of phenomena where causes make their effects more likely, regardless of any agency. That does not imply that the initial focus on agency was pointless. For it turned out that the asymmetry of causation could be understood largely in terms of the asymmetry of advancement, which essentially incorporates agency.

Focusing attention now on my explanation of causal asymmetry, we can once again see the methodology of empirical analysis in action. I started by ruminating about why we would ever come to think of causation as directed toward the future. Following the lead of others like J. L. Mackie (1973), I made an initial guess that *useful* causation or *exploitable* causation is demonstrably future-directed in the sense that if people are assigned tasks to do something in the past, any effort they exert will be futile. I then attempted to formulate an experimental schema to make this claim testable. My suggestion was to select some event-kind E to serve as a target and then randomly assign agents either the goal of making E happen or the goal of preventing E. The results of running such experiments will show that whenever E is situated in the past, E's occurrence does not depend statistically on what goal the agent was assigned.

My explanation of the expected results of the asymmetry experiment can be simplified into two components. The first component exploits several uncontroversial features of the (presumed) fundamental laws of physics to prove that any action an agent takes after receiving notification of the target goal will be effective only to the extent that it operates directly toward the past. The second component demonstrates that any effective influence going directly toward the past would require nature to exhibit a freakish pattern of coincidences where the output from randomizing devices like coin flips and dice are remarkably correlated with any given E the experimenter chooses to consider. Such coincidences do not exist in regions of the universe like ours where matter behaves (toward the future) in a way that does not systematically deviate from whatever fundamental or derivative chanciness it possesses. The novel features of my explanation include that it can apply to fairly small regions of space-time, it does not require the overly restrictive preconditions for the applicability of entropy, and it eschews the overly restrictive and explanatorily insufficient fork asymmetry.

Once again, it was important that the notion of agency invoked in the asymmetry experiment did not presuppose any metaphysics that went too far beyond what was needed to address the empirical phenomena. To that end, I ensured that the asymmetry of advancement held true even as one moved from human agents to the crudest agent-like objects around, like thermostats. Here, it proved convenient to draw on the distinction between the pattern-based aspects of causation—those that would be useful for a perceptive non-agent to have—and the influence-based aspects of causation—those that would be useful for a purposive agent to have. This delineation made room for an adequate explanation of the advancement asymmetry, which vindicates people's crude belief that the influence-based aspects of causation are directed only toward the future, without pronouncing on all the pattern-based aspects of causation in the material layout of the universe.

10.2 Future Directions

The topic of causation is vast, and limitations imposed on the length of this volume have required me to avoid digressing from the central task of sketching a large-scale blueprint for the conceptual architecture of causation. I will now survey a few neglected topics together with a brief explanation of why I chose not to discuss them in more detail. This section is included primarily to suggest ideas for any readers who find my overall conceptual structure layout compelling enough to warrant further research and wish to flesh out comparisons with other theories of causation. Its secondary purpose is to serve as a rebuttal to the charge that by not discussing every publication that could be considered "clearly relevant" to the subject of causation in the special sciences, I have revealed myself to be a solipsist, ignorant of the numerous penetrating insights into causation readily available at my local library.

First, my introductory chapter began to elucidate a certain conception of fundamental reality and how it differs from derivative reality, and this scheme certainly demands further exegesis. Although my preliminary characterization is sufficient, I think, for an initial formulation of the metaphysics of causation, I have provided further discussion elsewhere, (Kutach 2011b), and I will have more to say in the second volume of *Empirical Fundamentalism*. By attempting to clarify fundamentality, I have not tried to make any advance in our understanding of fundamental reality in the sense of providing a new theory of fundamental physics or redesigning space-time structure or anything remotely that sophisticated. Insofar as one is concerned with the character of (the actual) fundamental reality—whether it includes corpuscles or fields, whether it binds events together with determination or incorporates chanciness, what the arena is, or if there even is an arena—I have no insights to present here. Thus, much of the value of my theory hangs on the hope that enough principles like those identified in chapter 2 obtain in order for modified versions of my explanations of causal directness and the asymmetry of advancement to withstand future revisions to our best guesses at fundamental reality.

Second, my ability to contrast my own theory with alternative accounts of how causes are probability-raisers is greatly hindered by my current inability to comprehend the standard conception of probability as it applies to roughly human-scale coarse-grained events, what I have been calling 'mundane events'. I think I understand well enough the mathematical axioms that make a quantity count formally as a probability as well as the notion of probability-fixing that I invoked in my own account. Recall that probability is incorporated into my account in one of two ways: either as a metaphysical primitive—stochasticity built into the fundamental laws—or by being stipulated into existence as part of an abstraction away from fundamental events into (derivative) contextualized events. I recognize there are many unresolved questions about how best to understand the issues raised by these notions: insensitivity considerations, typicality, microconstancy, and generally why some ways of fuzzing the physics are better than others. Yet, I think these two basic forms of probability in my theory have been connected to fundamental reality in a way that makes their *metaphysical* status clear enough for the purpose of understanding causation.

By contrast, metaphysical theories employing probabilistic relations among mundane events (or types of such events) baffle me because I cannot figure out how their advocates intend these probabilities to be connected to fundamental reality if it includes fundamental physics. For example, the existence of reliable statistical correlations among germs and diseases is no refutation of determinism, so that the alleged probabilistic relations are presumably compatible with deterministic fundamental laws. Nor do such correlations imply that one's fundamental ontology must include germs and diseases and their probabilistic relations. In light of these two considerations, should the conditional probabilities that play a central role in many standard accounts of probability-raising be construed as fundamental? In that case, they seem to add an unnecessary ontological redundancy. Or should they be understood as merely derivative? In that case, it is mysterious to me at present how people who employ such relations think they should be connected with fundamental reality. I know the existing literature on interpretations of probability is not entirely silent on this topic, but in light of people's current unfamiliarity with my particular construal of the distinction between fundamental and derivative, it is not surprising that standard probability-raising accounts of causation such as Suppes (1970) do not connect their intended construal of probability to fundamental reality. As a result, I will postpone further discussion of theories that rely on probabilistic relations among mundane events or mundane event types. No one is obligated to adopt my framework for how to think about reality in terms of fundamental and derivative, but if this framework has been set aside, I cannot contribute much to the discussion of probabilistic relations.

Third, there are undoubtedly important connections between my account of causation and the nascent philosophical literature on causal modeling (Spirtes, Glymour, and Scheines 2000; Pearl 2000; Woodward 2003). Because this literature uniformly employs the standard conception of probability, all the considerations in the previous paragraph apply to forbid an adequate comparison or assessment. Moreover, to add some further distance between our approaches, I can report that regardless of how valuable causal modeling is to scientific practice and to scientific conceptions of causation, it is unclear to me how advocates of the causal modeling approach can reasonably expect it to illuminate the *metaphysics* of causation. From my perspective on the conceptual landscape, metaphysics is primarily aimed at fundamental reality while secondarily casting light on how fundamental reality relates to derivative reality. I have no idea of the extent to which philosophers who work in the causal modeling tradition would share this conception of metaphysics, but a common tack their research takes is to avoid commitment to the existence of a reduction of causation. For a good example, see Woodward (2003), where central concepts like 'intervention' and 'direct cause' are defined using words like 'causal' in a deliberate effort to insulate his theory from questions about the reducibility of causation. Furthermore, the directed graphs that play the key structural role in these sorts of accounts of causation do not appear to be particularly useful for understanding the asymmetry of causation because the direction of causation is just built in as a primitive in the graphs. So, the central metaphysical question about the direction of causation—whether it is fundamental—is unaddressed. Nor does the causal modeling approach explain which features of fundamental reality account for the utility of causal models and the utility of the concept of intervention. The connection between fundamental physics and the causal modeling approach has been discussed by Woodward (2007), where he finds no incompatibility between them but also no currently plausible reduction of the interventionist's conception to notions in fundamental physics. In the end, the causal modeling and interventionist literature as a whole has not yet adequately addressed the *metaphysics* of causation (in the sense of 'metaphysics' I invoked) because it has said little about how causation is connected to fundamental reality.

On a positive note, I think there are some promising avenues for bridging the existing work on causal modeling to my own. One hypothesis worth entertaining is that the rules defining causal models and their structural equations provide a means for abstracting away from fundamental reality in a way that does not presuppose nearly as much about the nature of fundamental reality as my auxiliary hypothesis that fundamental reality resembles models of the paradigm theories of fundamental physics. Any particular causal model on this construal could thus count as an alternative to my scheme of contrastive events fixing one another and prob-influencing coarse-grained events. Furthermore, a causal model need not be a *conflicting* alternative to my account because there is nothing wrong with having multiple schemes for abstracting away from fundamental reality, nor is there any need to insist that exactly one scheme is best. We sometimes have different explanatory goals that demand different criteria or perhaps different ways of weighting the relative importance of multiple criteria. A causal model of how a windmill generates electricity can be a better way of abstracting from the fundamental causation-like relations when we place a lot of emphasis on the explanatory value of readily existing vocabulary by treating as basic antecedently understood derivative entities and quantities such as blades, gears, wind-speed, and voltage. Understanding the interactions between the various components of a windmill is fantastically complicated in the language of contrastive events. The value of prob-influence lies not in its applicability to the explanatory needs of the practicing scientist but in its ability to encompass all causal generalities regardless of the scientific discipline that studies them. Causal modeling provides resources better tailored for specific applications of causal principles in the special sciences.

The connection between the causal modeling literature and my metaphysical scheme could be worked out in more detail by spelling out how causal models provide an abstraction more narrowly targeted at specific scientific domains. I suspect that if one sets aside the irrelevant nominal and structural differences between the causal modeling approach and mine, the empirical content of any causal model for general causation would turn out to be special cases of prob-influence. This could lead to further clarification of some of the key concepts in the causal modeling tradition. For example, the concept of intervention, which is defined only in terms of the structure provided by a causal model, can be explained as a reasonable fiction in light of my explanation of the unexploitability of pseudo-backtracking prob-influence. That is, my account of causal asymmetry explains why it is reasonable to consider hypothetically setting certain variables in the model and letting only the values of the variables that are "causally downstream" depend on these alterations. Notably, my account can make sense of interventions while remaining fully committed to the thesis that there are no fundamental interventions. The universe evolves according to fundamental laws, the story would go, and all the events we like to think of as interventions are fundamentally just more stuff evolving according to the same laws as everything else. Furthermore, the universal applicability of my account allows one to make sense of imperfect or approximate interventions and to provide deeper explanations of the conditions under which particular causal models lose accuracy.

It ought to go without saying that none of these comments constitutes an attack on the value of causal modeling. Instead, it is merely a conclusion (drawn from the considerations mentioned in the introductory chapter) to categorize the philosophical work on causal modeling and its concepts like "direct cause" and "variable" as a part of the conceptual organization work that rightly goes on in science but not necessarily in metaphysics. The utility of causal models has implications for metaphysics, no doubt; otherwise I would not have bothered to suggest that my metaphysical scheme is capable of explaining how its utility arises. But existing discussions of causation that use causal models have so far not said much about how causal models hold by virtue of how fundamental reality is structured and so are not metaphysical theories in the sense I understand 'metaphysics'.

Fourth, not only can my account help to support theories that appeal to interventions, it can also bolster related accounts of causation that emphasize the role of agency and manipulation, for example Collingwood (1940), von Wright (1971, 1975), Price (1991), Menzies and Price (1993). The central organizing principle in this tradition is that causes are means by which an agent can manipulate the world so as to bring about an effect. The standard criticisms of such accounts, Woodward (2012) for example, question whether this organizing principle is adequate for making sense of instances of causation where no agency or intervention is involved. My account was also initiated by the idea that causes are means for bringing about effects, but because I was able to model future-directed advancement of goals via manipulating causes as a special case of promotion, I demonstrated in effect that toward the future, causation involving agency is not importantly different from agency-free causation. The key role that agency played was to help make sense of causal asymmetry and to make sense of why we are unable to exploit common causes as a (superficially backtracking) route for goal-advancing influence. Because I have addressed these issues in the previous section and in more detail in §7.4, I will say no more.

Fifth, for emphasis, I will repeat some comments I made in §5.9. Some recent philosophical literature, summarized nicely by Phil Dowe (2009), associates causation with the transfer of some appropriate physical quantity or the presence of some appropriate process. Despite the invocation of physics in many of these theories and despite the use of the label 'empirical analysis' to describe them, these theories are not comparable to the account I have presented in this volume. None of these theories counts as an empirical analysis of causation according to the definition I laid out in §1.1 because none of them identify what empirical phenomena their theories are attempting to explain; none of them specify any experiments whose results constitute the target explananda. Furthermore, like Woodward (2003), I have not yet seen a plausible proposal for how these processes provide a comprehensive basis for making sense of the utility of our many practices that bear on causation, such as finding the causes of cancer or inventing gadgets. The strategy of supplementing transference or causal process theories with information about statistical relationships, (Salmon 1997) or counterfactuals concerning conserved quantities (Dowe 2000), if successful, would threaten to convert them into another variation on the many probability-raising or counterfactual accounts.[89]

Sixth, I have said little about causal explanation, and this omission might strike some readers as particularly troubling given that the point of empirical analysis is to facilitate explanations of empirical phenomena. I have left the notion of explanation unexplored because I think the imprecise cluster of explanatory principles that are routinely employed by scientists is sufficient for the purpose of understanding the *metaphysics* of causation and because many of the orthodox methods for identifying genuine explanations or for settling philosophical debates about the nature of explanation flout the spirit of empirical analysis.

My account is not silent on the explanatory role of singular partial causes because it provides two different ways to explain any effect. The first is a "complete story" explanation where any given fundamental event *e* is explained by citing all its contributors and how they fit together as terminants of *e*. The second focuses on the utility of our various rules of thumb for identifying culpable causes. In brief, because the concept of a successful inducer is so handy for quickly learning about promotion (and for other reasons), our psychology and culture have adopted the idea that we should treat as especially important those contributors we recognize as events whose inducement (or probability-raising) of the effect succeeds in leading to the occurrence of the effect. We call them "the causes" even though there is virtually always a much larger collection of events that play a non-superfluous role in bringing about the effect. Though we know it is sometimes unclear which events are culpable and though we know pragmatic factors sometimes motivate us to assign some culpable causes zero explanatory value, we philosophers still have a tendency to simplify reality by categorizing events as either cause or non-cause, and (as a first order approximation) to think of these (culpable) causes as prima facie explanatory.

[89] Related points are made by Hitchcock (1995) and Kim (2001).

Despite any prominent explanatory value culpable causes may have, my theory does not grant them any special metaphysical status. Fundamentally, all pure contributors are equally partial causes. By delegating concerns about the explanatory status of singular events to the special sciences themselves and to epistemology, the criteria for a successful *metaphysics* of causation are made much easier to satisfy. Orthodox theories, by contrast, are expected to deny the metaphysical status of 'cause' to contributors that are widely judged not to be causes, as illustrated in cases of preëmption and saved fizzles.

Because the methodology of empirical analysis makes a successful metaphysics much easier to develop, readers who have spent years struggling to solve the problem of preëmption may find it tempting to regard my empirical analysis of causation as some sort of cheat, a refusal to address the central issues of causation rather than a proper resolution of them. To that charge, I will offer three final comments.

First, my account does not forbid anyone from adding culpable causes to the metaphysics; it only says they are unnecessary for an empirical analysis that already includes comprehensive relations like terminance and prob-influence.

Second, when I clarified the boundary between metaphysics and the special sciences in order to make precise the scope of my metaphysics of causation, the fact that it ended up classifying as non-metaphysical some projects that philosophers routinely think of as metaphysical does not imply that these projects are unimportant. Nor does it imply that my refusal to provide a STRICT account of causal culpability constitutes a failure to provide an adequate account of the metaphysics of causation. There are many important topics bearing on causation that do not require a STRICT account. Finding the causes of cancer, for example, is undoubtedly important, but it does not require any philosophical reduction. The same can be said for the project of using assessments of preëmption to identify more efficiently what kinds of background conditions are likely to interfere with attempts to bring about a chosen effect. The same can be said for many other projects that address causation scientifically.

Third, the reason I drew the distinction between STRICT and RELAXED standards of adequacy was to facilitate the provision of causal explanations by freeing them from an unnecessarily stringent condition of theoretical adequacy. When we use culpable causes (or some variant of them) to explain effects in the special sciences, we can get by with theories that are "consistent enough" in the sense of having consistent subcomponents that provide whatever explanatory goods are needed without there being some overarching, comprehensive, principled, correct theory of culpable causes that addresses paradigm cases with a single, manageably simple, demonstrably consistent theory. As I announced at the very beginning of this volume, I have adopted standards for theoretical adequacy that are appropriate for the task of empirical analysis, and they turn out to be easier to satisfy than the orthodox standards. Many of the theoretical roles that philosophers want a theory of causation to play do not require a STRICT theory of culpable causes. If my goal strikes readers as all too easily achieved, remember that the point of providing an account of the metaphysics of causation is not to solve all the problems associated with causation that have traditionally been understood as part of metaphysics, nor is it to solve all the problems concerning causation that are important in science or ethics or the law. The goal I set out was to provide an explicitly consistent *foundation*, based in fundamental reality, for all our current evidence pertaining to causation. I believe it is a smart policy, when addressing any philosophical issue, to disavow any doctrine that makes success artificially difficult to achieve.

{ REFERENCES }

Adams, E. (1975). *The Logic of Conditionals: An Application of Probability to Deductive Logic.* Dordrecht: D. Reidel.

Adams, E. (1976). "Prior Probabilities and Counterfactual Conditionals," in W. L. Harper and C. A. Hooker (eds.), *Foundations of Probability Theory, Statistical Inference and Statistical Theories of Science* **1**, 1–21.

Albert, D. (2000). *Time and Chance.* Cambridge: Harvard University Press.

Alicke, M. (1992). "Culpable Causation," *Journal of Personality and Social Psychology* **63** (3), 368–78.

Allori, V., Goldstein, S., Tumulka, R., Zanghi, N. (2008). "On the Common Structure of Bohmian Mechanics and the Ghirardi-Rimini-Weber Theory," *The British Journal for the Philosophy of Science* **59**, 353–89.

Armstrong, D. (1983). *What Is a Law of Nature?* Cambridge: Cambridge University Press.

Austin, J. L. (1961). "A Plea for Excuses," in *Philosophical Papers*, 123–52. Oxford: Oxford University Press.

Bechtel, W. and Richardson, R. (1993). *Discovering Complexity: Decomposition and Localization as Strategies in Scientific Research.* Princeton, NJ: Princeton University Press.

Beebee, H., Hitchcock, C., and Menzies, P., eds. (2009). *The Oxford Handbook of Causation.* Oxford: Oxford University Press.

Beebee, H. (2004). "Causing and Nothingness," in J. Collins, N. Hall, and L. A. Paul (eds.), *Causation and Counterfactuals.* Cambridge, MA: MIT Press.

Beisbart, C. and Hartmann, S., eds. (2011). *Probabilities in Physics.* Oxford: Oxford University Press.

Bell, J. "Bertlmann's Socks and the Nature of Reality," *Journal de Physique*, Colloque C2, suppl. au numero 3, Tome 42 (1981) pp. C2 41-61; reprinted in Bell, J. (2004). *Speakable and Unspeakable in Quantum Mechanics*, 2nd ed. Cambridge: Cambridge University Press, 139–58.

Benton, M. (2005). *When Life Nearly Died: The Greatest Mass Extinction of All Time.* New York: Thames and Hudson.

Block, N. and Stalnaker, R. (1999). "Conceptual Analysis, Dualism, and the Explanatory Gap," *The Philosophical Review* **108** (1), 1–46.

Boltzmann, L. (1895). *Nature* **51**, 413.

Brewer, R. and Hahn, E. (1984). "Atomic Memory," *Scientific American* **251** (6), 36–50.

Campbell, J. K., O'Rourke, M., and Silverstein, H., eds. (2007). *Causation and Explanation.* Cambridge, MA: MIT Press.

Carroll, J. (1991). "Property-level Causation?" *Philosophical Studies* **63**, 245–70.

Carroll, J. (1994). *Laws of Nature.* Cambridge: Cambridge University Press.

Carroll, J. (2004). *Readings on Laws of Nature.* Pittsburgh: University of Pittsburgh Press.

Cartwright, N. (1979). "Causal Laws and Effective Strategies," *Noûs* **13**, 419–37. Reprinted in N. Cartwright, *How the Laws of Physics Lie* (Oxford: Clarendon Press, 1983, 21–43).

Cartwright, N. (1994). *Nature's Capacities and their Measurement* Oxford: Oxford University Press.

Chalmers, D. and Jackson, F. (2001). "Conceptual Analysis and Reductive Explanation," *The Philosophical Review* **110** (3), 315–60.

Collingwood, R. G. (1940). *An Essay on Metaphysics.* London: Oxford University Press.

Collins, J., Hall, N., and Paul, L. A., eds. (2004). *Causation and Counterfactuals.* Cambridge: MIT Press.

Dainton, B. (2001). *Time and Space.* Montreal: McGill Queen's University Press.

Diaconis, P. and Engel, E. (1986). "Some Statistical Applications of Poisson's Work," *Statistical Science* **1** (2), 171–74.

Dowe, P. (1992a). "An Empiricist Defence of the Causal Account of Explanation," *International Studies in the Philosophy of Science* **6**, 123–28.

Dowe, P. (1992b). "Wesley Salmon's Process Theory of Causality and the Conserved Quantity Theory," *Philosophy of Science* **59**, 195–216.

Dowe, P. (2000). *Physical Causation.* Cambridge: Cambridge University Press.

Dowe, P. and Noordhof, P. (2004). *Cause and Chance: Causation in an Indeterministic World.* London: Routledge.

Dowe, P. (2009). "Causal Process Theories," in H. Beebee, C. Hitchcock, and P. Menzies (eds.), *The Oxford Handbook of Causation.* Oxford: Oxford University Press.

Dretske, F. (1977). "Laws of Nature," *Philosophy of Science* **44** (2), 248–68.

Driver, J. (2008a). "Attributions of Causation and Moral Responsibility," in W. Sinnott-Armstrong (Ed.) *Moral Psychology (Vol. 2): The Cognitive Science of Morality: Intuition and Diversity* 423–39. Cambridge, MA: MIT Press.

Driver, J. (2008b). "Kinds of Norms and Legal Causation: Reply to Knobe and Fraser and Deigh," in W. Sinnott-Armstrong (ed.), *Moral Psychology (Vol. 2): The Cognitive Science of Morality: Intuition and Diversity* 459–61). Cambridge, MA: MIT Press.

Ducasse, C. J. (1926). "On the Nature and the Observability of the Causal Relation," *The Journal of Philosophy* 23 (3), 57–68.

Dummett, M. (1964). "Bringing About the Past," *The Philosophical Review* **73** (3), 338–59.

Earman, J. (1986). *A Primer on Determinism.* Dordrecht: Reidel.

Earman, J. (1995). *Bangs, Whimpers, Crunches, and Shrieks.* Oxford: Oxford University Press.

Edgington, D. (2004). "Counterfactuals and the Benefit of Hindsight," in P. Dowe and P. Noordhof (eds.), *Cause and Chance: Causation in an Indeterministic World.* London: Routledge.

Eells, E. (1991). *Probabilistic Causality.* Cambridge: Cambridge University Press.

Elga, A. (2007). "Isolation and Folk Physics," in H. Price and R. Corry (eds.), *Causation, Physics, and the Constitution of Reality.* Oxford: Oxford University Press.

Ernst, G. and Hüttemann, A., eds. (2010). *Time, Chance, and Reduction: Philosophical Aspects of Statistical Mechanics.* Cambridge: Cambridge University Press.

Fair, D. (1979). "Causation and the Flow of Energy," *Erkenntnis* **14**, 219–50.

Field, H. (2003). "Causation in a Physical World," in M. Loux and D. Zimmerman (eds.), *The Oxford Handbook of Metaphysics.* Oxford: Oxford University Press, 435–60.

Fisher, R. (1959). *Smoking: The Cancer Controversy.* London: Oliver & Boyd.

Friedman, M. (1983). *Foundations of Space-Time Theories: Relativistic Theories and Philosophy of Science.* Princeton: Princeton University Press.

Frigg, R. (2009). "Typicality and the Approach to Equilibrium in Boltzmannian Statistical Mechanics," *Philosophy of Science* **76**, Supplement, S997–1008.

Frigg, R. (2011). "Why Typicality Does Not Explain the Approach to Equilibrium," in M. Suarez (ed.), *Probabilities, Causes and Propensities in Physics*, Synthese Library, Vol. 347. Berlin: Springer, 77–93.

Frisch, M. (2005a). *Inconsistency, Asymmetry, and Non-Locality*. New York: Oxford University Press.

Frisch, M. (2005b). "Counterfactuals and the Past Hypothesis," *Philosophy of Science* **72**, 739–50.

Frisch, M. (2007). "Causation, Counterfactuals and Entropy," in H. Price and R. Corry (eds.), *Causation, Physics, and the Constitution of Reality*. Oxford: Oxford University Press.

Frisch, M. (2010). "Does a Low-Entropy Constraint Prevent Us from Influencing the Past?" in G. Ernst and A. Hüttemann (eds.), *Time, Chance, and Reduction*. Cambridge: Cambridge University Press.

Galilei, G. (1960). *De Motu* (I.E. Drabkin, Trans.). Madison: The University of Wisconsin Press. (Original work written 1590)

Ghirardi, G. C., Rimini, A., and Weber, T. (1986). "Unified Dynamics for Microscopic and Macroscopic Systems," *Physical Review D* **34**, 470–91.

Glennan, S. (1996). "Mechanisms and the Nature of Causation," *Erkenntnis* **44** (1), 49–71.

Glennan, S. (2002). "Rethinking Mechanistic Explanation," *Philosophy of Science* **69** (3): S342–S353.

Glennan, S. (2009). "Mechanisms," in H. Beebee, C. Hitchcock, and P. Menzies (eds.), *The Oxford Handbook of Causation*. Oxford: Oxford University Press.

Glymour, C. and Wimberly, F. (2007). "Actual Causes and Thought Experiments," in J. K. Campbell, M. O'Rourke, and H. Silverstein (eds.), *Causation and Explanation*. Cambridge, MA: MIT Press.

Godfrey-Smith, P. (2012). "Metaphysics and the Philosophical Imagination," *Philosophical Studies* **160** (1), 97–113.

Good, I. J. (1961). "A Causal Calculus I," *The British Journal for the Philosophy of Science* **11**, 305–18.

Good, I. J. (1962). "A Causal Calculus II," *The British Journal for the Philosophy of Science* **12**, 43–51.

Gold, T. (1962). "The Arrow of Time," *American Journal of Physics* **30**, 403–10.

Goodman, N. (1947). "The Problem of Counterfactual Conditionals," *The Journal of Philosophy* **44**, 113–28.

Goodman, N. (1954). *Fact, Fiction, and Forecast*. Cambridge, MA: Harvard University Press.

Goldstein, S. (2010). "Boltzmann's Approach to Statistical Mechanics," in J. Bricmont, D. Dürr, M.C. Gallavotti, G. Ghirardi, F. Petruccione, and N. Zanghì (eds.), *Chance in Physics: Foundations and Perspectives*, Springer, New York, 39–54.

Goldstein, S. "Bohmian Mechanics", The Stanford Encyclopedia of Philosophy (Spring 2013 Edition), Edward N. Zalta (ed.), URL = ⟨http://plato.stanford.edu/archives/spr2013/entries/qm-bohm/⟩.

Hall, N. (2000). "Causation and the Price of Transitivity," *Journal of Philosophy* **97**, 198–222.

Hall, N. (2004). "Two Concepts of Causation," in J. Collins, N. Hall, and L. A. Paul (eds.), *Causation and Counterfactuals*, 2004. Cambridge: MIT Press.

Hall, N. (2007). "Structural Equations and Causation," *Philosophical Studies* **132**, 109–36.

Hausman, D. (1998). *Causal Asymmetries*. Cambridge: Cambridge University Press.

Hesslow, G. (1981). "Causality and Determinism," *Philosophy of Science* **48**, 591–605.

Hitchcock, C. (1993). "A Generalized Probabilistic Theory of Causal Relevance," *Synthese* **97**, 335–64.

Hitchcock, C. (1995). "Discussion: Salmon on Explanatory Relevance," *Philosophy of Science* **62**, 304–20.

Hitchcock, C. 1996a. "Farewell to Binary Causation," *Canadian Journal of Philosophy* **26**, 267–82.

Hitchcock, C. 1996b. "The Role of Contrast in Causal and Explanatory Claims," *Synthese* **107** (3), 395–419.

Hitchcock, C. (2001). "The Intransitivity of Causation Revealed in Equations and Graphs." *Journal of Philosophy* **98**, 273–99.

Hitchcock, C. (2003). "Of Humean Bondage," *The British Journal for the Philosophy of Science* **54**, 1–25.

Hitchcock, C. (2004). "Do All and Only Causes Raise the Probabilities of Effects?" in J. Collins, N. Hall, and L. A. Paul (eds.), *Causation and Counterfactuals*. Cambridge, MA: MIT Press.

Hitchcock, C. (2007). "Three Concepts of Causation," *Philosophy Compass* 2/3: 508–16, 10.1111/j.1747-9991.2007.00084.x.

Hitchcock, C. (2009). "Structural Equations and Causation: Six Counterexamples," *Philosophical Studies* **144**, 391–401.

Horwich, P. (1987). *Asymmetries in Time*, Cambridge, MA: MIT Press.

Jackson, F. (1979). "On Assertion and Indicative Conditionals," *The Philosophical Review* **88**, 565–89.

Jackson, F. (1998). *From Metaphysics to Ethics: A Defence of Conceptual Analysis*. Oxford: Oxford University Press.

Keller, J. (1986). "The Probability of Heads," *The American Mathematical Monthly* **93** (3), 191–97.

Kim, J. (2001). "Physical Process Theories and Token-Probabilistic Causation," *Erkenntnis* **48**, 1–24.

Kistler M. (1999). *Causalité et lois de la nature*. Paris: Vrin.

Knobe, J. and Fraser, B. (2008). "Causal Judgment and Moral Judgment: Two Experiments," in W. Sinnott-Armstrong (ed.), *Moral Psychology (Vol. 2): The Cognitive Science of Morality: Intuition and Diversity* 441–47. Cambridge, MA: MIT Press.

Kutach, D. (2001). *Entropy and Counterfactual Asymmetry*, PhD. Dissertation, Rutgers.

Kutach, D. (2002). "The Entropy Theory of Counterfactuals," *Philosophy of Science* **69** (1), 82–104.

Kutach, D. (2007). "The Physical Foundations of Causation," in H. Price and R. Corry (eds.), *Causation, Physics, and the Constitution of Reality*. Oxford: Oxford University Press.

Kutach, D. (2010). "Empirical Analyses of Causation," in A. Hazlett, (ed.) *New Waves in Metaphysics*. Palgrave Macmillan, UK.

Kutach, D. (2011a). "Backtracking Influence," *International Studies in the Philosophy of Science* 25 (1), 55–71.

Kutach, D. (2011b). "Reductive Identities: An Empirical Fundamentalist Approach," *Philosophia Naturalis* **47–48**, 67–101.

Kutach, D. (2011c). "The Asymmetry of Influence," in C. Callender (ed.), *The Oxford Handbook of Philosophy of Time*. Oxford: Oxford University Press.

Kutach, D. (Forthcoming). "The Empirical Content of the Epistemic Asymmetry," in B. Loewer, B. Weslake, and E. Winsberg (eds.), *On Time and Chance*. Cambridge, MA: Harvard University Press.

Kvart, I. (2004). "Probabilistic Cause, Edge Conditions, Late Preemption and Discrete Cases," in P. Dowe and P. Noordhof (eds.), *Cause and Chance: Causation in an Indeterministic World*. London: Routledge.

Lange, M. (2002). *An Introduction to the Philosophy of Physics: Locality, Fields, Energy, and Mass*. Oxford: Blackwell Publishers.

Laraudogoitia, P. (1996). "A Beautiful Supertask," *Mind* **105**, 81–83.

Lewis, D. (1973a) *Counterfactuals*. Oxford: Blackwell.

Lewis, D. (1973b). "Causation," *The Journal of Philosophy* **70** 556–67, reprinted in *Philosophical Papers, Volume* 2, Oxford: Oxford University Press, 1986.

Lewis, D. (1986). "Postscripts to 'Causation'," *Philosophical Papers, Vol. II* Oxford: Oxford University Press.

Lewis, D. (2000). "Causation as Influence," *Journal of Philosophy* **97** 182–97, reprinted in J. Collins, N. Hall, and L. A. Paul (eds.), *Causation and Counterfactuals*. Cambridge, MA: MIT Press.

Lewis, P. (2006). "GRW: A Case Study in Quantum Ontology," *Philosophy Compass* **1** (2), 224–44.

Loewer, B. (2004). "Humean Supervenience," in J. Carroll, ed. *Readings on Laws of Nature*.

Loewer, B. (2007). "Counterfactuals and the Second Law," in H. Price and R. Corry (eds.), *Causation, Physics, and the Constitution of Reality*. Oxford: Oxford University Press.

Loewer, B. (2012). "Two Accounts of Laws and Time," *Philosophical Studies* DOI 10.1007/s11098-012-9911-x 1–23.

Machamer, P. and Wolters, G. (2007). *Thinking about Causes: From Greek Philosophy to Modern Physics*. Pittsburgh: University of Pittsburgh Press.

Machamer, P., Darden, L., and Craver, C. (2000). "Thinking about Mechanisms," *Philosophy of Science* **67** (1), 1–25.

Mackie, J. L. (1973). *The Cement of the Universe*. Oxford: Oxford University Press.

Maslen, C. (2004). "Causes, Contrast, and the Nontransitivity of Causation," in J. Collins, N. Hall, and L. A. Paul (eds.), *Causation and Counterfactuals*. Cambridge, MA: MIT Press.

Mather, J. and McGehee, R. (1975). "Solutions of the Collinear Four-Body Problem Which Become Unbounded in a Finite Time," in J. Moser (ed.), *Dynamical Systems, Theory and Applications*, New York: Springer-Verlag.

Maudlin, T. (2004). "Causation, Counterfactuals, and the Third Factor," in J. Collins, N. Hall, and L. A. Paul (eds.), *Causation and Counterfactuals*. Cambridge, MA: MIT Press. Reprinted in *The Metaphysics in Physics*, 2007. Oxford: Oxford University Press.

Maudlin, T. (2007a). *The Metaphysics in Physics*. Oxford: Oxford University Press.

Maudlin, T. (2007b). "What Could Be Objective About Probabilities?," *Studies in History and Philosophy of Modern Physics* **38**, 275–91.

Maudlin, T. (2011). "Three Roads to Objective Probability," in C. Beisbart and S. Hartmann (eds.), *Probabilities in Physics*. Oxford: Oxford University Press.

McCloskey, M. (1983). "Intuitive Physics" *Scientific American* **248** (4), 122–30.

Meheus, J. (2002). *Inconsistency in Science*. Dordrecht, Netherlands: Kluwer.

McDermott, M. (1995). "Redundant Causation," *The British Journal for the Philosophy of Science* **40**, 523–44.

Mellor, D. H. (1995). *The Facts of Causation*, New York: Routledge.

Menzies, P. (1989). "Probabilistic Causation and Causal Processes: A Critique of Lewis," *Philosophy of Science* **56**, 642–63.

Menzies, P. (1996). "Probabilistic Causality and the Pre-exemption Problem," *Mind* **105**, 85–117.

Menzies, P. and Price, H. (1993). "Causation as a Secondary Quality," *The British Journal for the Philosophy of Science* **44**, 187–203.

Mill, J. S. (1858). *A System of Logic: Ratiocinative and Inductive*, London: Longmans, Green and Co., 1930.

Ney, A. (2009). "Physical Causation and Difference-Making," *The British Journal for the Philosophy of Science* **60**, 737–64.

Northcott, R. (2008). "Causation and Contrast Classes," *Philosophical Studies* **139** (1), 111–23.

Northcott, R. (2010). "Natural Born Determinists: a New Defense of Causation as Probability-raising," *Philosophical Studies* **150**, 1–20.

Norton, J. (1987). "The Logical Inconsistency of the Old Quantum Theory of Black Body Radiation," *Philosophy of Science* **54**, 327–50.

Norton, J. (2003). "Causation as Folk Science," Philosophers' Imprint 3 (4), http://www.philosophersimprint.org/003004/. Reprinted in H. Price and R. Corry (eds.), *Causation, Physics, and the Constitution of Reality*, 2007. Oxford: Oxford University Press.

Norton, J. (2007). "Do the Causal Principles of Modern Physics Contradict Causal Anti-Fundamentalism?" in P. Machamer and G. Wolters (eds.), *Thinking about Causes: From Greek Philosophy to Modern Physics*. Pittsburgh: University of Pittsburgh Press.

Norton, J. (2008). "The Dome: An Unexpectedly Simple Failure of Determinism," *Philosophy of Science* **75**, 786–98.

Nozick, R. (1969). "Newcomb's Problem and Two Principles of Choice," in N. Rescher (ed.), *Essays in Honor of Carl G. Hempel*, 114–46. Dordrecht: Reidel.

Paul, L. A. (2000). "Aspect Causation." *Journal of Philosophy* **97**, 235–56, reprinted in J. Collins, N. Hall, and L. A. Paul (eds.), *Causation and Counterfactuals*. Cambridge, MA: MIT Press.

Paul, L. A. (2009) "Counterfactual Theories," in H. Beebee, C. Hitchcock, and P. Menzies (eds.), *The Oxford Handbook of Causation*. Oxford: Oxford University Press.

Pearl, J. (2000). *Causality: Models, Reasoning, and Inference*. Cambridge: Cambridge University Press.

Price, H. (1991). "Agency and Probabilistic Causality," *The British Journal for the Philosophy of Science* **42**, 157–76.

Price, H. (1996). *Time's Arrow and Archimedes' Point*. Oxford: Oxford University Press.

Price, H. and Corry, R., eds. (2007). *Causation, Physics, and the Constitution of Reality*. Oxford: Oxford University Press.

Price, H. and Weslake, B. (2009). "The Time-Asymmetry of Causation," in H. Beebee, C. Hitchcock, and P. Menzies (eds.), *The Oxford Handbook of Causation*. Oxford: Oxford University Press.

Putnam, H. (1975). "The Meaning of 'Meaning'," in K. Gunderson (ed.), *Language, Mind and Knowledge*, Minnesota Studies in the Philosophy of Science, VII. Minneapolis: University of Minnesota Press.

Ramachandran, M. (2004). "Indeterministic Causation and Varieties of Chance-raising," in P. Dowe and P. Noordhof (eds.), *Cause and Chance: Causation in an Indeterministic World*. London: Routledge.

Ramsey, F. (1928). "Universals of Law and of Fact," in *F. P. Ramsey: Philosophical Papers*, D. H. Mellor (ed.), Cambridge: Cambridge University Press, 1990, 140–44.

Reichenbach, H. (1956). *The Direction of Time*, Berkeley: University of California Press.

Roberts, J. (2009). *The Law Governed Universe*. Oxford: Oxford University Press.

Russell, B. (1913). "On The Notion of Cause," *Proceedings of the Aristotelian Society* **13**, 1–26.

Russell, B. (1948). *Human Knowledge*. New York: Simon and Schuster.

Salmon, W. (1977). "An 'At-At' Theory of Causal Influence," *Philosophy of Science* **44**, 215–24.

Salmon, W. (1984). *Scientific Explanation and the Causal Structure of the World*. Princeton: Princeton University Press.

Salmon, W. (1993). "Causality: Production and Propagation," in E. Sosa and M. Tooley (eds.), *Causation*. Oxford: Oxford University Press.

Salmon, W. (1997). "Causality and Explanation: A Reply to Two Critiques," *Philosophy of Science* **64**, 461–77.

Schaffer, J. (2000a). "Overlappings: Probability-Raising without Causation," *Australasian Journal of Philosophy* **78**, 40–46.

Schaffer, J. (2000b). "Causation by Disconnection," *Philosophy of Science* **67**, 285–300.

Schaffer, J. (2001). "Causes as Probability Raisers of Processes," *Journal of Philosophy* **98**, 75–92.

Schaffer, J. (2003). "Is There a Fundamental Level?" *Noûs* **37**, 498–517.

Schaffer, J. (2005). "Contrastive Causation," *The Philosophical Review* **114** (3), 327–58.

Schaffer, J. "The Metaphysics of Causation," *The Stanford Encyclopedia of Philosophy (Fall 2008 Edition)*, Edward N. Zalta (ed.), URL = ⟨ http://plato.stanford.edu/archives/fall2008/entries/causation-metaphysics/ ⟩.

Sellars, W. (1962). "Philosophy and the Scientific Image of Man," in R. Colodny (ed.), *Frontiers of Science and Philosophy*. Pittsburgh: University of Pittsburgh Press. Reprinted in *Science, Perception and Reality*, 1963.

Skyrms, B. (1981). "The Prior Propensity Account of Subjunctive Conditionals," in W. L. Harper, R. Stalnaker, and G. Pearce (eds.), *Ifs*. Dordrecht: D. Reidel, 259–65.

Sklar, L. (1977). *Space, Time, and Spacetime*. Berkeley, CA: University of California Press.

Sober, E. (1985). "Two Concepts of Cause," in P. Asquith and P. Kitcher (eds.), *PSA 1984, vol. 2*, 405–24. East Lansing: Philosophy of Science Association.

Spirtes, P., Glymour, C., and Scheines, R. (2000). *Causation, Prediction, and Search*. Cambridge: MIT Press.

Stalnaker, R. (1968) "A Theory of Conditionals," in N. Rescher (ed), *Studies in Logical Theory, American Philosophical Quarterly Monograph Series, No. 2*, Oxford Basil Blackwell, 98–112, reprinted in E. Sosa (ed.), *Causation and Conditionals*, Oxford: Oxford University Press, 165–79, and in W. L. Harper, R. Stalnaker, and G. Pearce (eds.), *Ifs*, Dordrecht: D. Reidel, 107–28.

Steglich-Petersen, A. (2012). "Against the Contrastive Account of Singular Causation," *The British Journal for the Philosophy of Science* **63**, 115–43.

Strevens, M. (1998). "Inferring Probabilities from Symmetries," *Noûs* **32**, 231–46.

Strevens, M. (2003). *Bigger than Chaos: Understanding Chaos through Probability*. Cambridge, MA: Harvard University Press.

Strevens, M. (2011). "Probability Out Of Determinism," in C. Beisbart and S. Hartmann (eds.), *Probabilities in Physics*. Oxford: Oxford University Press.

Suppes, P. (1970). *A Probabilistic Theory of Causality*. Amsterdam: North-Holland.

Swoyer, C. (1982). "The Nature of Natural Laws," *Australasian Journal of Philosophy* **60** (3), 203–23.

Tahko, T. (2009). "The Law of Non-Contradiction as a Metaphysical Principle," *Australian Journal of Logic* 7, 32–47.

Talmy, L. (1988). "Force Dynamics in Language and Cognition," *Cognitive Science* **12**, 49–100.

Tooley, M. (1977). "The Nature of Laws," *Canadian Journal of Philosophy* **7** (4), 667–98.

Tumulka, R. (2006). "A Relativistic Version of the Ghirardi-Rimini-Weber Model," *Journal of Statistical Physics* **125** (4), 821–40.

Volchan, S. (2007). "Probability as Typicality," *Studies In History and Philosophy of Modern Physics* **38** (4), 801–14.

Wald, R. (1984). *General Relativity*. Chicago: The University of Chicago Press.

Ward, B. (2001). "Humeanism without Humean Supervenience: A Projectivist Account of Laws and Possibilities," *Philosophical Studies* **107** (3), 191–218.

Weslake, B. (2006). "Common Causes and the Direction of Causation," *Minds and Machines* **16** (3), 239–57.

Wolff, P. and Zettergren, M. (2002). "A Vector Model of Causal Meaning," *Proceedings of the twenty-fifth annual conference of the Cognitive Science Society*. Hillsdale, NJ: Erlbaum.

Wolff, P. (2007). "Representing Causation," *Journal of Experimental Psychology: General* **136** (1), 82–111.

Woodward, J. (2003). *Making Things Happen: A Theory of Causal Explanation*. Oxford: Oxford University Press.

Woodward, J. (2007). "Causation with a Human Face," in H. Price and R. Corry (eds.), *Causation, Physics, and the Constitution of Reality*. Oxford: Oxford University Press.

Woodward, J. (2012). "Causation and Manipulability", *The Stanford Encyclopedia of Philosophy (Winter 2012 Edition)*, Edward N. Zalta (ed.), URL = ⟨http://plato.stanford.edu/archives/win2012/entries/causation-mani/⟩.

von Wright, G. (1971). *Explanation and Understanding*. Ithica, New York: Cornell University Press.

von Wright, G. (1975). *Causality and Determinism*. New York: Columbia University Press.

Wüthrich, C. (2011). "Can the World be Shown to be Indeterministic After All?" in C. Beisbart and S. Hartmann (eds.), *Probabilities in Physics*. Oxford: Oxford University Press.

Yablo, S. (1992). "Mental Causation," *The Philosophical Review* **101**, 245–80.

Yablo, S. (1997). "Wide Causation," *Philosophical Perspectives: Mind, Causation, and World* **11**, 251–81.

Yablo, S. (2002). "De Facto Dependence," *The Journal of Philosophy* **99** (3), 130–48.

{ INDEX }

abstreduction, 34–36, 123; of general causation, 36–37, 138–40, 186–87
acceleration, 88
actual causation, 43, 266
actual coarse-grained event, 62
advancement asymmetry, 17, 226
agency, 143, 171, 198, 229, 253, 316
Albert, D., 250
amelioration of conflicts, 39–42
antecedent, 124
apparent conflict, 39–42
arbitrarily fast contribution, 88, 212
arena, 57; quantum-mechanical, 112–13; space-time, 79–82
aspect promotion, 154
asymmetry, of advancement, 16–17, 226; of bizarre coincidences, 184–86, 192, 239, 247–48; of causation, 105, 122, 230–39, 244–46, 249–62, 289–90; of entropy, 179, 244–47
asymmetry experiment, 226
attribute, 24

background of a contrastive event, 137
backtracking, causation, 214; fixing, 203; influence, 214; nomic connection, 16, 203; prob-influence, 204, 207–10, 212, 214, 234–36; routes of influence, 214, 216, 240
backtracking experiment, 216–18
backtracking experiments B1, B2, B3, 240–42
Bell, J., 237
Bernoulli distribution, 181
bizarre, coincidences, 184–86, 192, 233, 247–48; development, 129, 172; evolution, 129, 180, 181, 233; world, 129, 132, 192
Bohmian mechanics, 113–14, 116, 260
bold prediction, 228
Boltzmannian universe, 251–52
bottom conceptual layer of causation, 20, 22, 36, 37, 48–49, 53, 266, 278
brain correlator, 219–20, 240–42
branch systems, 244–46
broad prob-influence, 189–92, 237

c-connected, 96
c-path, 96
Cartesian spirit, 109, 219
Cartwright, N., 14, 122, 223, 274n, 307
causal, asymmetry, 105, 122, 230–39, 244–46, 249–62, 289–90; backtracking, 214; chain, 287, 300; contributor, 75 (*see also* contribution); culpability, 22 (*see also* culpability); eliminativism, 53–54; explanation, 18–19, 303; irreflexivity, 289; law, 53; line, 54; modeling, 19, 239, 314–16; pluralism, 197; process, 199, 297; relata, 21; variable, 316
causal contribution. *See* contribution
causal directness, 16, 17, 32, 49, 203–04, 234, 238, 239, 241; proof of, 207–11
causal process accounts, 3, 200
causation, actual, 43, 266; by omission, 155; culpable, 22 (*see also* culpability); direction of, 105, 122, 230–39, 244–46, 249–62, 289–90; egalitarian, 46, 266; general, 22, 121, 140, 144, 284; metaphysics of, 13, 17, 19–21, 44, 48, 152, 171, 194, 215, 225, 265, 277, 317; non-metaphysical aspects, 13, 17–20, 37, 44, 48, 267, 268, 306; singular, 22, 121, 265–66, 284
causation-like relation, 36
chain, causal, 287, 300
chaotic influence, 141
charge, 92–93
classical gravitation, 56, 82–88; sparse interpretation, 85; standard interpretation, 82
classical mechanics, simple theory of, 25–26
classical unified field theory, 92–94
the clever, 235
closed time-like curves (CTCs), 111, 211, 229
coarse-grained event, 60, 62, 110; trivial, 61
coarse-graining, 201–02
Collins, J., 5
common-cause pattern, 207, 214–16, 223, 239, 242, 243
conceptual analysis, 4; empirical, xi, 1–20, 317; orthodox, 9–10
conceptual engineering, 3, 5, 8
conceptual insulation, 5
conditional, 123; contrary-to-fact, 123–24; corner, 125, 127; counterfactual, 123–25; nomic, 128–35, 138, 176, 251; ordinary language, 134–35
configuration space, 112
conflict, 39–42, 302

conflict amelioration, delegation of, 49, 267, 283, 302
conjunctive fork, 231, 254
consequent, 124
conserved quantity (CQ) accounts, 3, 6, 7n, 200
conspiratorial coincidences, 87
content completeness, 93–94
content independence, 85n, 94–98, 110
contextualization, 69; trivial, 69
contextualized event, 69
continuity, of causal culpability, 303; of fixing, 161; of prob-influence, 161; of probability-fixing, 160; of terminance, 98–100, 211–12
contrary-to-fact conditional, 123–24
contrast, 136
contrastive causation, 155–57, 286
contrastive event, 136–38, 157; maximal, 137
contrastivity, 155–57
contrastivization, irregular, 165; regular, 162
contribution, 75–76, 109, 121, 265–66; arbitrarily fast, 88, 212; instantaneous, 88; pure, 76, 77, 121; space-like pure, 88
contributor, 75, 109 (*see also* contribution)
control, 188, 250, 252
corner conditional, 125, 127
corpuscle, 25, 80
cotenability, 125
counterfactual, 123–24; mundane, 128n, 130; vs contrary-to-fact conditional, 124
counterfactual dependence, 37, 123, 125–26, 135, 138
culpability, 22, 46, 266, 270, 272, 284, 285; as an explanatory device, 277; as proxy for promotion, 282, 284; for learning about promotion, 275; tension in concept of, 284; transitivity of, 160, 303
culpability$_1$, 286, 287, 291–92
culpability$_2$, 287, 292–99
culpability$_3$, 287, 299–302, 305
culpability$_4$, 287, 300–05
culpable causation, 22 (*see also* culpability)

Dainton, B., 245, 246
dappled worlds, 150
decision theory, 223, 243
dependence, counterfactual, 37, 123, 125–26, 135, 138
derivative, event, 60; future, 205; law, 66; past, 205; reality, 24–25
derivative quantity, 27
determinant, 67
determination, 23, 36, 67–68, 84, 110; by constitution, 77–78; trivial, 76; ubiquitous, 108, 176, 191
determinism, 84, 101–04, 108, 111; local, 101; possible-worlds, 102; unique-propagation, 102–03
deterministic law, 84, 102, 129, 173, 191
development, 129
difference-making, 37, 121, 135, 138, 142, 164
dimension, 80
direct cause, 314, 316
direction of causation, 105, 122, 230–39, 244–46, 249–62, 289–90
direction of influence, 142, 205–07, 257–62
team DO, 227
domain of contribution, 95
domain of dependence, 97; maximal, 103, 108
domain of influence, 95, 141
domain of terminance, 95, 97
team DONT, 227
Dowe, P., 2–4, 6, 200, 316, 317
Dummett, M., 230
duration, 80
dynamical law, 38

e-wards domain of influence, 98
Earman, J., 97, 102
effective strategies, 14–17, 170–71, 197–99, 222–24, 226
egalitarian cause, 46, 266
electromagnetic charge, 93
electromagnetic field, 64, 89
electromagnetism, relativistic, 56, 89–92
eliminativism, causal, 53–54
emergence, 55
empirical analysis, xi, 1–20, 317; of the metaphysics of causation, 13–17, 44, 48, 152, 171, 194, 215, 225, 277, 317; of the non-metaphysical aspects of causation, 13, 17–20, 37, 44, 48, 267, 268, 306; two forms of, 12–13, 19–20, 48
Empirical Fundamentalism, xi, 31
enabling cause, 304
energy, in ecology, 45; kinetic, 27, 30; mechanical, 28–30; thermal, 28–30
enion probability, 173, 174
entropy, 179, 184; asymmetry of, 179, 244–47
event, 58–64; actual, 59; coarse-grained, 60, 62, 110; contextualized, 69; contrastive, 136–38, 157; derivative, 60; fine-grained, 59; full, 63; fundamental, 59; impossible, 164; intermediate, 98; maximal, 70; maximal contrastive, 137; mundane, 46; possible, 59; temporally extended, 150–51; turnaround, 203; world, 64
Everettian quantum mechanics, 116
evolution, 129
example, acid rain, 151; airbags, 153; anti-thermodynamic ocean, 245–46; beach, 145;

bear box, 159; being born, 228–29; bird perches, 182–83, 237; blue roses, 189; bridge collapse, 288; cannonball in pond, 169, 170; college admissions, 194–96; cranberries, 154; dartboard, 172–73; duck reproduction, 304–05; electrolysis of water, 202; falling rock, 172; fire-starter, 148; fizzled fuse, 277, 298; flag-raising, 220–22, 240–42; food/nutrient, 3–4; gas in box, 177; gas in box in tank, 177, 184; hungry dog, 155; intelligent plants, 230; itch and fever, 207, 239; light switch, 144; lightning, 146–47; lit panel button, 289; match striking, 287; melting marshmallow, 158; moon, 141, 206; mouth, 188; nuclear-spin echo, 190–91; P-T extinction, 279–81; paint balls, 291–99; pizza, 228; power plants, 151; purple fire, 158; quark mass, 188–90; rock thrown in lake, 169; rock-throwers, 275; rust, 144; salty dessert, 155; salty dish, 156–57; silver coin, 189; smoking gene, 242–43; spicy dish, 156; super-light cone, 212; thermostat, 229; thrombosis, 305; thunder, 147–48; thwarting ubiquitous determination, 191–92; track switching, 300–04; tree falling on track, 301; volcanos and meteors, 279–81; voluntary stay in locked room, 261; wind farms, 151; windmill, 315
excision, 75
existent, 26–27
experiments B1, B2, B3 (backtracking), 240–42
experiments in empirical analysis, 11–12
explanation, complete story, 17–18, 271, 317; of effective strategies, 17
explicitly true, 7–8, 10, 48, 309

fibrillation, 178, 184
field, electromagnetic, 89; gravitational, 86–87, 91; physical, 80; strong, 92–93; weak, 92–93
Field, H., 307
Fisher, R., 242
fixed intermediate, 98, 160
fixing, 69–72; backtracking, 203; continuity of, 161; probability 70–71; shielding of, 161; strong transitivity of, 158; trivial, 137; unidirectional transitivity of, 158–59; weak transitivity of, 157, 211
fizzle, 297–98; saved, 298, 318
flashy quantum mechanics, 115
flow of time, 70, 205
focal set, 287
foliation, 82, 91
food/nutrient example, 3–4
force dynamic model of causation, 285n
foreground of a contrastive event, 137
fork, asymmetry, 231, 255–57; conjunctive, 254; open, 254
forking model of counterfactuals, 258
fragment of history, 271
free will, 143
full cause, 67, 75
full event, 63
functional relations, 68
fundamental; constants, 198; direction of influence, 142; event, 59; existence, 21; future, 205, 225, 257, 260; law, 66; level, 34; past, 205; reality, 24–25; volition, 109, 219
fundamental physics, paradigm theories of, 56
fundamentally arbitrary parameter, 27, 33–35
future, 205
future-bizarre, 182–83
future-typicality, 182, 184, 233, 238, 256
fuzzing, 49, 59

Galilean space-time, 79, 82–83, 91, 112, 259
general causation, 22, 121, 140, 144, 284
general relativity, 56, 109–11
generic state, 84
genuine conflict, 39–42
global state, 81; of suitable shape, 101, 103
Godfrey-Smith, P., 11n
Gold universe, 248–49
Goldstein, S., 181
Goodman, N., 124–32
graceful degradation, 5–6, 15, 187, 229, 251
gravitation, classical theory of, 82–88; universality of, 94, 212
gravitational field, 86–87, 91
gremlin, 108
growing block, 205
GRW interpretation of QM, 114

Hall, N., 5, 266, 300
Hitchcock, C., 10, 268
hole argument, 103
Horwich, P., 255
Humeanism, 65, 231

imposed contrastive effect, 296, 299
impossible event, 164
indeterminant, 72
indetermination, 72–73, 106
indeterminism, Newtonian, 107–08; non-stochastic, 107–09; quantum, 114; stochastic, 104–06, 114
inducement, 285–86, 294
inextendible, c-path, 97; space-like surface, 81
inferring prob-influence, 193, 196

influence, 122, 140–41, 167; backtracking, 214; chaotic, 141; direction of, 142, 205–07, 257–62; partial, 162–66; past-directed, 206
influence-based set of platitudes, 230
initial value problem, 109
insensitivity considerations, 172–74, 198
instantaneous causation, 87–88
instantaneous contribution, 88
interaction linkage, 93–94, 212
intermediate event, 98
intermediate region, 98
intervention, 239, 314–16
inus account of causation, 36–37, 76
inus condition, 76
irregular contrastivization, 164–65, 222

kinetic energy, 27, 30
Kistler, M., 200

Lange, M., 31
lattice, stochastic, 104–05
law, 64–67; derivative, 66; dynamical, 38; fundamental, 66; special science, 66–67
law facts, 64–66
law of non-contradiction, 38
law-like vs. accidental regularities, 16
level of reality, 34
Lewis, D., 65, 156
light cone, 90; of an event, 91
light-like, 90
link, fuzzy, 187; precise, 187
local determinism, 101
locality, 166
locality, relativistic, 91
Locke, J., 261
Loewer, B., 250–52
Lorentz force law, 38
Lorentzian space-time, 109

Mackie, J. L., 36, 76, 311
macroperiodicity, 174
macroscopic condition, 179
macrostate, 177, 179, 251
managed inconsistency, 38, 45, 47, 283, 318
manifold, 79
manipulability accounts, 316
Markov process, 99n
mass, 90
material contents, 57
material facts, 64–66
matter, 57
Maudlin, T., 181
maximal, arena, 102; domain of dependence, 103, 108; event, 70; space-time, 101–02
Maxwell's laws, 38, 65, 91
mechanical energy, 28–30
mechanism, causal, 199–201
metaphysics, 43, 267; of causation, 13, 17, 19–21, 44, 48, 152, 171, 194, 215, 225, 265, 277, 317
metric, Minkowski, 89; spatial, 83; spatio-temporal, 89; temporal, 80, 83
microcanonical distribution, 177, 181
microconstancy, 174
middle conceptual layer of causation, 20, 22, 36–37, 49, 186, 266, 278
Mill, J. S., 36, 118, 124
minimal terminant, 84
Minkowski, metric, 89; space-time, 79, 89–91
modal interpretations of QM, 116
morality and causation, 288
mundane counterfactual, 128n, 130
mundane event, 46

naked singularity, 110–11
narrow prob-influence, 189–91, 237, 249
neuron diagram, 53
Newcomb's problem, 219
Newton's second law of motion, 84
Newtonian indeterminism, 107–08
Newtonian space-time, 259
Ney, A., 164
nomic conditional, 128–35, 138, 176, 251
nomic connection, backtracking, 16, 203
non-locality, 165
non-maximal space-time, 101
non-metaphysical aspects of causation, 13, 17–21, 37, 44, 48, 267–68, 306
non-spatiality, 81, 212–13
Norton's dome, 108
Norton, J., 108
Nozick, R., 219
nuclear-spin echo, 190–91
null cone, 90
nutrient/food example, 3–4

$O = P$ prediction, 169–70, 190
objectivity, 23, 49, 54, 134, 268, 308; of causal direction, 249, 253–54; of probability distributions, 149, 174, 180, 181
omission, promotion by, 155
ontological vagueness, 150
open fork, 254
orthodox (conceptual) analysis, 9–10
overdetermination, 85–87

paradigm theories of fundamental physics, 56
parterminance, 72, 106, 115
partial cause, 75
partial influence, 162–66; space-like, 166

particle decay, toy theory of, 105–06
past, 205
past hypothesis, 185
past-bizarre, 184
past-directed influence, 206
pattern-based set of platitudes, 230
Paul, L. A., 5
perception, 281
phase space, 175
physicalism, xi, 55
pixelation, 179, 184
platitudes, 9–10
point-like, 83, 89
possibility, 60
preëmption, 1, 150, 310–11, 318; early cutting, 298; late cutting, 298
Price, H., 233, 252–54
protrast, 136
prob-dependence, 122, 135–38, 140, 142
prob-influence, 140, 142–44; backtracking, 204, 207–10, 212, 214, 234–36; broad and narrow, 189–91, 237; continuity of, 161; pseudo-backtracking, 204, 222–23, 239–44; shielding of, 162
probability, 74, 274; enion, 173–74
probability distribution, objective, 149, 174, 180–81
probability-fixing, 23, 36, 70–71; continuity of, 160; shielding of, 161
probability-raising, 24, 149
problem of relevant conditions, 127–28
production, 37, 199
prominent background, 234
prominent foreground, 234, 236
promotion, 144–46, 152–53; aspect, 154; by omission, 155; contrastive, 155–57; trivial, 145
promotion experiment, 169–71
protrast, 160
pseudo-backtracking prob-influence, 204, 222–23, 239–44
psychology of causation, 17–18, 282–84
pure contribution, 76–77, 121

quantum field theory, 56n
quantum mechanics, 56, 112–16; Bohmian, 113–14, 116, 260; Everettian, 116; flashy, 115; GRW, 114; modal interpretations, 116; spontaneous collapse interpretations, 114–16, 166
quantum non-locality, 166
quantum state, 112
quinergy, 35–36

real basis, 25, 32–33
real cause, 36, 67
realism, 308
realistic circumstance, 39
reality, 24
reduction, 34; abstreduction, 34–37, 123, 138–40, 186–87
reductive explanation, 34
region, 58
regular contrastivization, 162
regularities, law-like vs. accidental, 16
Reichenbach, H., 231, 254–55, 257
relationism, 58
relativistic electromagnetism, 56, 89–92
relativistic locality, 91, 166
relativity, 56
RELAXED standards of theoretical adequacy, 40–44, 302
routes of influence, 214; backtracking, 216, 240
Russell, B., 53–54, 68, 307

salient promoter, 287
Salmon, W., 200, 317
saved fizzle, 298, 318
Sellars, W., 8
settledness of the past, 262
shielding; of fixing, 161; of prob-influence, 162; of probability-fixing, 161; of terminance, 99–100, 211–12
significant promotion, 290
simple theory of classical mechanics, 25–26
simplon, 77
Simpson's paradox, 152, 194–96
Simpson's reversal, 197
singular causation, 22, 121, 265–66, 284
singularity, 110–11
Sklar, L., 88
SM-conditional, 250–51
space invaders, 107
space-like, 83, 90; dimension, 80; partial influence, 166; pure contribution, 88; surface, 80; terminance, 81
space-time, arena, 79–82, 213; connected, 79; Galilean, 79, 82–83, 91, 112, 259; homogeneous, 79; Lorentzian, 109; maximal, 101; Minkowski, 79, 89–91; Newtonian, 259
sparse interpretation of classical gravitation, 85
spatial metric, 83
spatio-temporal metric, 89
special science law, 66–67
spin echo experiment, 190–91
spontaneous localization, 114
standard interpretation, of classical gravitation, 82, 83; of electromagnetism, 89
START, 226
state, 79, 81; generic, 84; global, 81
statistical mechanics, 175–80
stochastic lattice, 104–05

stochasticity, 68, 73, 74, 104–06
Strevens, M., 173–74
STRICT standards of theoretical adequacy, 40–44; for metaphysics, 43, 47–48, 267
strong charge, 92
strong field, 92–93
strong transitivity of fixing, 158
stuff, 57
subevent, 63
subjective chance, 149
subjectivity, 308; of causal direction, 253–54
superevent, 63
superluminal influence, 91

temporal asymmetry of disjunctive deliberation, 252–53
temporal metric, 80, 83
temporal passage, 70, 205
terminance, 72, 116–17; and time, 81; continuity of, 98–100; in classical gravitation, 83–85; in relativistic electromagnetism, 91–92; intermediate, 98; minimal, 84; shielding of, 99–100; space-like, 81; transitivity of, 100; trivial, 76–78
terminant, 72 (*see also* terminance)
thermal energy, 28–30
thermodynamics, 175
time, 79, 81
time slice, 80, 83
time-like, 83, 90; dimension, 80
top conceptual layer of causation, 20, 22, 37, 50, 265, 269, 282
transference accounts, 3, 200
transitivity, of causal culpability, 160, 303; of fixing, 157–59, 211; of terminance, 100
trivial, contextualization, 69; determination, 76; promotion, 145; terminance, 76–78
Tumulka, R., 166
turnaround event, 203
typicality, 181–83

ubiquitous determination, 108, 176, 191
unidirectional transitivity of fixing, 158–59
unified field theory, classical, 92–94

vagueness, ontological, 150
volition, fundamental, 109, 219

wave function, 112
weak charge, 92–93
weak field, 92–93
weak transitivity of fixing, 157, 211
Weslake, B., 252–54, 256
Woodward, J., 200, 314–17
world event, 64
world line, 80
wormhole, 100

Yablo, S., 289